T0351797

American Agriculture, Water Resources, and Climate Change

National Bureau of Economic Research

Conference Report

NATIONAL BUREAU of ECONOMIC RESEARCH

American Agriculture, Water Resources, and Climate Change

Edited by **Gary D. Libecap and Ariel Dinar**

The University of Chicago Press

Chicago and London

The University of Chicago Press, Chicago 60637
The University of Chicago Press, Ltd., London
© 2024 by National Bureau of Economic Research
Published 2024
Printed in the United States of America

33 32 31 30 29 28 27 26 25 24   1 2 3 4 5

ISBN-13: 978-0-226-83061-2 (cloth)
ISBN-13: 978-0-226-83062-9 (e-book)
DOI: https://doi.org/10.7208/chicago/9780226830629.001.0001

Library of Congress Cataloging-in-Publication Data

Names: Libecap, Gary D., editor. | Dinar, Ariel, 1947– editor.
Title: American agriculture, water resources, and climate change /
    edited by Gary D. Libecap and Ariel Dinar.
Other titles: National Bureau of Economic Research conference report.
Description: Chicago : The University of Chicago Press, 2024. | Series:
    National Bureau of Economic Research conference report | Includes
    bibliographical references and index.
Identifiers: LCCN 2023016767 | ISBN 9780226830612 (cloth) | ISBN
    9780226830629 (ebook)
Subjects: LCSH: Crops and climate—United States. | Crops and
    water—United States. | Water in agriculture—United States. |
    Irrigation farming—Climatic factors—United States.
Classification: LCC S600.6 .A44 2023 | DDC 630.2/515—dc23/
    eng/20230512
LC record available at https://lccn.loc.gov/2023016767

♾ This paper meets the requirements of ANSI/NISO Z39.48-1992
(Permanence of Paper).

Relation of the Directors to the Work and Publications of the NBER

1. The object of the NBER is to ascertain and present to the economics profession, and to the public more generally, important economic facts and their interpretation in a scientific manner without policy recommendations. The Board of Directors is charged with the responsibility of ensuring that the work of the NBER is carried on in strict conformity with this object.

2. The President shall establish an internal review process to ensure that book manuscripts proposed for publication DO NOT contain policy recommendations. This shall apply both to the proceedings of conferences and to manuscripts by a single author or by one or more co-authors but shall not apply to authors of comments at NBER conferences who are not NBER affiliates.

3. No book manuscript reporting research shall be published by the NBER until the President has sent to each member of the Board a notice that a manuscript is recommended for publication and that in the President's opinion it is suitable for publication in accordance with the above principles of the NBER. Such notification will include a table of contents and an abstract or summary of the manuscript's content, a list of contributors if applicable, and a response form for use by Directors who desire a copy of the manuscript for review. Each manuscript shall contain a summary drawing attention to the nature and treatment of the problem studied and the main conclusions reached.

4. No volume shall be published until forty-five days have elapsed from the above notification of intention to publish it. During this period a copy shall be sent to any Director requesting it, and if any Director objects to publication on the grounds that the manuscript contains policy recommendations, the objection will be presented to the author(s) or editor(s). In case of dispute, all members of the Board shall be notified, and the President shall appoint an ad hoc committee of the Board to decide the matter; thirty days additional shall be granted for this purpose.

5. The President shall present annually to the Board a report describing the internal manuscript review process, any objections made by Directors before publication or by anyone after publication, any disputes about such matters, and how they were handled.

6. Publications of the NBER issued for informational purposes concerning the work of the Bureau, or issued to inform the public of the activities at the Bureau, including but not limited to the NBER Digest and Reporter, shall be consistent with the object stated in paragraph 1. They shall contain a specific disclaimer noting that they have not passed through the review procedures required in this resolution. The Executive Committee of the Board is charged with the review of all such publications from time to time.

7. NBER working papers and manuscripts distributed on the Bureau's web site are not deemed to be publications for the purpose of this resolution, but they shall be consistent with the object stated in paragraph 1. Working papers shall contain a specific disclaimer noting that they have not passed through the review procedures required in this resolution. The NBER's web site shall contain a similar disclaimer. The President shall establish an internal review process to ensure that the working papers and the web site do not contain policy recommendations, and shall report annually to the Board on this process and any concerns raised in connection with it.

8. Unless otherwise determined by the Board or exempted by the terms of paragraphs 6 and 7, a copy of this resolution shall be printed in each NBER publication as described in paragraph 2 above.

Contents

Acknowledgments

This volume was made possible by two generous grants from the Economic Research Service (ERS) at the US Department of Agriculture (grants #58-6000-7-0114 and #59-3000-7-0102/3) to the National Bureau of Economic Research (NBER). It is part of a multiyear research initiative at the NBER on agricultural economics. We are grateful to Spiro Stefanou, the administrator of ERS, and James Poterba, the president of the NBER, for their support of this initiative. We would also like to thank Helena Fitz-Patrick for outstanding assistance in the editorial process and in shepherding the papers toward final publication, Denis Healy for expert oversight of the USDA research grants to NBER, and Carl Beck and the NBER Conference Department staff for supporting the meeting at which the papers in this volume were presented.

We acknowledge the invaluable input and feedback from the six dedicated discussants of the papers presented in the conference leading to this book: Andrew Ayres, Research Fellow, Water Policy Center, Public Policy Institute of California; Tamma Carleton, Assistant Professor of Economics, Bren School of Environmental Science and Management, University of California Santa Barbara; Zeynep Hansen, Professor of Economics, Boise State University; Lynne Lewis, Elmer W. Campbell Professor of Economics, Bates College; Prabhu L. Pingali, Professor of Applied Economics and Director of the Tata-Cornell Institute, Cornell University; and Karina Schoengold, Associate Professor of Economics, University of Nebraska–Lincoln.

Introduction

Gary D. Libecap and Ariel Dinar

Introduction

As access to water is altered due to climate change, there will be new challenges that face agriculture. Traditional locations and production practices for crops will be affected. The collected works in this volume explore the various margins of adjustment available to farmers in the US in light of these conditions.

Broadly the papers focus on four main areas at the intersection among agriculture, water, and climate change: the movement of water (drainage and irrigation); the potential for negative externalities from private responses (use of conservation easements, fertilizer runoff, groundwater overextraction, depletion, and surface stream flow interaction); institutional adaptation to solve collective action problems (movement of water, groundwater conservation); and adjustments following exposure to extreme conditions (irrigation technologies, fertilizer use, crop mix changes, cover crops). There are complex incentives facing farmers as they respond to conditions where water supplies are less reliable.

The chapters demonstrate the various margins of adjustment, some of

Gary D. Libecap is Emeritus Distinguished Professor at the Bren School of Environmental Science and Management and the Economics Department at the University of California, Santa Barbara, and a research associate of the National Bureau of Economic Research.

Ariel Dinar is a Emeritus Distinguished Professor of Environmental Economics and Policy at the School of Public Policy at the University of California, Riverside.

For acknowledgments, sources of research support, and disclosure of the authors' material financial relationships, if any, please see https://www.nber.org/books-and-chapters/american -agriculture-water-resources-and-climate-change/introduction-american-agriculture-water -resources-and-climate-change.

which are informed by past experiences with drought and intense precipitation, as well as production across the continent where conditions vary. The adjustment margins include changes in crop mix, new fertilizer intensity, shifts in capital investment for irrigation, and more reliance on groundwater. Farmers can invest in draining swampy fields, reducing water loss in transport (by either lining ditches or installing pipes), adopting more efficient central pivot or drip irrigation technology, or shifting to drought-tolerant crop varieties.

Many of these investments, however, have the potential to create externality problems that may require institutional solutions to internalize the external costs. For instance, increased fertilizer runoff can contaminate the watershed downstream, leading to hypoxia zones. Drainage can create water flows across neighboring properties, and increased central pivot irrigation can overburden aquifers. Importantly, this volume not only documents such problems but also highlights possible solutions. Responses also are conditioned by past subsidy policies that affect crop selection, even water-intensive crops. These issues, while important, are beyond the scope of this volume.

The central theme of this volume is that agricultural production in the US relies critically on having water available for production, a reliance that historically had not been an issue. That relationship is now challenged, not only in the semiarid US West, but in the Midwest and East, where historically, water access was more uniform and predictable. Innovations in technology, production practices, irrigation, crop mixes, and institutions will be needed to deal with the challenges of climate change in droughts, extreme precipitation, and aquifer depletion futures. Fortunately, many insights can be obtained from the varied conditions that exist across the continental US, and point to their use in the broad range of topics covering the country.

This collection of papers addresses a diverse set of problems and potential solutions to challenges of water in agriculture. Chapters include topics from improving institutions for water drainage to paying for land easements for conversion to wetland. The studies about the water/agriculture nexus address crop mix, irrigation, groundwater, fertilizer, and related externalities.

To begin, we highlight the role of agriculture in the American economy and society over time; point to farmer historical and contemporary responses to varying climatic conditions; indicate the importance of water as an input to agricultural production; identify possible impacts of climate change on access to water; and briefly summarize 11 papers on these topics presented at a conference organized by Gary Libecap and Ariel Dinar on May 12–13, 2022. The conference was supported by the National Bureau of Economic Research and the US Department of Agriculture, Economic Research Service.

Role of Agriculture in the American Economy and Society and Historical Responses to Changes in Access to Water

Agriculture has been critical in the development of the American society and economy. It was a pathway for immigrant settlement; was a basis for employment and community formation; and has provided critical foodstuffs, fibers, and other sources of industrial production. Critical inputs have been land, labor, capital equipment, nutrients, and water. Until the late 19th century, agriculture was centered in the eastern part of the country, where precipitation was frequent, as was general access to water. The western part of the country always has been drier, and water supplies more limited and costly, leading to differences in water institutions and infrastructure. Even so, except in parts of the US West, water access has not been a critical constraint in agriculture. But this is changing.

With climate change, water supplies are apt to be much more problematic in most parts of the country, affecting agricultural production and rural populations. Fortunately, the wide range of spatial climatic conditions affecting water access that were encountered as settlement and production moved across the continent provides valuable insights to contemporary climate change. In the research briefly summarized below, focus is on farmer interpretation of available climatic data; their reactions and related investments; potential externalities; and institutional/coordination challenges posed by efforts to secure water.

In terms of the overall impact of agriculture on American economic development, access to agricultural land was a primary driver of migration to temperate North America. Large-scale migration, mainly from Europe, of entire families in the colonial and subsequent federal periods resulted in dense population settlements and internal market development from the East Coast through the 98th meridian (Wilcox 1929). Thousands of small landowning farmers became the decision makers regarding farm size, input use, production, and responses to various climatic signals.

Small farms, organized under federal land laws, such as the Homestead Act of 1862 and the rectangular survey of the 1785 Northwest Ordinance (Libecap and Lueck 2011), relied upon family labor with minimal agency problems (Allen and Lueck 1998). Midwestern farm populations, in particular, invested in education, leading to high levels of human capital, perhaps the highest in the world by the early 20th century (Goldin 1998, 2001). The turnover of farmlands via very active land markets encouraged the development of capital markets (Hartnett 1991). The capital gains from land sales, in turn, were a major source of wealth creation (Kearl, Pope, and Wimmer 1980; Steckel 1989; Ferrie 1993; Stewart 2009). Overall, easy access to farmland resulted in a relatively egalitarian society in rural US areas compared to urban centers in the 19th and early 20th centuries (Pope 2000, 118).

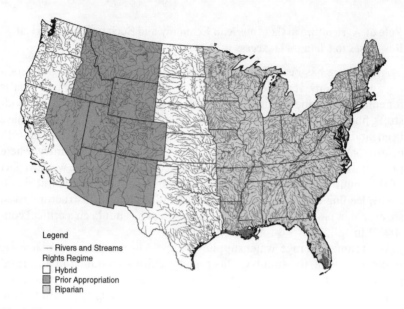

Fig. 0.1   Stream density
Source: Modified from Leonard and Libecap (2019), fig. 2.

The role of water for agricultural settlement and production was stressed early. Thomas Jefferson commented in 1811 that farmers wished for "a rich spot of earth, well watered, and near a good market . . ." (Atack, Bateman, and Parker 2000, 245). In the eastern US, farmers relied upon rainfed agriculture possible from relatively reliable precipitation and absence of serious drought (Libecap and Hansen 2002, 91–92). Irrigation was uncommon, and drainage primarily was aimed at shifting swamplands into farm production. Because water was available locally, there was little large-scale water movement, which would have posed significant coordination problems under the riparian doctrine. Riparian water rights granted use of water to all adjacent landowners, and collective agreement was required to transfer any water from its source.

Figure 0.1 shows stream densities in the US, along with the three major water rights practices by state (riparian, prior appropriation, and joint or hybrid practices). The figure clearly shows that local surface water sources for agriculture were far more prevalent east of the 98th meridian, running from North Dakota through Texas.

To improve yields and profits, farmers adopted innovative management practices, technologies, and varieties, such as novel seed types in corn, wheat, other grains, and cotton as increased aridity, lower mean and more variable temperatures, and insect pests were encountered (Griliches 1957; Olmstead and Rhode 2011; Sutch 2011). Research on new seeds and agricultural

Table 0.1   **Farm population**

Year	Farm Population	Percent of Total US Population
1910	32,077,000	34.9
1912	32,210,000	33.9
1950	23,048,000	15.3
1960	15,652,000	8.7
1970	9,712,000	4.8
1980	6,051,000	2.7
1990	4,591,000	1.9

Source: Agriculture. Farms and Farm Structure Alan L. Olmstead and Paul W. Rhode 2006. Historical Statistics of the United States, Volume Four Part D, Series Da 1-13, 4-39.

practices was provided by private companies, such as DeKalb and Pioneer; by land-grant colleges under the Morrill Act of 1862; and by the USDA experiment stations, Agricultural Research Service (established 1953), and the Economic Research Service (established 1961). Additionally, farmers invested in innovative capital equipment introduced by Ford, McCormick-Deering, and Farmall, including mechanized reapers and threshers, tractors riding plows, seed drills, and balers (Olmstead and Rhode 1995). Farmers also incorporated new chemical fertilizers and changes in tillage practices to raise yields.

Between 1870 and 1990 farm productivity grew by nearly six times (Olmstead and Rhode 2000, 701). At the same time, however, farm populations and their share of US total population fell dramatically, as shown in table 0.1. As farm sizes grew, farming became more capital intensive, and rural-to-urban migration increased.

The data in table 0.1, however, understate the continuing economic, social, and political role of agriculture in the US. In addition to farm populations, urban centers based on agricultural research and development, marketing, processing, manufacturing, and shipment emerged in Minneapolis, Chicago, Kansas City, Cincinnati, Fort Worth, Omaha, Stockton, and elsewhere. The value of agricultural output and processing remain key element of overall state GDPs as indicated in table 0.2. Moreover, figure 0.2 shows agricultural exports as major elements of US trade between 1970 and 2020, as well as critical sources of food worldwide. Beef and beef products exports approached $8 billion in 2020, and among commodities, soybean exports grew to over $25 billion by 2020.

Figure 0.3 reveals the role of agricultural production in providing relatively low-cost domestic food supplies. The figure reveals a continuous decline in the share of household disposable personal income spent on food from 1920 through 2020.

As noted above, through the 19th century, agriculture largely was centered

Table 0.2   Agricultural output and processing share of state GDP for selected states 2020

State	Share of State GDP (%)
California	2.8
Colorado	2.3
Idaho	12.5
Illinois	2.9
Indiana	2.5
Iowa	9.3
Kansas	6.8
Montana	4.9
Nebraska	21.6
Ohio	3.2
Oregon	13.0
Washington	12.0
US	5.0

Source: USDA ERS.

east of the 98th meridian with rich soil, flat terrain, dense streams (figure 0.1) and abundant precipitation. After that time, however, the area west of the 98th meridian, especially the Pacific region, became a major source of domestic food production and exports, as well as employment in processing. Agriculture in the Pacific region, however, relied upon far different sources of water supply.

The region is more drought prone; generally, is drier; depends upon water storage in surface reservoirs and aquifers (Libecap and Hansen 2002); utilizes canal and ditch networks for water delivery; and applies irrigation more than elsewhere in the US. As such, these experiences are indicative of future conditions with climate change that suggest greater prevalence of drought along with alternating very wet and dry periods, with more reliance upon irrigation, longer distance of water transport from storage sites, and need to dispose of drainage.

The western region has always been recognized as more arid. John Wesley Powell in his 1878 *Report on the Lands of the Arid Region of the United States* quite accurately illustrated the dramatic change in precipitation beyond the 98th meridian.

Drought led to Homestead farm failure (Hansen and Libecap 2004a, 2004b). Most of the region's more limited and variable precipitation comes as snow in higher elevations. Snowpack melt has fueled stream flows, often with water stored in reservoirs. Arable land generally is remote from streams, requiring water movement for irrigation. Water transport, however, has required a change in water rights from riparian to prior appropriation (Leonard and Libecap 2019).

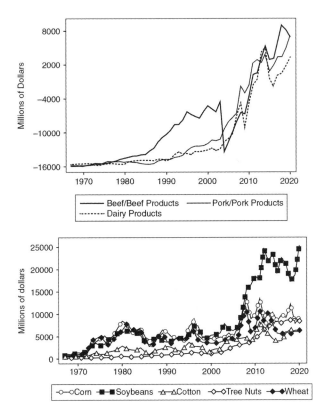

Fig. 0.2 US exports 1970–2020 in chain-weighted $
Source: https://apps.fas.usda.gov/gats/default.aspx.

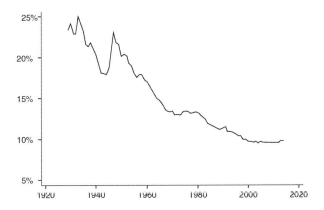

Fig. 0.3 Share of household disposable personal income spent on food
Source: USDA ERS, Food Expenditure Series.
Disposable Personal Income = Post-Tax Income

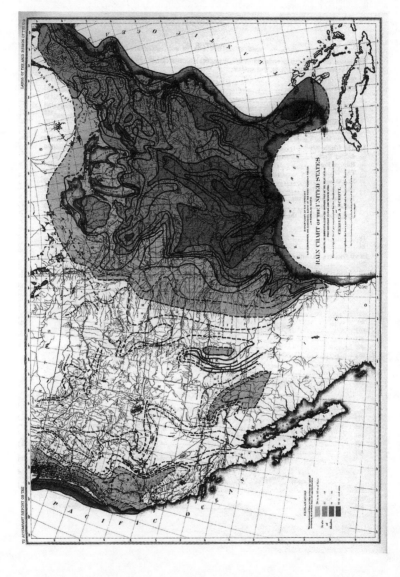

Fig. 0.4   John Wesley Powell, 1878's indication of increased aridity
Source: J. W. Powell, frontispiece. *Report on the Lands of the Arid Region of the United States* (1878).

Prior appropriation water rights are an institutional innovation that allowed water to be separated from the source and moved to the site of agricultural production. It was first introduced in California and Colorado and then spread to all western states and Canadian provinces either in full or as a hybrid with riparian systems. Irrigation districts were formed to coordinate diversion dam construction on streams, canal investments, ditch maintenance, and to protect the priority of diversion. Dams and irrigation systems initially were private, but followed the 1902 Reclamation Act with large-scale federal government investment, particularly after 1940 (Wahl 1989; Pisani 2002). By storing and moving water in an otherwise arid region, dams, related reservoirs, and water infrastructure smoothed supplies during annual summer dry periods and droughts (Hansen, Libecap, and Lowe 2011).

As shown in figure 0.5a, there are many dam sites in the western region of the US, and most are small for local stream water diversion and storage for irrigation. Larger dams, such as Shasta and Oroville in California, American Falls and Palisades in Idaho, Grand Coulee and Tieton in Washington, Canyon Ferry and Tiber in Montana, for example, may have multiple uses with reservoirs to support irrigation, hydroelectric power generation, and flood control.

Figure 0.5b details irrigation projects and networks in the western US that include dams, reservoirs, and extensive canal systems to deliver water to irrigated farmland, and acreage covered. The largest projects are associated with construction and operation by the Bureau of Reclamation (the agency name is indicative of the primary objective), while smaller, earlier developments are private (see details in the 1890, 1900, 1910, 1920 Agricultural Irrigation Censuses). In the most arid regions where arable lands were remote from streams and lacked sufficient precipitation, agriculture would not have been feasible without such supplemental projects.

Irrigation from snowmelt and reservoir storage and shipment was augmented after 1940 with groundwater pumping. Aquifer access became feasible with greater access to electricity, more powerful combustion engines and turbine pumps, deeper wells, and new pumping technologies. Advances in irrigation with new dam construction and groundwater delivery provided new water and led to major increases in agricultural production and higher productivity in the US West, especially in the Pacific region (Edwards and Smith 2018).

Figure 0.6 maps aquifers, primarily for the US West and Midwest, by surrounding geologic formation. These formations bound the subterranean basin; determine its size, depth, and uniformity; influence conductivity or movement of water within the aquifer; and affect recharge and leakage. As such, geology helps determine how much groundwater is available for pumping in various parts of the aquifer and for how long, and extraction costs. Although aquifers appear to cover much of the region, they are extremely

(A) Dam site locations

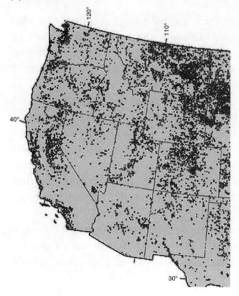

(B) Irrigation infrastructure projects

Fig. 0.5   Western dams, reservoir, and irrigation projects
Source: A. Novak et al. (2016); B. Library of Congress, https://hdl.loc.gov/loc.gmd/g4051c
.ct011656.

Fig. 0.6   Aquifers (primarily western) by surrounding geologic formation

Source: US Geological Survey (2000), Ground Water Atlas of the United States, Introduction and National Summary, figure 4a, https://pubs.usgs.gov/ha/ha730/ch_a/A-text1.html.

heterogeneous in structure, leading to important differences within and across groundwater basins in the stock of water, qualities, and linkages between recharge and extraction.

These variations make modeling and aquifer management difficult. The basins are not like uniform bathtubs as early discussions had assumed to simplify approaches (Gisser and Sanchez 1980). They also have varying surface growing conditions and farming practices. Moreover, groundwater pumping occurs for a variety of uses—urban (especially in the southern San Joaquin valley and near Los Angeles in California), as well as for annual

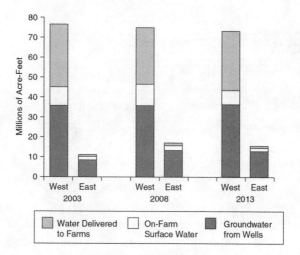

Fig. 0.7   Sources of irrigation water, 2003–2013
Source: Stubbs (2016), fig. 4.

crops, such as hay, grains, and vegetables, and permanent crops, such as fruit and nut orchards and vineyards.

These geologic and user differences, as well as the open access nature of groundwater, compound problems of coordinating pumping among users to address depletion and implement any sustainability objectives. Unlike surface water and prior appropriation, groundwater lacks clear water rights, making it subject to competitive withdrawal and associated externalities (Ayres, Edwards, and Libecap 2018). For larger and more varied aquifers with more heterogeneous pumpers, the challenges in controlling rent dissipation are formidable. As climate change leads to greater reliance upon groundwater for irrigation, these issues are likely to increase in severity.

Figure 0.7 details differences in irrigation water delivery for farms in the eastern and western US between 2003 and 2013. Western farms rely far more on irrigation water, including water conveyed from reservoirs via canals and ditches and groundwater pumping, than do those in the eastern US. With climate change, these distinctions may become less apparent.

Figure 0.8 illustrates the path of irrigation water use from 1984 through 2013. The data underlying the figure reveal that in addition to changes in crop varieties and management practices, US agriculture has witnessed a swift overhaul in irrigation technologies that not only saved water but also increased yields and allowed for more efficient use of fertilizers (Stubbs 2016). The figure shows the decline in total irrigation water despite an increase in total irrigated acres. This is due mainly to the steady increase in pressure-based irrigation technologies replacing gravity-based irrigation technologies. Although on-farm surface water and water delivered to farms

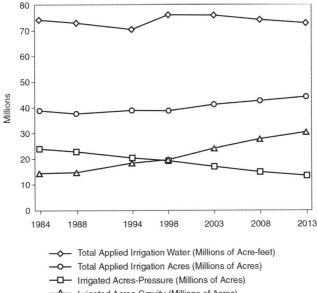

Fig. 0.8   Irrigated acres and applied irrigation water, western states 1984–2013
Source: Stubbs (2016), fig. 3.

in the West for irrigation have declined by 2.5 million acre-feet and 1.2 million acre-feet, respectively since 2003, groundwater withdrawals have risen by 740,000 acre-feet (an acre-foot equals approximately 326,000 gallons and 1,235 cubic meters).

Figure 0.9 shows the percent of market value of crops sold from irrigated farms by state in the US in 2012. Generally, western states have the largest share of crops produced by irrigation to provide water.

As climate change leads to greater reliance upon irrigation, especially in previously rainfed agricultural regions, the techniques, institutional responses, and other innovations observed in the drier western US will provide important laboratories for new learning (Schoengold and Zilberman 2007; Libecap 2011; Hornbeck and Keskin 2014).

Water Use in Agriculture

Agriculture is the largest single user of water in the US. It accounts for approximately 80 percent of the consumptive use of water, and of that, irrigation amounts to about 42 percent.[1] As we have indicated, productivity advances have resulted in declines in water use per irrigated acre, while

1. See https://www.ers.usda.gov/topics/farm-practices-management

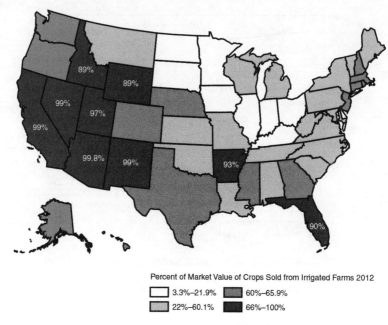

Fig. 0.9   Percent of total market value of all crops from irrigated farms, 2012
Source: Stubbs (2016), fig. 1.

the area irrigated has increased. These trends likely will continue as water becomes scarcer and more costly, forcing farmers to further adapt.

Figure 0.10 traces the growing trend in irrigated land in the US over the period 1890–2018 and reductions in water use per acre over the period 1975–2018. There also are noticeable gains in productivity as water use per acre has fallen, often with a shift from gravity surface flow onto fields from ditches with an associated extravagant delivery of water.

Climate Change Projections and Agricultural Water

For crops to grow and be economically productive, several inputs, such as sunlight, water, carbon dioxide, nutrients, and limited weeds, diseases, and insects, have to be present at optimal amount (Mendelsohn and Dinar 2009). An optimal growing process of agricultural crops requires a certain distribution of dry matter within each plant, especially the reproductive components (in non-weed crops), that lead to yield increases, compared with the green matter components that are non-marketable. Climate change as it impacts temperature, CO_2, and water availability may alter this distribution and related productivity.

The effect of climate change on US agriculture (with focus on irrigated

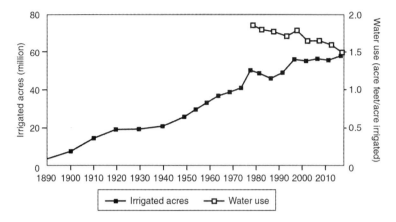

Fig. 0.10   US irrigated acres and water use per acre, 1890–2018

Source: USDA-ERS, n.d. Irrigation and Water Use, https://www.ers.usda.gov/topics/farm
-practices-management/irrigation-water-use/.

agriculture) has been examined in multiple studies, and estimation results have varied (Mendelsohn and Dinar 2003; Deschenes and Greenstone 2011; Massetti and Mendelsohn 2011). In part, these differences reflect the underlying uncertainty and complexity of climate change projections, as well as the variables examined. Deschenes and Greenstone (2011) estimate that the average present value (in 2005 dollars) of an annual decline in agricultural profit across 2,256 counties in the US is $38.7 billion. Alternatively, Massetti and Mendelsohn (2011) found that depending on the severity of climate change, the agricultural sector of the US could benefit (due to CO_2 effects on crop yields) from mild impacts.

Mendelsohn and Dinar (2003) used a 1997 census of 2,863 counties in the US and provide estimates of the role of adaptation—specifically, adoption of irrigation technologies—in reducing damage from climate change. They found that the value of irrigated cropland is not sensitive to precipitation changes, and values increase with temperature. They also found that new sprinkler systems are used primarily in wet cool sites, whereas gravity and especially drip irrigation systems, help compensate for higher temperatures. These results underscore the importance of irrigation in adapting to increased water scarcity.

Drought is a major indicator of potential patterns of increased aridity associated with climate change. As indicated in figure 0.11, over a 20-year period, drought has become increasingly more intense, covering a larger area, especially in the central and western US. Agriculture is very sensitive to drought, as precipitation and water access for irrigation are disrupted. When drought persists, the hydrological cycle can be altered, affecting agricultural productivity (Hayes et al. 2011).

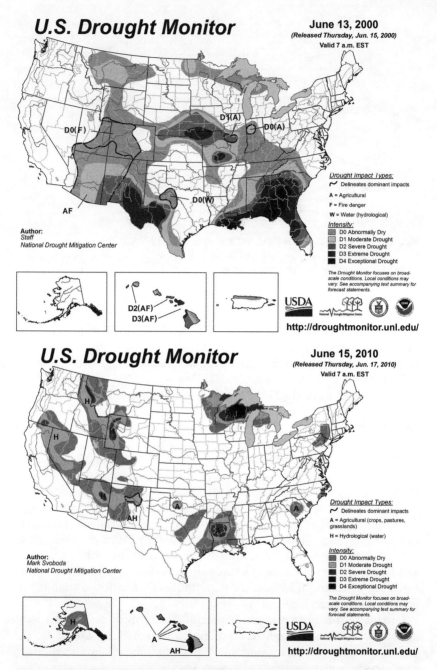

Fig. 0.11   Drought intensity changes in the US 2000–2022

Note: The U.S. Drought Monitor is jointly produced by the National Drought Mitigation Center at the University of Nebraska-Lincoln, the United States Department of Agriculture, and the National Oceanic and Atmospheric Administration. Map courtesy of NDMC.

Source: National Drought Mitigation Center, University of Nebraska-Lincoln, https://drought monitor.unl.edu/.

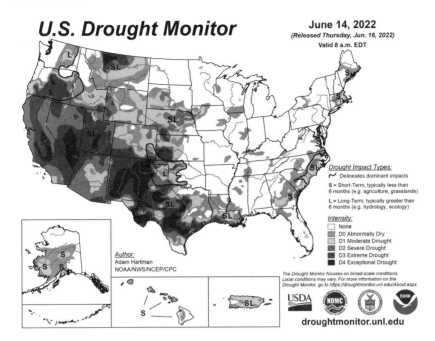

Fig. 0.11   (cont.)

What is the role of adaptation in securing the agricultural sector's profitability from climate change–induced water scarcity? Gollin (2011) analyzes the role of various science-related technological innovations such as plant breeding for climate adaptation, modifications of farm management practices, water control and improved water use efficiency, mechanical innovations, and chemical use to compensate for yield losses, including the negative effects of pollution externalities from increased intake of chemicals and fertilizers.

Overall, farmer adaptations range from new crops, especially drought-tolerant varieties; intermediate fallowing during dry periods (if climate change results in times of increased water availability followed by drought); permanent withdrawal of marginal production areas; use of cover crops and tillage practice to conserve water; addition of fertilizer and other inputs; greater reliance upon irrigation, particularly in the eastern US, as well as adoption of new irrigation technologies in both regions of the US; greater movement of water from storage sites for irrigation and for drainage; increased reliance of marginal water sources such as recycled wastewater; and reliance upon more groundwater pumping. Many of these responses will require institutional arrangements to coordinate groundwater extraction and water movement, and to address other potential externalities associated with fertilizer runoff (Saleth, Dinar, and Aapris Frisbie 2011). In addition, adjustments in crop insurance programs may assist farmers in responding to uncertainty associated with assessing climatic variability and crop yields (Garrido et al. 2011).

New Research on Water, Agriculture, and Climate Change

Agriculture is practiced in the US under a variety of climatic conditions, with wetter and humid climates in the eastern part and drier and semiarid to arid climates in the western part of the nation. The research outlined below addresses the role of water in irrigated agriculture from snowmelt and groundwater west of the 98th parallel and supplemental water to the east. The effects of too little or too much water resulting from climate change; the adaptations needed to address them; farmer interpretation of past droughts and their responses; adoption of new irrigation practices; institutional adjustments required to promote cooperation; as well as any negative externalities from efforts to maintain yields are examined in the research summarized below.[2]

The first group of research papers refers to agricultural adaptation in the eastern part of the US, dealing with rainfed agriculture and/or supplemental irrigation and the need to remove excess water. Edwards and Thurman analyze the role of drainage under the increasing likelihood of extreme precipitation events across the entire US due to climate change. Alongside with technical innovations to be introduced in drainage tile technologies required for collection and disposal of excess water, the research highlights the relevance of institutional innovation necessary for efficient coordination of drainage reduction, and its associated costs. The chapter begins with the observation that all US regions (even arid and semiarid regions) are projected to see periodic heavier rainfall events under climate change. Poorly drained soils see excess water in the root zone of cultivated crops, leading to waterlogging and salinity, which in turn create aeration deficits and productivity losses, both of which drastically reduce yields or eliminate production.

The ability of farmers to remove excess water from fields is crucial for ensuring secure and reliable food supply. Legislation for establishing local institutions (drainage districts) has been essential in successful drainage-management adaptation. The analysis suggests that after the enactment of drainage district legislation, poorly drained counties realized a rise in improved-drainage acres, resulting in increase in land value. Estimated increases in the value of land in the worst-drained counties of the eastern

2. During the May 12–13, 2022 conference, the authors benefitted from comments provided by Andrew Ayers (Public Policy Institute of California); Tamma Carleton (University of California, Santa Barbara); Zeynep Hansen (Boise State University); Lynne Lewis (Bates College); Prabhu Pingali (Cornell University); and Katrina Schoengold (University of Nebraska, Lincoln).

US after adaptation of improved drainage increased by 13.5 percent to 30.3 percent, with a combined increase in land value after the enactment of drainage district legislation of between $7.4B and $16.6B in 2020 dollars. This finding suggests an important role to adaptation of drainage institutions.

Karwowski adds another adaption angle to climate change in humid regions by analyzing the value of the land easement program. Large agricultural areas in the eastern US exist in regions that were reclaimed on wetlands and floodplains but which now are subject to flooding risks under increased precipitation. Easements might promote removal of some of these areas from production. Approximately 3 million acres of eased wetlands and 185,000 acres of floodplain easements existed in the US in 2020. The easements program impacts agricultural production both directly, by reducing planting on marginal land, and indirectly, by changing flood patterns that improve yields on surrounding cropland. The easement program provides payments to farmers who withdraw inundated cropland from production and restore it to its natural condition.

Karwowski analyzes data on crops (corn, soybean, and wheat) in 1,700 rainfed and non-irrigated counties east of the 100th meridian. She finds that easements can be an effective adaptation strategy. For example, a 100 percent increase in wetland easement land share increases county yields by 0.34, 0.77, and 0.46 percent for corn, soybeans, and wheat, respectively. Doubling of wetland easement land share reduces losses by $3.59, $6.07, and $11.23 from excess moisture, heat, and disease for each dollar of soybean liability, respectively. In the case of corn, the same change in easement leads to reduction in insect losses by $8.50 per dollar of liability. All in all, the results suggest that increasing land share in floodplain and wetland easements leads to reduced risk of loss for all three crops.

Other research addresses the roles of off-farm water conveyance and on-farm irrigation technologies in response to shifting precipitation. Hrozencik, Potter, and Wallander focus on the value of water savings in the conveyance of water from the source to farms, as opposed to most water conservation efforts that have focused on farm-level improved irrigation efficiency. Given that more than one-third of the applied agricultural irrigation in the US originated from off-farm sources, improvements in delivery and conveyance efficiency have the potential to significantly reduce water losses. These improvements include lining of canals and converting open canals to pipes.

Using a data set of irrigation water delivery organizations in the western US, the authors estimate the impact of lining and piping of conveyance infrastructure on water losses. The potential resource savings are large. On average, reported conveyance losses are nearly 15 percent of the delivered water in 2019. The findings of the study indicate that at the margin, an increase of 1 percent in the share of conveyance piped infrastructure leads to an expected 0.16 percent reduction in conveyance losses. Using a simulated water-conservation supply curve, the authors suggest that nearly 2.3 percent of all water delivered to farms could remain in the system, rather than lost through evaporation or leakage at a private capital cost lower than $10,000 per acre-foot of delivered water.

Cooley and Smith add to understanding of the role of irrigation technologies in adapting to water scarcity in the US Midwest, a humid region that actually faces relative water scarcity due to climate change. Irrigated agriculture in the state of Illinois saw increased irrigation-equipped cropland by threefold since 1978, mainly by a rise in center pivot irrigation systems (CPIS) a decade later. CPIS adoption came in certain locations with monetary benefits in terms of annual crop yield, greater irrigated acreage, new crop selection, and reduction in drought-related insurance payments. The authors demonstrate the value of CPIS adoption by using a data set that includes CPIS locations during drought years and the remaining control variables of crop type, yield levels, and insurance payments. The results of the statistical analysis suggest that in drought years CPIS presence has a significant positive effect on corn yield and a significant negative effect on indemnity payments for both soybeans and corn.

The results provide insights into an emerging trend of irrigation in humid regions, and the role of irrigation in replacing crop insurance. CPIS adoption has reduced drought indemnity for both corn and soybeans. Namely, an increase of 1 percent in cropland equipped with a CPIS decreases insurance payments for corn by approximately 6.34 percent and for soybeans by about 2.81 percent. In addition, CPIS presence during a drought year has a significant effect on corn yield but no significant effect on soybeans yield. Findings suggest that during a drought year, increase in 1 percent of cropland equipped with CPIS yields nearly 0.46 percent increase in corn yield per acre across the state.

Adoption of costly new irrigation technologies and cropping patterns by farmers depends upon their perception of future drought. Blumberg, Goemans, and Manning examine how farmers interpret past droughts in implementation of new irrigation technologies. Their theoretical framework suggests that farmers facing possible reductions in surface water availability will be more likely to adopt water-efficient irrigation systems. Using data on corn production from one water region in Colorado (corn is considered more sensitive to water stress than are other popular crops) over seven observation years during 1976–2015, the authors identify a change in beliefs arising from past droughts about the reliability of farmers' water supply. Water access is reduced through a curtailment of water supplies through an administrative system of "calls." Past drought and associated calls on water allow the authors to observe shifts in beliefs and infer their impact on the adoption of water-saving sprinkler irrigation technology at the field level to replace older flood irrigation. Several important findings include that by the year 2015, there was on average a 11.2 percent increase in land converted from flood to sprinkler irrigation; further, generalizing to the entire water-supply region, the reduction in water availability from increased "calls" brought an increase of over 52,000 sprinkler-irrigated acres; and finally, a reduction in surface water availability led to more groundwater use to augment existing corn irrigation practices.

In addition to on-farm adoption of new irrigation technologies, farmers can also turn to new seed varieties that are more tolerant to drought and related climate-induced effects; they can also introduce new management practices, such as planting cover crops to conserve water. McFadden, Smith, and Wallander investigate the determinants of farmer adoption of drought-tolerant corn varieties in response to an increased frequency of drought in the US. Given that corn is a water-intensive crop and given corn's economic importance due to its large share in US agricultural value, adaptation of drought tolerant corn might have significant economic benefits. The authors used 2016 data from a survey of corn operations in the US and a sample covering over 73.3 million acres, representing nearly 78 percent of 2016 US corn acreage and where drought-tolerant corn was grown on non-irrigated land in 2016.

Their analysis suggests that the duration and severity of recent droughts do not appear to affect adoption of drought-tolerant plants, but that higher average temperatures and variability of rainfall instead lead to higher adoption rates, although temperature variability is statistically insignificant. In addition, higher adoption rates occur on lower quality, more highly erodible land. Predictably, increased rainfall leads to lower adoption rates. These findings suggest that irrigation could be increasingly important to support adoption of drought-tolerant corn under changing long-term climate conditions.

The studies provided by Hrozencik et al., Cooley and Smith, Blumberg et al., and McFadden et al., illustrate the factors affecting adoption of new water conveyance investments, irrigation technologies, and crop mixes. At the same time, they raise the problem of groundwater depletion and interruption of surface stream flows as farmers resort to pumping from aquifers. Crop mix decisions, especially continued reliance upon water-intensive corn, suggests the influence of historic crop subsidies. These policies will require reconsideration if more flexible and climatic responsive crop adjustments are to take place.

Dong investigates adoption of cover cropping to improve resilience to drought. Cover crops include grasses; legumes, including annual cereals, such as rye, wheat, barley, oats; annual or perennial forage grasses, such as ryegrass; and warm-season grasses, such as sorghum. Cover crops can protect and improve soil between periods of regular crop production through control of erosion, weeds, and pests; addition and recycling of nutrients; provision of habitat for beneficial organisms; and greater water efficiency by reducing evaporation from bared soil. Trade-offs associated with cover cropping include incremental costs of soil preparation, seeds, and labor, as well as difficulties in implementation and management of rotating cover crops with major cash crops.

With such background and data available for soybean production in the US, Dong explores factors influencing farmer's adoption of cover crops and examines the impact of cover crops on soybean yield and risk. She finds regional differences in adoption, likely the result of hedonic effects, such as soil types and quality, landscape, and climate. She also finds that cover crops adoption was affected by farmers' concerns regarding production outcomes. Farmers who had concerns over wind-driven erosion, soil compaction, water quality, or other concerns were more likely to adopt cover crops than were those who did not have such concerns. Still, she finds that the voluntary adoption rate of cover crops is relatively low. Financial support, however, increased cover crop acres enrolled in government programs from 312.6 thousand acres in 2009 to 2,443.1 thousand acres in 2020.

Greater use of fertilizers along with new cropping patterns can be strategies for farmers to maintain or improve yields under more uncertain water supply conditions. While local agricultural production can benefit from such adaptation practices, downstream costs can be inflicted from runoff. Elbakidze, Xu, Gassman, Arnold, and Yen present a valuable analysis of the unintended consequences of greater use of fertilizers and associated nitrate concentrations in runoff from farmland upstream on water quality downstream. They use a set of models applied to the Mississippi River basin to estimate the costs of externalities in the Gulf of Mexico. The estimated increase in nitrate runoff to the gulf is in the range of 0.4–1.58 percent compared to the baseline. The effects vary because changes in production, including nitrate use, are spatially heterogeneous. In some counties, nitrate use will intensify, while in others it will decrease.

Similarly, Metaxoglou and Smith explore the extent of nutrient pollution in US agriculture associated with climate change responses, also using the Mississippi River basin as their study area. They apply their econometric approach to a long-term data set and introduce an analytical framework for nutrients, corn production, and precipitation in estimating and interpreting their results. If corn yield is not affected by overapplying nitrate fertilizer, farmers overapply as insurance against yield reduction arising from reduced precipitation, common in many locations in the basin. Any residual nitrogen from overdoses remains in the soil. With precipitation, the nitrogen leaches into lakes, rivers, and streams as nutrient pollution. Therefore, less rainfall leads to more nitrogen applied by farmers, increasing yields and expanding acreage, whereas more rainfall leads to more nutrient leakage into waterways. Under this framework, increases in corn acreage are expected to increase nitrogen concentration in the soil and downstream waterways.

The authors use data on changes in corn acres planted for counties east of the 100th meridian (excluding Florida), precipitation patterns, Mississippi stream flow for 1970–2017, along with secondary estimates of the median potential damage costs of nitrogen increases in the Gulf of Mexico from declines in fisheries and estuarine/marine life at $15.84 per kg of nitrate disposed to the gulf (in 2008 values). They estimate that an additional 50,800 metric tons of nitrogen in the Gulf of Mexico yield an estimated damage of nearly $805 million per year.

Increased climate change–induced surface water scarcity will direct more investment in groundwater pumping to support irrigation. As noted above, increased reliance upon pumping and competitive extraction, however, depletes subsurface stocks in a nonoptimal manner, raises pumping costs, generates surface land subsidence, and reduces water quality. These effects will be intensified if precipitation and groundwater recharge are reduced following climate change. The assignment of tradable groundwater rights or implementation of other regulatory controls will be required to reduce rent dissipation in such a critical resource. While seemingly obvious, despite their benefits these institutional changes are complex and costly due to heterogeneities across groundwater resources and among the pumpers who draw from them, as well as to the many external constituencies who also seek groundwater claims.

The research by Bruno, Hagerty, and Wardle demonstrates the importance of new institutional arrangements to regulate groundwater withdrawal in California with consequences for both long-term water levels and farmland values in the vicinity of the regulated aquifers. Despite the growing importance of the issue of groundwater overextraction, its impact, and the range of mitigating institutions, the literature on the topic is scant. To help address this, the authors use the case of the Sustainable Groundwater Management Act (SGMA) enacted in California in 2014 as an example of the benefits and costs, and hence complexity, of legislative policy intervention. The law identified local groundwater sustainability agencies (GSAs) as key for negotiation and implementation of pumping controls among members to achieve sustainable withdrawals. Despite advertised benefits of locally higher land values and enhanced groundwater stocks, SGMA adoption has been controversial, with opposition from many pumpers and their irrigation districts. Pumpers bear direct costs as they cut back on water extractions. These costs vary. The impact of reductions is immediate and generally predictable, while the benefits are longer term and more uncertain.

Using data for all 343 groundwater agencies (GSAs) formed following the enactment of SGMA, the authors estimate the gross cost of agricultural groundwater regulation through the changes in land values across GSA boundaries before and after the SGMA enactment. Their findings suggest that although SGMA encouraged a move from the previous *status quo* of open access to a joint management regime, the high costs of reduced pumping are significant. Their estimates suggest that, on average, a reduction of 1 acre-foot per acre of expected future water pumping from an aquifer reduces land values of farms within the borders of the GSA by 55 percent in the post

SGMA period. The study suggests that although institutional changes to address common-pool extraction of critical groundwater resources may have broad public good benefits, localized private net costs may be significant and not be Pareto improving. The implication nationwide is that groundwater extraction controls may be resisted, slow, and incomplete.

In another study in a region with precipitation variability, Kovacs and Rider develop an approach to quantify how the demand for in situ groundwater can help identify the value of groundwater to farmers who experience climatic change effects. Using detailed field-level data and data from land markets in eastern Arkansas overlaying the Mississippi River alluvial aquifer, the authors provide empirical evidence of decreases in the value of agricultural land due to increased overdraft of groundwater. Levels fall as farmers use more water from the aquifer to compensate for reduced availability of surface sources.

As part of their analysis, the authors estimate a willingness to pay for a foot increase in saturated water thickness of $4.70 and $24.80 for all farms and rice farms, respectively, when current thickness is between 100 to 120 feet. The authors also show that the demand for in situ groundwater is more elastic for rice farmers than for all other farm landowners. The main finding is that in all regional land markets analyzed, a decrease in saturated thickness by 20 feet from 120 feet to 100 feet (as was experienced in the region in the past 30 years) would decrease the per acre property value by $148 for all farms and $296 for rice farms. The analysis in this study demonstrates that declining precipitation patterns and related groundwater withdrawals can have a significant impact on the profitability of the agricultural sector and land values in the presence of interacting natural capital stocks, such as surface and groundwater.

Conclusions

As described, the research summarized here provides valuable insights into how American agriculture responds to changes in water access as climate change unfolds. Fortunately, there is abundant historical and contemporary experience for analyzing farmer reaction to greater drought and intense short-term precipitation. These responses include expanded use of irrigation and related technologies, extensive water transport and drainage, introduction of new drought-tolerant crop types and cover crops, shifts to greater reliance upon groundwater to augment surface water reductions, and intensified application of nutrient fertilizers to maintain yields. At the same time, new institutional arrangements, consistent with local farmer incentives, could mitigate the losses of open access in groundwater, promote use of easements, and reduce downstream negative externalities from upstream fertilizer runoffs.

Because climate change is a global process with significant international collective action impediments to mitigation (Libecap 2014), its further unfolding is likely to be inexorable. The research included here, however, indicates that US agriculture and the food stocks, fibers, other outputs and exports, as well as related employment and viability of rural communities, are likely to be resilient. There are many margins for adaptation, and farmers have incentives to exploit them.

The studies focus on a subset of adaptation options and provide examples of possible directions available for varying farm types, regions, and water situations. Overall, the research indicates that the responses examined lead to positive changes in the performance of the agricultural sector at the region or state level analyzed either in terms of yield or net revenue. A complete benefit-cost assessment of farmer adaptation strategies, however, would include any external costs associated with new crop and seed varieties, water efficient irrigation technologies, resort to common groundwater, investment in water conveyance systems, and design and implementation of new institutional arrangements.

In the case of groundwater, where property rights are relatively complete, such as with tradable extraction rights to Southern California's Mojave Aquifer (Ayres, Meng, and Plantinga 2021), or where management institutions exist, such as in groundwater management districts in Nebraska (Edwards 2016), the losses may be minimal. Externalities are more significant where these conditions are lacking. Increased fertilizer application and associated downstream runoff are an example, and when costs are not privately internalized, fertilizer use may be excessive within a cost/benefit framework. Alternatively, where farmers adopt easements with downstream benefits, not all gains are privately captured, resulting in under-adoption. In these respects, the research can be seen as part of an emerging and critical agenda for analysis of adaptation in the agricultural sector to greater water scarcity resulting from climate change.

References

Allen, Douglas W., and Dean Lueck. 1998. "The Nature of the Farm." *Journal of Law and Economics* 41: 343–86.

Atack, Jeremy, Fred Bateman, and William N. Parker. 2000. "The Farm, the Farmer, and the Market." In *The Cambridge Economic History of the United States, Volume II, The Long Nineteenth Century*, 245–285. New York: Cambridge University Press.

Ayres, Andrew B., Eric C. Edwards, and Gary D. Libecap. 2018. "How Transaction Costs Obstruct Collective Action: The Case of California's Groundwater." *Journal of Environmental Economics and Management* 91: 46–65.

Ayres, Andrew B., Kyle C. Meng, and Andrew J. Plantinga. 2021. "Do Environmental Markets Improve on Open Access? Evidence from California Groundwater Rights." *Journal of Political Economy* 129 (10): 2817–2860.

Deschenes, Olivier, and Michael Greenstone. 2011. "Using Panel Data Models to Estimate the Economic Impacts of Climate Change on Agriculture." Chapter 7 in *Handbook on Climate Change and Agriculture*, edited by Ariel Dinar and Robert Mendelsohn. Cheltenham: Edward Elgar Publishing.

Edwards, Eric C. 2016. "What Lies Beneath? Aquifer Heterogeneity and the Economics of Groundwater Management." *Journal of the Association of Environmental and Natural Resource Economists* 3 (2): 453–91.

Edwards, Eric C., and Steven M. Smith. 2018. "The Role of Irrigation in the Development of Agriculture in the United States." *Journal of Economic History* 78 (4): 1103–141.

Edwards, Eric C., Martin Fiszbein, and Gary D. Libecap. 2022. "Property Rights to Land and Agricultural Organization: An Argentina–United States Comparison." *Journal of Law and Economics* 65 (1) Part 2: S1–S34.

Ferrie, Joseph P. 1993. "'We Are Yankeys Now:' The Economic Mobility of Two Thousand Antebellum Immigrants to the United States." *Journal of Economic History* 53: 388–91.

Garrido, Alberto, María Bielza, Dolores Rey, M. Inés Minguez, and M. Ruiz-Ramos. 2011. "Insurance as an Adaptation to Climate Variability in Agriculture." Chapter 19 in *Handbook on Climate Change and Agriculture*, edited by Ariel Dinar and Robert Mendelsohn. Cheltenham: Edward Elgar Publishing.

Gisser, Micha, and David A. Sanchez. 1980. "Competition versus Optimal Control in Groundwater Pumping." *Water Resources Research* 16 (4): 638–42.

Goldin, Claudia. 1998. "America's Graduation from High School: The Evolution and Spread of Secondary Schooling in the Twentieth Century." *Journal of Economic History* 58: 345–74.

———. 2001. "The Human-Capital Century and American Leadership: Virtues of the Past." *Journal of Economic History* 61: 263–92.

Gollin, Douglas. 2011. "Climate Change and Technological Innovation in Agriculture: Adaptation through Science." Chapter 17 in *Handbook on Climate Change and Agriculture*, edited by Ariel Dinar and Robert Mendelsohn. Cheltenham: Edward Elgar Publishing.

Griliches, Zvi. 1957. "Hybrid Corn: An Exploration in the Economics of Technological Change." *Econometrica* 25 (4): 501–22.

Hansen, Zeynep K., and Gary D. Libecap. 2004a. "The Allocation of Property Rights to Land: US Land Policy and Farm Failure in the Northern Great Plains." *Explorations in Economic History* 41: 103–29.

Hansen, Zeynep K., and Gary D. Libecap. 2004b. "Small Farms, Externalities, and the Dust Bowl of the 1930s." *Journal of Political Economy* 112 (3): 665–94.

Hansen, Zeynep K., Gary D. Libecap, and Scott E. Lowe. 2011. "Climate Variability and Water Infrastructure: Historical Experience in Western United States." In *The Economics of Climate Change: Adaptations Past and Present*, edited by Gary D. Libecap and Richard H. Steckel, 253–80. Chicago: University of Chicago Press.

Hartnett, Sean. 1991. "The Land Market on the Wisconsin Frontier: An Examination of Land Ownership Processes in Turtle and LaPrairie Townships, 1839–1890." *Agricultural History* 65: 38–77.

Hayes, Michael, Donald A. Wilhite, Mark Svoboda, and Miroslav Trnka. 2011. "Investigating the Connections between Climate Change, Drought and Agricultural Production." Chapter 5 in *Handbook on Climate Change and Agriculture*, edited by Ariel Dinar and Robert Mendelsohn. Cheltenham: Edward Elgar Publishing.

Hornbeck, Richard, and Pinar Keskin. 2014. "The Historically Evolving Impact of the Ogallala Aquifer: Agricultural Adaptation to Groundwater and Drought." *American Economic Journal: Applied Economics* 6 (1): 190–219.

Kearl, J. R., Clayne L. Pope; and Larry T. Wimmer. 1980. "Household Wealth in a Settlement Economy: Utah, 1850–1870." *Journal of Economic History* 40: 477–96.

Leonard, Bryan, and Gary D. Libecap. 2019. "Collective Action by Contract: Prior Appropriation and the Development of Irrigation in the Western United States." *Journal of Law and Economics* 62 (1): 67–115.

Libecap, Gary D., and Zeynep Kocabiyik Hansen. 2002. "'Rain Follows the Plow' and Dry-farming Doctrine: The Climate Information Problem and Homestead Failure in The Upper Great Plains, 1890–1925." *Journal of Economic History* 62 (1): 86–120.

Libecap, Gary D. 2011. "Institutional Path Dependence in Climate Adaptation: Coman's 'Some Unsettled Problems of Irrigation.'" *American Economic Review* 101: 64–80.

———. 2014. "Addressing Global Environmental Externalities: Transaction Costs Considerations." *Journal of Economic Literature* 52 (2): 1–57.

Libecap, Gary D., and Dean Lueck. 2011. "The Demarcation of Land and the Role of Coordinating Property Institutions." *Journal of Political Economy* 119 (June): 426–467.

Massetti, Emanuele, and Robert Mendelsohn. 2011. "The Impact of Climate Change on US Agriculture: A Repeated Cross-Sectional Ricardian Analysis." Chapter 8 in *Handbook on Climate Change and Agriculture*, edited by Ariel Dinar and Robert Mendelsohn. Cheltenham: Edward Elgar Publishing.

Mendelsohn, Robert, and Ariel Dinar. 2003. "Climate, Water and Agriculture, Land Economics." 79 (3): 328–41.

Mendelsohn, Robert, and Ariel Dinar. 2009. *Climate Change and Agriculture*. Cheltenham: Edward Elgar Publishing.

Novak, R., J. G. Kennen, R. W. Abele, C. F. Baschon, D. M. Carlisle, L. Glugolecki, D. M. Eignor, J. E. Flotemersch, P. Ford, J. Fowler, R. Galer, L. P. Gordon, S. E. Hansen, B. Herbold, T. E. Johnson, J. M. Johnston, C. P. Konrad, B. Leamond, and P. w. Seelbach. 2016. "Final EPA-USGS Technical Report: Protecting Aquatic Life from Effects of Hydrologic Alteration." Technical Report 822-R-156-007. U.S. Geological Survey, U.S. Environmental Protection Agency. https://www.researchgate.net/publication/311846500_Final_EPA-USGS_Technical_Report_Protecting_Aquatic_Life_from_Effects_of_Hydrologic_Alteration.

Olmstead, Alan L., and Paul W. Rhode. 1995. "Beyond the Threshold: An Analysis of the Characteristics and Behavior of Early Reaper Adopters." *Journal of Economic History* 55 (1): 17–57.

———. 2000. "The Transformation of Northern Agriculture, 1910–1990." In *The Cambridge Economic History of the United States*. Volume 3, *The Twentieth Century*, edited by Stanley L. Engerman and Robert E. Gallman, 693–742. New York: Cambridge University Press.

———. 2006. "Agriculture. Farms and Farm Structure." In *Historical Statistics of the United States, Earliest Times to the Present. Millennial Edition, Volume Four*, Part D. 4–39.

———. 2011. "Responding to Climatic Challenges: Lessons from US Agricultural Development." In *The Economics of Climate Change: Adaptations Past and Pres-*

ent, edited by Gary D. Libecap and Richard H. Steckel, 169–94. Chicago: University of Chicago Press.

Pisani, Donald J. 2002. *Water and the American Government: The Reclamation Bureau, National Water Policy and the West, 1902–1935*. Berkeley: University of California Press.

Pope, Clayne L. 2000. "Inequality in the Nineteenth Century." In *The Cambridge Economic History of the United States*. Volume 2, *The Long Nineteenth Century*, edited by Stanley L. Engerman and Robert E. Gallman, 118. New York: Cambridge University Press.

Powell, John Wesley. 1878. *Report on the Lands of the Arid Region of the United States*. Washington, DC: Government Printing Office.

Saleth, R. Maria, Ariel Dinar, and J. Aapris Frisbie. 2011. "Climate Change, Drought and Agriculture: The Role of Effective Institutions and Infrastructure." Chapter 21 in *Handbook on Climate Change and Agriculture*, edited by Ariel Dinar and Robert Mendelsohn. Cheltenham: Edward Elgar Publishing.

Schoengold, Karina, and David Zilberman. 2007. "The Economics of Water, Irrigation, and Development." *Handbook of Agricultural Economics* 3: 2933–77.

Steckel, Richard H. 1989. "Household Migration and Rural Settlement in the United States, 1850–1860." *Explorations in Economic History* 26: 190–218.

Stewart, James I. 2009. "Economic Opportunity or Hardship? The Causes of Geographic Mobility on the Agricultural Frontier, 1860–1880." *Journal of Economic History* 69: 238–68.

US Geological Survey. 2000. *Ground Water Atlas of the United States*. Reston, VA: Office of Ground Water, US Geological Survey.

Stubbs, Megan. 2016. "Irrigation in US Agriculture: On-Farm Technologies and Best Management Practices." Congressional Research Service 7–5700 R44158. https://sgp.fas.org/crs/misc/R44158.pdf.

Sutch, Richard. 2011. "The Impact of the 1936 Corn Belt Drought on American Farmers' Adoption of Hybrid Corn." In *The Economics of Climate Change: Adaptations Past and Present*, edited by Gary D. Libecap and Richard H. Steckel, 195–223. Chicago: University of Chicago Press.

Wahl, Richard W. 1989. *Markets for Federal Water: Subsidies, Property Rights, and the Bureau of Reclamation*. Washington, DC: Resources for the Future.

Wilcox, Walter F. 1929. *International Migrations, Volume I: Statistics*. New York: National Bureau of Economic Research.

1

The Economics of Climatic Adaptation
Agricultural Drainage in the United States

Eric C. Edwards and Walter N. Thurman

1.1 Introduction

Climatic shocks like heat stress and drought can dramatically reduce agricultural yields (Ortiz-Bobea et al. 2019). In the western United States, anthropogenic climate change is expected to increase the frequency of high temperature days and drought (Strzepek et al. 2010). Accordingly, there has been considerable focus on the role irrigation will play in adaptation (Schlenker, Hanemann, and Fisher 2005; Smith and Edwards 2021). In the eastern United States, however, projections of climate change suggest a need to adapt to more precipitation, not less (Rosenzweig et al. 2002).

Water is a key agricultural input, but its marginal product depends on local availability. In water-scarce areas, the marginal product of water is high and investments are made in irrigation. In water-abundant areas, the marginal product of water can be negative, especially at certain times of the year.[1]

Eric C. Edwards is an assistant professor of agricultural and resource economics at North Carolina State University.

Walter N. Thurman is the William Neal Reynolds Professor Emeritus in the Agricultural and Resource Economics Department at North Carolina State University.

The authors thank Carl Kitchens, Karina Schoengold, David Hennessy, Michael Castellano, Dale Hoover, and Bill Eagles, as well as participants at the 2022 ASSA Meetings, the 2022 Cliometric Conference, the 2022 NBER Summer Institute, and the NBER/USDA Conference on Economic Perspectives on Water Resources, Climate Change, and Agricultural Sustainability for their helpful comments. For acknowledgments, sources of research support, and disclosure of the authors' material financial relationships, if any, please see https://www.nber.org/books-and-chapters/american-agriculture-water-resources-and-climate-change/economics-climatic-adaptation-agricultural-drainage-united-states.

1. In irrigated areas, drainage is also used to reduce the salinization of soil that results from long-term irrigation. Irrigation increases evapotranspiration, which can effectively wick salts at deeper soil levels upward into the root zone. Drainage flushes salts from the root zone (Castellano et al. 2019).

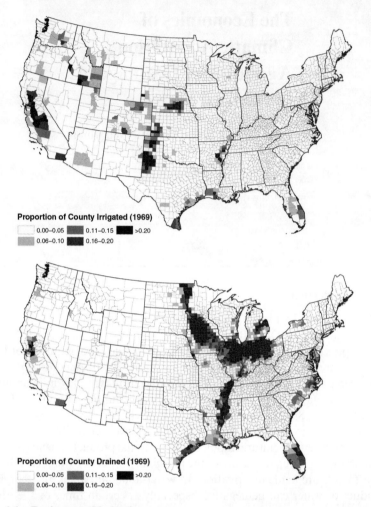

Fig. 1.1   Drainage and irrigation

Note: Percentage of county area irrigated (top) or drained (bottom) from the 1969 Census of Agriculture. Counties are scaled to 1910 borders for consistency with our later analysis.

In wet and poorly drained soils, excess water in the root zone of cultivated crops can create waterlogging, preventing the absorption of oxygen and drastically reducing yields or killing the plants entirely. Wet soils prevent field access by heavy equipment like tractors, limiting the ability of farmers to plant crops at optimal times during wet years. To address issues of excess soil moisture, water tables can be artificially lowered via within-soil flow if nearby drainage provides a pathway for water out of the plant root zone. Figure 1.1 shows the relative intensity of irrigation and drainage across the

US historically, illustrating adaptation to water scarcity in the West and excess water in the East.

Across the US, all regions are projected to see heavier rainfall events under climate change, even those regions that see less precipitation overall (Easterling et al. 2017). The left panel of figure 1.2 shows the change in precipitation for the period 1987–2007 relative to the earlier part of the 20th century (1900–1992) and shows significant increases in precipitation for most areas east of the Mississippi. Most field crops experience no long-term damage from rainfall events if water tables fall to 6 inches below ground surface within one day and continue falling to 18 inches from ground surface within three days (Hofstrand et al. 2010). The ability of farmers to rapidly remove excess water from fields will be crucial to ensuring a secure and reliable food supply.

The current location and intensity of agricultural drainage remains similar to that shown in figure 1.1 and was largely the result of the dramatic investment of 19th- and early 20th-century farmers in the Midwest and eastern US in technologies to move water off saturated lands. (See for example Bogue 1951, 1963.) In fact, a significant portion of the eastern US, including the upper Midwest, Mississippi River basin, and eastern Coastal Plain, would not be suitable for agricultural production absent such drainage. The physical fact of extensive tiled acreage in the Midwest is an important determinant of the costs of future adaptation that involves drainage.

The right panel of figure 1.2 shows a similar map for changes in average temperature. The Dakotas, Wisconsin, and northern Minnesota have already seen increased temperatures. As climate change shifts growing regions north, we expect investment in drainage to occur in currently swampy areas. Looking forward to changing climatic conditions, agronomists, engineers, and economists are discussing the technical issues associated with intensifying drainage to deal with increasing precipitation and shifting growing regions while, at the same time, addressing the environmental costs associated with drainage, including nutrient runoff and reduced wetland acreage (Castellano et al. 2019).

Historical context on drainage is needed to understand potential future climate adaptation. The locations where investments in drainage tiling, field leveling, and deswamping have been made are inherited from historic decisions—starting as early as the late 1800s. Further, adaptation is not just a technical problem about plant physiology and hydrology. It is an economic problem with a complex set of priced and unpriced costs and benefits, with important externalities and transaction costs into which historic analysis can provide key insight.

Using data from the agricultural census on improved agricultural acres and farm value spanning the period of agricultural drainage across the eastern US, we compare counties with poorly drained soils to those that

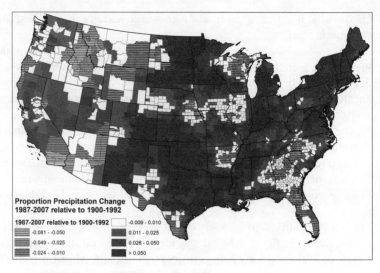

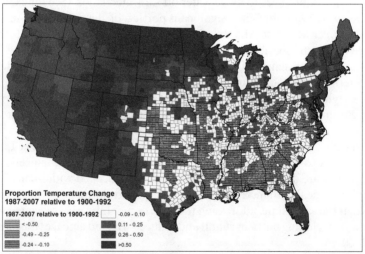

Fig. 1.2   Recent changes in temperature and precipitation
Note: Change in the average 1987–2007 precipitation (top) and temperature (bottom) relative to the 1900–1992 average using PRISM data: PRISM Climate Group, Oregon State University, https://prism.oregonstate.edu.

were well drained within the same state. We explain how economic factors determined the timing and locations of drainage adoption and tie these factors to lessons for modern adaptation to climate change. We provide a framework for understanding benefits and costs of drainage in historical context and describe the historical performance of drainage.

Agricultural drainage in the eastern US is in many ways analogous to irrigation in the West, in that both are fixed investments in location specific assets that greatly increase agricultural production or reduce variability in production due to climatic shocks. In the modern day, both have important implications for farm adaptability to a changing climate. Changing patterns of precipitation and shifting growing regions could make agricultural production less suitable in the locations that in the past have made drainage investments, opening the possibility of new investment in drainage elsewhere. Meyer and Keiser (2018) project increased tile drainage in the northern Midwest—North Dakota, Minnesota, Michigan, and Wisconsin—and suggest that absent drainage, losses due to climate change will be larger.

Today, agricultural drainage is perceived to have had much higher social costs than initially anticipated, both in terms of reducing the amount of wetlands and providing a more direct conduit for agricultural nutrients to make their way into rivers, lakes, and oceans. The region projected to see increases in drainage would undoubtedly include important undeveloped wetlands, such as the Prairie Pothole Region (Dahl 2014). New institutional innovations that coordinate over wetland protection and nutrient pollution, as well as agricultural production, can enhance the net value of such an adaptation.

1.2   Empirical Setting

1.2.1   US Drainage History

Open ditches for drainage were dug throughout the US from its founding. The earliest attempts at drainage in the Midwest, in 1818, were of this type (Prince 2008, 205). Ditches, typically three to five feet deep, were labor intensive, and because they bisected fields at regular intervals, they reduced the available land surface area and made planting and harvesting difficult. The technology that ultimately replaced open ditches was the laying of drain tile. Installing drain tile involved digging a trench in which flat clay tiles were laid end to end and covered with a second, inverted-V, layer of tile, creating a porous water channel. The tile was covered again with soil. The resulting subterranean channel drained water above it down to its level, typically four feet below the surface. Unlike open ditching, installed tile drainage was invisible and allowed farming above it.[2] The advent and diffusion of clay drain tile, first used in the US in Seneca County, New York, in 1835, transformed American agriculture (McCrory 1928).

The natural wetlands of the US were viewed by federal government policy as "unproductive and an economic waste" until at least 1956 (Palmer 1915; McCorvic and Lant 1993). To encourage their development via drainage,

2. Modern land drainage follows the same principle, but involves the burying of perforated, corrugated, plastic tubing using advanced drilling and trenching machines. While still called "tile drainage," the technology bears little superficial resemblance to its ancestor, and no longer involves clay tile.

Table 1.1   **Swamp Land Acts**

Year	State	Acres
1849	Louisiana	9,493,456
1850	Alabama	441,289
	Arkansas	7,686,575
	California	2,192,875
	Florida	20,325,013
	Illinois	1,460,184
	Indiana	1,259,231
	Iowa	1,196,392
	Michigan	5,680,310
	Mississippi	3,347,860
	Missouri	3,432,481
	Ohio	26,372
	Wisconsin	3,360,786
1860	Minnesota	4,706,503
	Oregon	286,108
TOTAL		84,895,415

Source: Fretwell (1996).

Congress passed a series of Swamp Land Acts in 1849, 1850, and 1860 that granted wetlands to the states. The lands made available under the acts are shown in table 1.1. The acts were a first step, but the land still required investment for reclamation. The initial belief that states would simply use land sale funds to finance drainage proved incorrect because the funds raised were insufficient and because state governments were disinclined to fund these types of public works. Responsibility for the investment in reclamation passed from the states to counties, which subsequently divested the lands in the hopes that private investors would drain them (Prince 2008). The task of improving drainage fell to individual landowners and was achieved over time through local investment, private and public, not federal or state support.

1.2.2 Classifying Drained Land

The focus of our analysis is on 24 states in which deposits of soil created flat areas with relatively poor drainage. Palmer (1915) articulated two categories of such areas, "glacial swamps" and "tidewater or delta overflowed lands." We follow by segmenting states into "Coastal Plain" and "Midwest Tile" categories, as shown in figure 1.3. Coastal Plain states follow the definition of the Atlantic Coastal and Mississippi River alluvial plain in the map created by Fenneman and Johnson (1946): Virginia, North Carolina, South Carolina, Georgia, Florida, Mississippi, Louisiana, Texas, Tennessee, and Arkansas. Poor natural soil drainage in these states results from sedimentary deposits from rivers near the coast and around the Mississippi River.

The area of glacial swamps described by Palmer (1915) coincides with

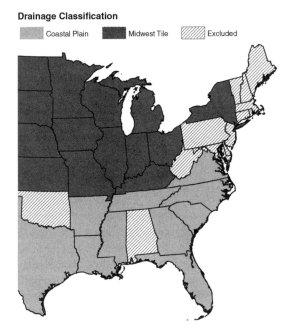

Drainage Classification

Coastal Plain   Midwest Tile   Excluded

Fig. 1.3   **Drainage state classification**

the areas subject to glaciation during the Pleistocene, where the retreating ice sheets left flat soil deposits. This area roughly coincides with the Upper Midwest, and our definition of "Midwest Tile Drainage" includes North and South Dakota, Nebraska, Iowa, Minnesota, Wisconsin, Illinois, Indiana, Michigan, and Ohio. To this list we add Kentucky, Missouri, and Kansas, portions of which share drainage similarities with these states, and New York, which was the initial location of tile drainage in the US and shares a similar geologic history.[3]

1.2.3   Public and Private Benefits and Costs

Laying tiles constituted a private investment in agricultural production, and the incentives to invest were meaningful. Prince (2008) suggests that in the Upper Midwest prior to 1880, unimproved wetland sold for an average of $7 per acre (ranging from $2–$12), but that the sale price once drained could increase by a factor of five. Such investment, however, was generally not effective on a small scale. Drainage projects required coordination across hundreds or thousands of acres as well as new ditches, levees, and embank-

3. Excluded states may lack much need for agricultural drainage, or in the case of New Jersey, Maryland, and Delaware, differ from most other states in the institutions managing drainage. See Edwards and Thurman (2022) for additional discussion of drainage categorization.

ments on private lands (Wright 1907; Prince 2008). Thus, drainage incurred not only the costs of construction and maintenance, but also the contracting costs associated with collective investment and action.

Because drainage investment was generally not effective on a small scale, coordination was essential. Institutional innovation in the form of drainage districts, enabled by state law, reduced the costs of coordination by facilitating, and compelling, coordination. Although they varied in specifics, drainage districts were generally legislated to be formed via a petition from landowners residing in a region, then requiring a vote by the majority of land area and/or landowners (McCorvie and Lant 1993). District decisions were typically made by locally elected boards. Their power was restricted to investments that met some definition of public benefit, which courts often interpreted as requiring public health benefits (Prince 2008). Another key feature of the districts was financial, with districts able to issue low-interest bonds to secure cash for investment (McCrory 1928).

Additional benefits of drainage in the late 19th and early 20th centuries accrued due to the improved health outcomes of surrounding communities through the reduction of malarial infections.[4] Throughout the 1800s, malaria affected most of the populated regions of the US, was one of the country's leading causes of death (in 1850 45.7 of every 1,000 deaths were caused by malarial fevers), and had long-term health impacts including stunting and chronic conditions in later life (Hong 2007). The reduction of malaria was cited explicitly as the key benefit that warranted governmental intervention to create drainage legislation.

At the time of initial construction of drainage in the late 19th century, environmental costs were not well understood or relevant to farmers, but they have become important today. Of the 215 million acres of wetlands estimated to have existed in the contiguous US at colonization, 124 million have been drained, 80–87 percent for agricultural purposes. Tile systems and associated drainage canals increase the transport of nitrogen and phosphorous from inland farms to waterways, leading to environmental damage. The leading example is the Mississippi water basin and the hypoxic zone in the Gulf of Mexico.

Some benefits of drainage were also not well understood and diminished over time. Drainage increases the availability of soil nitrogen and reduces the optimal application of N-containing fertilizer. Once drained, soils sponsor greater microbial activity, which decomposes organic matter already in the soil. This releases nitrogen into inorganic forms, which are then available for uptake by crops. The initial increase in microbial decomposition of soil organic matter effectively mines the finite amount of organic matter in the soil to begin with, setting in motion a dynamic adjustment in available nitro-

4. Malaria had been linked to marshy areas through the 19th century and was definitively linked to mosquitoes near the end of the 19th century.

gen that reduces benefits in future years. Eventually a lower steady state of available N is established in the drained soil, with farmers relying more on the application of N-containing fertilizer than they did immediately after the installation of drain tile. Castellano et al. (2019) say that the time it takes for soils to reach the new post-drainage steady state in the Midwest is on the order of 15 to 30 years.

Although all 24 states in our sample (see figure 1.3) adopted similar drainage district laws, there were important differences in the development of drainage between the glaciated Midwest and the alluvial coastal plain. Drain tile was well suited for use across the glaciated regions but was not as successful on the coastal plains, where the need for additional investment in levees and pumping, as well as other challenges, limited its effectiveness. Coastal Plain states developed drainage using a combination of in-field ditching, levee systems, pump houses, and tile in select areas.

The magnitude of the drainage investment problem in the glaciated (Midwest Tile) regions is illustrated by Blue Earth County, Minnesota, and Story County, Iowa. Blue Earth County saw 92 districts form between 1898 and 1952, with around 100,000 acres in drainage districts in total in the county (Burns 1954). Districts ranged in size from 320 to 7,202 acres, with an average district consisting of around 1,200 acres and 20 farms. Similarly, in Story County there were 95 districts by 1920 with around 60 percent of the county in districts (Hewes and Frandson 1952). The districts average around 2,100 acres and 20 farms.

In the Coastal Plain states, single drainage districts require coordination over areas orders of magnitude larger than the glaciated regions, and they are larger and fewer in number. The Ross Drainage District in Arkansas is around 40,000 acres (Deaton 2016), and the Cypress Creek District in Arkansas is around 285,000 acres, the later facing decades of litigation resulting from opposition to its formation (Harrison and Kollmorgen 1948). In 1920, North Carolina had 81 districts draining 543,000 acres, making the average district size about 6,700 acres (O'Driscoll 2012). These districts were already substantially larger than those in the glaciated regions, and by 1985 drained acres had increased by an order of magnitude to 5.4M acres, but the number of districts had decreased to 53. How regional differences might continue to be important in understanding response to climate change remains an outstanding question.

Drainage districts provided local landowners with the tools to undertake collective investment suggested by Olson (1965) in a form consistent with the nested structure described by Ostrom (1990). Drainage district laws provided sufficient legal structure to coordinate investment in drainage infrastructure, through local taxing authority. In addition to facilitating public investment, eminent domain authority solved the problem of neighbors preventing drainage onto or across their land. Bogue (1951) describes "violent opposition" from neighboring landowners to drainage projects in Illinois, but under drainage district law these types of issues were resolved in the courts and generally in favor of the public good, i.e., draining land.[5]

1.2.4   Trial and Error

The historic drainage experience suggests trial and error will play a large role in future climate adaptation (more generally see Ridley 2020). Institutions needed time to adapt and evolve, and property rights were refined over time through case law (see Tovar 2020). Successes and failures were not obvious prior to investment. The historical record offers several examples of drainage failures—projects deemed ex ante to be profitable, but ex post proven to be not. Two large-scale examples come from physiographically distinct areas—the marshy region of central Wisconsin and Lake Mattamuskeet in the coastal plain of North Carolina.

Prince (1995) develops the chronology for Wisconsin. Beginning around 1900, soon after the period of drainage district formation and successful drainage in nearby Midwestern states, Wisconsin marsh lands were organized into drainage districts. The attempts to improve agricultural land through drainage followed by several decades the economically successful efforts in Iowa and Illinois, but they failed in central Wisconsin. By the 1930s, lands that had been drained largely reverted to public ownership and became recreational havens. In a section titled "A Chronicle of Repeated Failure," Prince places Wisconsin in the context of a global history of land reclamation and drainage:

> Most accounts of changing geographies of marshlands have happy endings. The reclamation of the Dutch polders, the draining of the English fenlands and Somerset levels, the draining of the wet prairies in Ohio, Indiana, Illinois, and Iowa were success stories. The marshland chronicle of central Wisconsin is unusual in that it records a succession of short-lived unsuccessful economic ventures. (page 17)

Another example of project failure and reversion of drained land to its prior state is Lake Mattamuskeet, the largest natural lake in North Carolina. The shallow freshwater bay near the Atlantic coast drains into the Pamlico Sound. In many ways, the 19th- and 20th-century history of the lake parallels that of central Wisconsin.

By the mid-19th century, the fertility of the area surrounding Lake Mattamuskeet suggested that draining the several-foot-deep lake could create thousands of acres of productive land. Partial draining of the lake took place as early as 1837. Larger-scale draining took place in the early 20th

5. From a modern governance perspective, a drainage district is one of many examples of the special district, commonplace today and encompassing varied responsibilities that include mosquito abatement and the operation of airports, mass transit, and libraries. Special districts allow landowners to retain rights to operate their properties at the scale and for the purposes that economic factors dictate, while ceding one property right "stick" to a local elected body.

century (see Forrest 1999). Private investors fully drained the lake in 1916 by dredging 130 miles of canals and building water control dams and a large coal-fired pumping station. Soon afterward the investing firm failed, due in part to low commodity prices, and the pumping station was abandoned. The lake refilled. Twice more between 1916 and 1926, the lake was drained but then abandoned and allowed to refill. The privately owned lake eventually was sold to the federal government in 1934 to become the Lake Mattamuskeet Wildlife Refuge.

Just as Prince (1995) casts the history of central Wisconsin as a series of failed attempts to claim partially submerged land for agriculture, Lake Mattamuskeet followed a similar trajectory. In both cases, the land ultimately reverted to its initial state and the primary land use became recreation, hunting, and fishing. The similar timing of drainage efforts in Wisconsin and North Carolina suggests that drainage was not simply a technology waiting for discovery and then application to different landforms; it was also a product of the state of knowledge about drainage and land use at certain times, conditions in agricultural markets at those times, and possibly a contagion of ideas deemed at the time to be suitable and promising by investors.

Like investment in agricultural production generally, the development of drainage was shaped by the fertility and climate of each county, as well as changes in input and output prices. For instance, the panic of 1873 and subsequent fall in farm prices would have reduced demand for drainage, while emerging transportation networks would have lowered the cost of moving tile, increasing the cost effectiveness of drainage investment.

1.3   Data and Empirical Strategy

We construct a 109-year panel from 1850–1969 on *Improved Acres* and *Total Farm Value* from United States Censuses of Agriculture collected once per decade and digitized by Haines, Fishback, and Rhode (2019).[6] We focus on counties east of the 100th meridian, generally the dividing point between the humid and semiarid portions of the US. Areas east of this line can be farmed without irrigation and were generally settled or being settled during the entire panel.[7] We scale county data to 1910 county boundaries using area-weight crosswalks constructed by Ferrara, Testa, and Zhou (2021). The USDA also conducted drainage censuses in 1920, 1930, and 1969, which recorded the number of *Drained Acres* in a county. We construct measures

6. After 1920 Improved Acres is not provided as a standalone variable. We construct this measure using other available variables after 1920 (see Edwards and Smith 2018 for details).

7. The removal of Indigenous groups from these states generally preceded drainage by several decades or more, starting with the Indian Removal Act of 1830.

of *Percent of County Improved* and *Percent of County Drained* by dividing by total county area.

We use a Soil Drainage Index (DI) to represent the natural wetness of soil in a given county (Schaetzl et al. 2009). The DI is an ordinal measure of long-term soil wetness ranging from 0 to 99. Soils with a DI of around 60 are generally termed "somewhat poorly drained," while higher DI values represent more poorly drained up to 99, open water. The DI is derived from the soil classification and slope and so is not affected by investment in drainage or irrigation.[8]

Using a 240-meter cell resolution raster, we extract the median DI value for each county. We then categorize poorly drained counties in two ways. First, we create a variable *High DI* (poor natural drainage), for counties with a median DI greater than 60. Second, within each state we can separate median DI into *DI Quartile*, and then can compare outcomes from Q4 (poorly drained) to Q1 (well-drained). Figure 1.4 shows the original drainage index raster (top left) as well as our two measures (bottom panels).

The figure also shows a measure of land productivity, the Soil Productivity Index (PI), developed by Schaetzl, Krist Jr., and Miller (2012). The PI is an ordinal measure of how advantageous the soil is to crop production based on soil taxonomy. The index ranges from 0 to 19, with 19 being the most productive. Because soil PI is correlated with DI, and PI is also potentially affected by practices like drainage, we generally do not include PI as a control, although our results are not driven by its inclusion or exclusion.[9] A comparison of the top panels of figure 1.4 demonstrates that lands with a high drainage index value are more productive once drained, on average, than low DI land.

Figure 1.5 shows the relationship between median DI and the observed percent of a county drained in 1969 by quartile. The DI=60 cutoff is shown on each plot. The two measures of poor drainage are highly correlated. There are no *High DI* counties in Q1 while most counties in Q4 are classified as *High DI*. The *DI Quartile* measure ensures a more balanced comparison within each state. Within poorly drained states, however, the comparison may be between counties that are, in absolute terms, all poorly drained.

Finally, we collect USDA Agricultural Census data from 2017 on land

8. A soil's taxonomic classification is not initially affected by on-farm investments like irrigation or artificial drainage and so the DI does not change unless these investments change the classification of the soil in the long run. "Instead, the DI reflects the soil's *natural* wetness condition. Each soil *series* has, in theory, its own unique DI." (Schaetzl et al. 2009).

9. "Soil productivity can be easily and rapidly amended by human activities. Thus, no index of productivity can accurately assess current soil productivity where soils have had a long history of cropping, erosion, and/or additions of soil amendments. Particularly, irrigation and drainage practices impact soil fertility/productivity and, therefore, any index of productivity is only an estimate; it is always affected by land-use practices, both current and those in the past. Thus, we focus on natural native soil productivity, as expressed in a soil's taxonomic classification and recognize that such an estimate is, at best, a good starting point." (Schaetzl, Krist Jr., and Miller 2012).

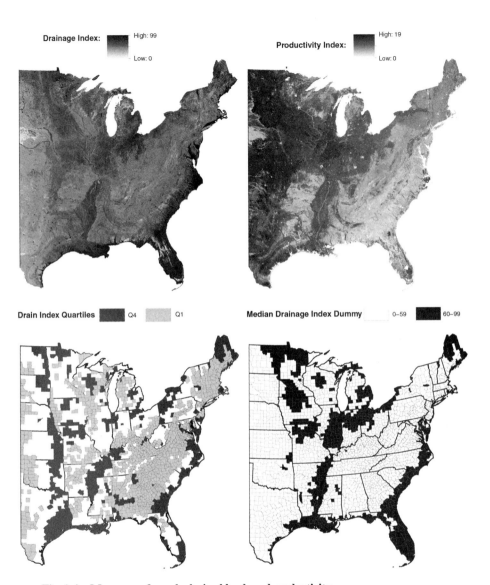

Fig. 1.4   Measures of poorly drained lands and productivity

Note: The top panels show the Drainage Index and Productivity Index rasters used to create county-level measures. The bottom left panel shows the *High Drainage* variable, which is the counties with median drainage index greater than 60. The bottom right panel shows the best and worst drained quartile of counties in each state.

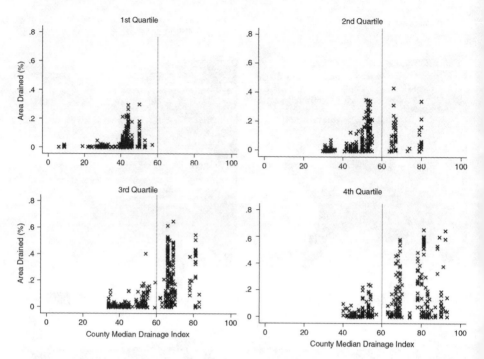

Fig. 1.5   Drainage index and drainage

Note: This figure depicts, for each county in our sample, the relationship between the median drainage index and the percent of county area drained in 1969, separating the data into state-specific drainage quartiles.

production practices from the National Agricultural Statistics Service Quick Stats. These data provide area (in acres) drained by artificial ditches and area drained by tile as two separate categories. To get total drainage acres we add these together.

1.4   Results

In this section we examine the contribution of drainage to agricultural production in the US east of the 100th meridian. Conditional summary statistics are provided in table 1.2. This table offers a comparison of area with high potential need for drainage relative to others in 1880 and 1920.

We begin by examining outcomes across states in our two groups, Midwest Tile and Coastal Plain, to provide a comparison of counties likely to be treated with drainage relative to others. We regress two variables, percentage of a county with improved agricultural land and total county farm value (logged) on a flexible set of controls—state by year and county fixed effects—and then group counties in each state by drainage index quartile.

Table 1.2   **Conditional summary statistics**

Variable	Drainage Index < 60		Drainage Index > 60	
	1880	1920	1880	1920
Total Value in Farms (2020$ millions)	103.68	265.91	124.02	413.22
	(149.00)	(277.84)	(144.60)	(403.80)
Farm Value per Acre (2020$)	448	1,054	496	1,708
	(844)	(1,714)	(405)	(6,284)
Pct. of County Improved	0.34	0.45	0.36	0.51
	(0.23)	(0.22)	(0.29)	(0.30)
Total Farms	1,717	2,405	1,705	2,410
	(1,202)	(1,343)	(1,300)	(1,510)
Total Acres in Farms	225,923	279,310	207,490	263,111
	(127,343)	(157,627)	(129,705)	(149,661)
Median Drainage Index	43.65		72.32	
	(6.38)		(7.75)	
Median Productivity Index	8.29		10.25	
	(3.96)		(3.41)	

Note: Summary statistics conditional on high/low drainage index (DI > 60) for 1880 and 1920. All values are the mean value of all the counties in that treatment status for the variable described on the left. Standard deviations are reported in parentheses.

We exclude Q2 and Q3 and then plot the yearly mean for each quartile group. Comparing the best- and worst-drained quartiles shows the changing trends over time.

The Midwest Tile states generally drained land via small districts, on the order of around 1,000 acres with around 20 farms. Coastal Plain districts were orders of magnitude larger and thus required more coordination. Because the transaction and overall costs of forming and draining larger areas are higher, we expect the Midwest Tile states to begin draining first. The top panels of figure 1.6 show how the percentage of a county in improved acres changes over time for Q1 and Q4 counties. After controlling for state by year and county specific variation, counties in Q4 see less improved agricultural land than counties in Q4, in both the Midwest Tile states (left) and Coastal Plain states (right). By 1900 in the Midwest Tile states, Q4 counties have as much improved land as Q1 counties, which we attribute to the rapid rollout of drain tile. In contrast, Coastal Plain states do not see success in drainage, bringing high-DI counties to parity in improved acres until 1940.

The bottom panels use the same approach with agricultural land value per acre as the variable of interest. Midwest Tile states see Q4 land values initially far below Q1 counties, but increase as buyers and sellers begin to anticipate drainage success. By 1880—a time when tile was being successfully deployed and land buyers could reasonably be assured that drainage could successfully improve their poorly drained lands—Q4 and Q1 counties

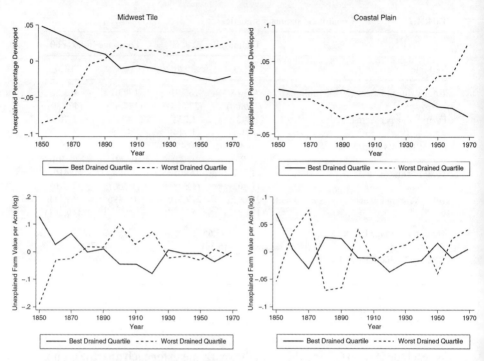

Fig. 1.6   Drainage quartile comparisons

Note: This figure depicts the unexplained variation, the residuals of a regression on county fixed effects and state-by-year fixed effects, of percentage of county improved (top panels) and total farm value (bottom panels) for counties in the Midwest (left panels) and Coastal Plain (right panels). Outcomes are bifurcated by drainage index quartile, Q4 relative to Q1.

have similar per acre land values. Thus capital markets appear to anticipate on-the-ground improvements. The story is much less clear for Coastal Tile states, with fluctuations in the relative prices of land in Q1 and Q4 counties. We attribute some of this to the uncertainty surrounding drainage in these states. Land markets may have alternatingly anticipated successful and unsuccessful drainage over time, while on-the-ground implementation ultimately took longer.

We perform a similar analysis using our alternative measure of *High DI*, and these results are shown in figure 1.7. The results are quite similar to figure 1.6, with *High DI* counties having improved acreage levels similar to low DI counties by 1900 for Midwest Tile states, and around 1940 for Coastal Plain states. The per acre land value trends, shown in the bottom panels, are clearer using the *High DI* measure. For Midwest Tile states, per acre land value for high- and low-DI counties is equal by around 1890, and by about 1900 for Coastal Plain states.

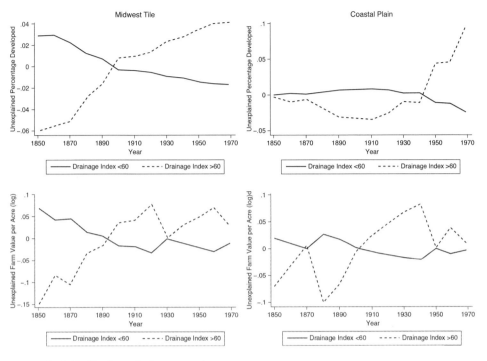

Fig. 1.7   Drainage index comparisons

Note: This figure depicts the unexplained variation, the residuals of a regression on county fixed effects and state-by-year fixed effects, of percentage of county improved (top panels) and total farm value (bottom panels) for counties in the Midwest (left panels) and Coastal Plain (right panels). Outcomes are bifurcated by drainage index.

1.5   Lessons for Future Adaptation

Our historical analysis offers insight into the real barriers associated with institutional change for drainage. A key conclusion of our work is that institutional development to solve drainage coordination problems, including trial and error and legislative revision, took time. The problem of adaptation via drainage was not known in advance and was discovered over time through the development process. The particular circumstances of time and place as in Hayek (1945) played an important role in adaptation. It took time for local agents to discover and learn the areas that needed to be drained, what technology to apply, and to coordinate on a solution. Hewes and Frandson (1952) discuss the evolution in Iowa:

> Within Story County, the pattern of small, discontinuous wet tracts intermingled with well drained land is the general rule except in the northeastern one-half of Lafayette Township, where the one extensive continuous poorly-drained prairie portion of the county is found. Although as an

early settler put it, 'only the higher laying lands could be broken, wet prairie land was necessarily included in most prairie farms.' The wet areas, if used at all, served for pasture or wild hay, or for open range grazing into the 1880's.

Projected increases in temperature and precipitation suggest that Midwestern agriculture will shift north, likely leaving behind some of the previously dominant Corn Belt.[10] The sequence of 19th-century American settlement revealed the important coincidence of wet and fertile soils in the Midwest and the resulting importance of land drainage technology. More extensive agricultural development in northern areas like North Dakota, Minnesota, Michigan, and Wisconsin will similarly incentivize land drainage (Meyer and Keiser 2018). It is useful to consider what can be learned from the settlement of the Corn Belt (read Drainage Belt) and how the context of investment in drainage is and will be different from what it was historically.

From a strictly agricultural perspective, the opportunity to drain wet land and increase its productivity is a welcome aid to climate adaptation. Opportunities to drain and improve soil productivity are likely to be broadly similar to those taken advantage of by earlier generations of farmers. Further, drainage technology, like virtually all agricultural technology, has dramatically improved. Terra-cotta tile is no longer the foundation of drainage, having long since been replaced by corrugated PVC pipe, installed with modern trenching machinery. Market incentives can be expected to lead to extensive investment in drainage of wet, but productive, farmland. Institutions such as the drainage district are well suited to coordinate communities of farmers with related interests in drainage.

Different now from then—beyond technology—are the perceived external effects of draining land. There are two main categories: loss of wetland ecosystems (and wildlife habitat) and off-farm transport of nutrients. With respect to wildlife, for example, the benefits from seasonal habitat for migratory ducks extend well beyond the borders of individual farms. But especially far reaching are the effects of nutrient runoff, abetted by the drainage systems put in place over the past 150 years. Nitrogen transported from Midwestern farms—all in the Mississippi watershed—collect in the Gulf of Mexico, creating a hypoxic dead zone (see Dale et al. 2010). External effects in both categories extend beyond the boundaries of farms as well as existing drainage districts, or any districts devised to deal with agricultural production.

There are both regulatory and market approaches to incorporating offsite effects into the calculations of farmers investing in drainage. Direct regu-

10. Changes in climate will not make North Dakota agronomically identical to Illinois because day lengths still will differ. See Olmstead and Rhode (2011) on adoption of wheat varieties from similar latitudes.

lation of land use, including drain tile installation, is a natural regulatory suggestion. Private mechanisms are available as well. Conservation easements placed on farms restrict future development and specify management methods, including drainage measures, that promote wildlife habitat and other off-site environmental benefits. Parker and Thurman (2019) discuss how private land trusts serve to aggregate public demands for environmental goods and incentivize their provision through the easement mechanism. Government policy plays a significant role in this private provision by granting substantial tax benefits to landowners who restrict their rights with easements (see Parker and Thurman 2018).

The path of human ingenuity to adapt to changing conditions is impossible to predict. We can say, though, that both institutional and technical innovation will play important roles. We argue here and elsewhere (Edwards and Thurman 2022) that institutional innovation in the form of drainage districts was vital to 19th-century agricultural development. Further institutional innovation will be key if we are to effectively discover and account for the full spectrum of benefits and costs to future adaptation. Technological innovation will continue to play an important role. Castellano et al. (2019) discuss and promote advanced drainage technologies that allow improvements in agricultural value at lower external cost.

Finally, in addition to the land use changes likely to result from Corn Belt agriculture moving north, important broad scale changes are likely to result in lands left behind—lands now less suitable to the types of agriculture relied on in the past. Much of this land is already drained, and the future private and public benefits from its use will be conditioned by the existing network of drain tile and ditches.

Figure 1.8 looks at the changes in agricultural drainage made over the last 50 years, from the last year in our data analysis, 1969, to 2017. Intensive investment in drainage continued to be made in Minnesota, Iowa, Illinois, the Dakotas, and Michigan. This region includes important undeveloped wetlands, such as the Prairie Pothole Region, where the use of agricultural subsurface drainage systems continues to increase (Tangen and Wiltermuth 2018). Meyer and Keiser (2018) project that under climate change, states in the northern Midwest—the eastern Dakotas and northern areas of Minnesota, Michigan, and Wisconsin—will see additional drainage development, continuing the trends that have emerged over the last 50 years.

In contrast, the Coastal Plains states have seen relative declines in the amount of drainage. Florida, Texas, Louisiana, South Carolina, and Mississippi all have numerous counties where share of drainage has decreased by more than 2.5 percentage points. This distinct difference between glaciated and plain areas suggests the importance of understanding the history of the challenges facing drainage enterprises. The costs and benefits of future adaptation via drainage are likely to be heterogeneous across space. Where coordination among large numbers of landowners is needed, transaction

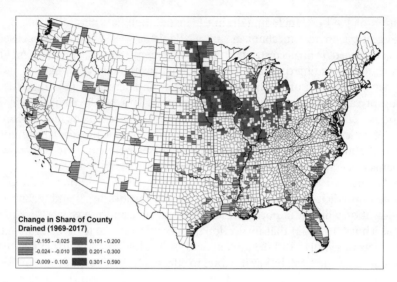

Fig. 1.8   Change in drainage

costs create challenges to adaptation that cannot be assumed away or solved ex ante.

While we can learn about how to solve drainage coordination problems of adjacent landowners from historical experience, the coordination problems involved in adapting to climate change will be different. One important difference is that third-party interests in nutrient runoff and wetland protection should, and will, influence the drainage solutions that groups of landowners will select as they adapt. Today, agricultural drainage is perceived to have had high external costs due to the reduction in wetland acreage (McCorvie and Lant 1993), an increase in water pollution and sedimentation (Skaggs, Brevé, and Gilliam 1994), and degradation of soil quality (Castellano et al. 2019). New institutional innovations that address these impacts as well as agricultural production will need to emerge.

1.6   Conclusion

In this paper, we document and analyze how local drainage enterprises invested in tile drainage, ditches, and drainage works. After federal and state funding for these projects failed to materialize, drainage management districts formed to locally finance drainage investment over wetlands spanning thousands to hundreds of thousands of acres. Of the 215 million acres of wetlands estimated to have existed in the contiguous US at colonization, today 124 million have been drained.

While modern drain tile is no longer baked clay, much original drain tile is still in place, representing an important long-term investment in agricultural adaptation. Meyer and Keiser (2018) suggest that this adaptation was related to climate, with the probability of adopting tile drainage increasing with precipitation. Once installed, drainage eliminates the inverted "U" relationship between precipitation and land values, making areas with excess precipitation more profitable to farm. For the eastern US, drainage is perhaps the most important climatic adaptation, despite its low profile, perhaps as a result of being buried out of sight.

References

Bogue, A. G. 1963. *From Prairie to Corn Belt: Farming on the Illinois and Iowa Prairies in the Nineteenth Century*. Chicago: University of Chicago Press, 310.

Bogue, M. B. 1951. "The Swamp Land Act and Wet Land Utilization in Illinois, 1850–1890." *Agricultural History* 25 (4): 169–80.

Burns, B. E. 1954. "*Artificial Drainage in Blue Earth County, Minnesota.*" PhD thesis, University of Nebraska-Lincoln.

Castellano, M. J., S. V. Archontoulis, M. J. Helmers, H. J. Poffenbarger, and J. Six. 2019. "Sustainable Intensification of Agricultural Drainage." *Nature Sustainability* 2 (10): 914–21.

Dahl, T. E. 2014. *Status and Trends of Prairie Wetlands in the United States 1997 to 2009*. Washington, DC: US Fish and Wildlife Service.

Dale, V. H., D. Wright, C. L. Kling, W. Boynton, J. L. Meyer, K. Mankin, J. Sanders, J. Opaluch, D. J. Conley, H. Stallworth, et al. 2010. *Hypoxia in the northern Gulf of Mexico*. New York: Springer.

Deaton, R. 2016. "A History of Terre Noire Creek and the Ross Drainage District." *The Arkansas Historical Quarterly* 75 (3): 239–54.

Easterling, D., K. Kunkel, J. Arnold, T. Knutson, A. LeGrande, L. Leung, R. Vose, D. Waliser, and M. Wehner. 2017. "Precipitation Change in the United States." *Climate Science Special Report: Fourth National Climate Assessment 1*(GSFC-E-DAA-TN49608).

Edwards, E. C., and S. M. Smith. 2018. "The Role of Irrigation in the Development of Agriculture in the United States." *The Journal of Economic History* 78 (4): 1103–141.

Edwards, E. C., and W. N. Thurman. 2022. "The institutional costs of adaptation: Agricultural Drainage in the United States." Technical report, National Bureau of Economic Research.

Ferrara, A., P. Testa, and L. Zhou. 2021. "New Area-and Population-Based Geographic Crosswalks for US Counties and Congressional Districts, 1790–2020." Available at SSRN 4019521.

Fenneman, N. M. and D. W. Johnson. 1946. *Physical division of the United States: US geological survey*. Physiography Committee Special Map.

Forrest, L. C. 1999. *Lake Mattamuskeet, New Holland and Hyde County*. Charleston, SC: Arcadia Publishing.

Fretwell, J. D. 1996. *National Water Summary on Wetland Resources*, Volume 2425. US Government Printing Office.

Haines, M., P. Fishback, and P. Rhode. 2019. United States Agricultural Data, 1840–2012 (icpsr 35206).

Harrison, R. W., and W. M. Kollmorgen. 1948. "Socio-economic History of Cypress

Creek Drainage District and Related Districts of Southeast Arkansas." *The Arkansas Historical Quarterly* 7 (1): 20–52.

Hayek, F. 1945. "The Use of Knowledge in Society." *The American Economic Review* 35 (4): 519–30.

Hewes, L., and P. E. Frandson. 1952. "Occupying the Wet Prairie: The Role of Artificial Drainage in Story County, Iowa." *Annals of the Association of American Geographers* 42 (1): 24–50.

Hofstrand, D. et al. 2010. "Economics of Tile Drainage." *Ag Decision Maker Newsletter* 14 (9). Ames, IA: Iowa State University.

Hong, S. C. 2007. "The Burden of Early Exposure to Malaria in the United States, 1850–1860: Malnutrition and Immune Disorders." *The Journal of Economic History* 67 (4): 1001–1035.

McCorvie, M. R., and C. L. Lant. 1993. "Drainage District Formation and the Loss of Midwestern Wetlands, 1850–1930." *Agricultural History* 67 (4): 13–39.

McCrory, S. H. 1928. "Historic Notes on Land Drainage in the United States." *Transactions of the American Society of Civil Engineers* 92 (1): 1250–1258.

Meyer, K., and D. Keiser. 2018. "Adapting to Climate Change through Tile Drainage: Evidence from Micro Level Data." Working Paper.

O'Driscoll, M. A. 2012. "The 1909 North Carolina Drainage Act and Agricultural Drainage Effects in Eastern North Carolina." *Journal of North Carolina Academy of Science* 128 (3–4): 59–73.

Olmstead, A. L., and P. W. Rhode. 2011. "Adapting North American Wheat Production to Climatic Challenges, 1839–2009." *Proceedings of the National Academy of Sciences* 108 (2): 480–85.

Olson, M. 1965. *The Logic of Collective Action: Public Goods and the Theory of Groups.* Cambridge, MA: Harvard University Press.

Ortiz-Bobea, A., H. Wang, C. M. Carrillo, and T. R. Ault. 2019. "Unpacking the Climatic Drivers of US Agricultural Yields." *Environmental Research Letters* 14 (6): 064003.

Ostrom, E. 1990. *Governing the Commons: The Evolution of Institutions for Collective Action.* Cambridge, UK: Cambridge University Press.

Palmer, B. W. 1915. *Swamp Land Drainage with Special Reference to Minnesota.* Volume 5, *Bulletin of the University of Minnesota.* Minneapolis, MN: University of Minnesota Press.

Parker, D. P., and W. N. Thurman. 2018. "Tax Incentives and the Price of Conservation." *Journal of the Association of Environmental and Resource Economists* 5 (2): 331–69.

Parker, D. P., and W. N. Thurman (2019). "Private Land Conservation and Public Policy: Land Trusts, Land Owners, and Conservation Easements." *Annual Review of Resource Economics* 11 (1): 337–54.

Prince, H. 1995. "A Marshland Chronicle, 1830–1960: From Artificial Drainage to Outdoor Recreation in Central Wisconsin." *Journal of Historical Geography* 21 (1): 3–22.

Prince, H. 2008. *Wetlands of the American Midwest.* Chicago: University of Chicago Press.

Ridley, M. 2020. *"How Innovation Works: And Why It Flourishes In Freedom.* New York: HarperCollins.

Rosenzweig, C., F. N. Tubiello, R. Goldberg, E. Mills, and J. Bloomfield. 2002. "Increased Crop Damage in the US from Excess Precipitation under Climate Change." *Global Environmental Change* 12 (3): 197–202.

Schaetzl, R. J., F. J. Krist, K. Stanley, and C. M. Hupy. 2009. "The Natural Soil

Drainage Index: An Ordinal Estimate of Long-Term Soil Wetness." *Physical Geography* 30 (5): 383–409.

Schaetzl, R. J., F. J. Krist Jr., and B. A. Miller. 2012. "A Taxonomically Based Ordinal Estimate of Soil Productivity for Landscape-Scale Analyses." *Soil Science* 177 (4): 288–99.

Schlenker, W., W. M. Hanemann, and A. C. Fisher. 2005. "Will US Agriculture Really Benefit from Global Warming? Accounting for Irrigation in the Hedonic Approach." *American Economic Review* 95 (1): 395–406.

Skaggs, R. W., M. A. Brevé, and J. W. Gilliam. 1994. "Hydrologic and Water Quality Impacts of Agricultural Drainage." *Critical Reviews in Environmental Science and Technology* 24 (1): 1–32.

Smith, S. M., and E. C. Edwards. 2021. "Water Storage and Agricultural Resilience to Drought: Historical Evidence of the Capacity and Institutional Limits in the United States." *Environmental Research Letters* 16 (12): 124020.

Strzepek, K., G. Yohe, J. Neumann, and B. Boehlert. 2010. "Characterizing Changes in Drought Risk for the United States from Climate Change." *Environmental Research Letters* 5 (4): 044012.

Tangen, B. A., and M. T. Wiltermuth. 2018. "Prairie Pothole Region Wetlands and Subsurface Drainage Systems: Key Factors for Determining Drainage Setback Distances." *Journal of Fish and Wildlife Management* 9 (1): 274–84.

Tovar, K. 2020. "Iowa Drainage Law: A Legal Review." Iowa State University Center for Agricultural Law and Taxation. Online Resource. https://www.calt.iastate.edu/article/iowa-drainage-law-legal-review.

Wright, J. O. 1907. *Swamp and Overflowed Lands in the United States*. Circular 76, US Department of Agriculture Office of Experiment Stations. Washington, DC: US Government Printing Office.

Estimating the Effect of Easements on Agricultural Production

Nicole Karwowski

2.1 Introduction

Agricultural systems and food production are vulnerable to climate. Excess water poses a particular risk for agricultural production. In 2019, when above average precipitation inundated the eastern half of the country, the US experienced its record-wettest year to date (NOAA 2020). The central US experienced a series of severe storms, preventing farmers from planting; flooding crops; and accruing debilitating losses in the billions for agrarian communities across the Corn Belt and Mid-South (English et al. 2021). Heavy precipitation and floods have caused catastrophic damage to US crop production and profits (Rosenzweig et al. 2002; NOAA 2023). The scientific literature has identified that regional rainfall patterns are already changing, and that we can expect more frequent occurrences of climate extremes, and ultimately, higher flood risk in certain regions (Urban et al. 2015). Studies consistently show lower crop yields and higher losses attributed to a changing climate, and that these losses are expected to increase in frequency and severity (Schlenker and Roberts 2009; Deschênes and Greenstone 2012; IPCC 2012; Rosenzweig et al. 2014; NOAA 2023; Perry et al. 2020). Finding strategies to deal with increased precipitation and flooding under future climate change is critical for mitigating climate risks.

Here, I evaluate the adaptation benefits of some of the largest conservation programs in the United States. The Natural Resources Conservation

Nicole Karwowski is a PhD candidate in Agricultural and Applied Economics at the University of Wisconsin-Madison.

For acknowledgments, sources of research support, and disclosure of the author's material financial relationships, if any, please see https://www.nber.org/books-and-chapters/american-agriculture-water-resources-and-climate-change/estimating-effect-easements-agricultural-production.

Services (NRCS) of the US Department of Agriculture (USDA) offers voluntary buyouts through the Wetlands Reserve Program (wetland easements) and Emergency Watershed Protection Program (floodplain easements). In 2020, there were approximately 3 million acres of eased wetlands and 185,000 acres of eased floodplains in the US. These programs buy out land from farmers through easements contracts. The farmer retains ownership of the land and receives a lump-sum transfer to forgo the right to plant crops on that field in perpetuity. Eased land is then restored to its natural floodplain or wetland state. Restoration includes planting native species, breaking or removing tiling, and building topographical features (for example, creating a berm or filling a ditch) to redirect water onto the eased land. Land restoration is hypothesized to provide flood protection by storing water and acting as natural buffers for nearby developed land.

Using over 30 years of national data and a two-way fixed effects strategy, I quantify the impacts of the wetland and floodplain easement programs on agricultural production at the county level. I focus on rainfed, non-irrigated counties producing corn, soybeans, and wheat. I discover that a 100 percent increase in wetland easement land share increases county yields by 0.34 percent, 0.77 percent, and 0.46 percent for corn, soybeans, and wheat. I find that easements decrease risk for soybeans: doubling wetland easement land share reduces indemnities from excess moisture by $3.59, from heat by $6.07, and from disease by $11.23 for each dollar of soybean liability. Corn crops also see less insect losses by $8.50 per dollar of liability. To better understand the drivers of these effects, I interact easement acreage with measures of precipitation and degree days to understand the weather pathways through which easements provide adaptive benefits. Wetland easement land share attenuates the impact of extreme degree days for soybeans and excess precipitation for corn. My results indicate that these easement conservation programs can serve a critical role in mitigating climate risk.

I also identify the potential channels through which easements impact agricultural outcomes. I estimate the effect of easements on acres planted, acres failed, and acres prevented planted to understand the underlying mechanisms changing yields and risk. Easements impact agricultural production in three main ways: removing marginal land from production (direct effect), improving yields on surrounding cropland (indirect effect), and by changing the cultivation choices of producers (slippage effect).

Easements lead to the retirement of cropland from production permanently. Easements also include non-cropland to create a more effective habitat. Easing cropland mechanically improves the average county-level yields for commodities, since the lowest yielding land is no longer cropped. There is also some evidence of a positive yield externality: wetland and floodplain habitats improve yields on surrounding croplands. I parse out the direct and indirect effect in my data by estimating how cropland and non-cropland easement land share impacts yields. Doubling cropland in the wetland program directly improves soybean yields by 0.82 percent and wheat yields by 0.33 percent, while doubling non-cropland indirectly improves corn yields by 0.22 percent and soybean yields by 0.29 percent. There is also evidence of an indirect floodplain yield effect: doubling non-cropland in the floodplain program increases corn, soybean, and wheat yields by 0.14 percent, 0.06 percent, and 0.09 percent. Easement habitats offer flood buffer protection to surrounding agricultural fields. It may also be the case that producers re-optimize their inputs and production strategies on their non-eased land and this improves producer-level yields.

Producers switch their production away from soybean and wheat toward corn. Easing land may encourage farmers to continuously crop corn on their remaining fields or alternatively convert non-cropland into corn cropland. There is a 2 percent decrease in soybean acres planted and 1 percent decrease in wheat acres planted as expected with a doubling of wetland easement land share. Surprisingly, planted acreage for corn increases by 3 percent after a 100 percent increase in wetland easement land share. A similar slippage effect has been found for the Conservation Reserve Program (CRP) (Wu 2000; Fleming 2014; Uchida 2014) and other conservation programs (Lichtenberg and Smith-Ramírez 2011; Pfaff and Robalino 2017). Learning that easements impact cultivation choices for producers may have implications for the sustainability benefits of the program. Continuous corn cropping tends to be more profitable for farmers but also associated with yield penalties and worse environmental outcomes (Seifert, Roberts, and Lobell 2017). This slippage effect may offset some of the ecosystem benefits of easements.

I find mixed results regarding how easements impact acres failed to harvest, failed, and prevented planted. The slippage story may help explain why easements have an insignificant or even positive impact at times on harvest failure and prevented planting. Based on National Agricultural Statistics Survey (NASS) data analysis, conditional on failure occurring, increasing wetland acres by 100 percent in a county is associated with a 1.67 percent increase in corn harvest failure, and −1.19 percent and −1.38 percent change in soybean and wheat acres failed to harvest. Meanwhile, doubling floodplain easement acreage results in a 1.66 percent increase in acres failed to harvest for soybean crops and 0.83 percent decrease for wheat crops. Using a decadal panel of data from the Farm Service Agency (FSA), I find that wetland acres decrease acres failed for soybeans by over 10 percent and for wheat by over 21 percent. Floodplain easements during this time reduce corn acres failed by 5.77 percent yet increase wheat acres failed by 5.45 percent. When examining incidences of acres prevented planted, wetland acres actually increase corn acres prevented planted by 43 percent. On the other hand, floodplain easement doubling reduces corn acres prevented planted by 14 percent and soybean acres prevented planted by 8.94 percent. These mixed findings suggest that further examination of how easements impact acreage outcomes is warranted.

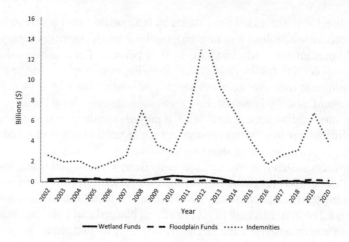

Fig. 2.1   Spending comparison of NRCS easement programs and crop indemnities

NRCS floodplain and wetland easements account for only 0.01 percent of land in the US, while 40 percent of land is used for agricultural purposes. From 2002 to 2020, the NRCS spent US$4.9 billion and US$3.4 billion on the wetland and floodplain programs respectively (USDA 2021). In comparison, indemnity spending for corn, soybean, and wheat losses in that same period reached over US$85.9 billion. Figure 2.1 emphasizes the difference in NRCS and indemnity spending over time. Putting land into easement may be a cost-effective adaptation strategy for agricultural resiliency.

Although the acreage of land under easement seems minimal, easements impact agricultural economic production through a number of pathways. These easement programs eliminate the moral hazard associated with insured farmers planting on marginal fields, decrease indemnities and tax-payer spending on agricultural losses, and offer other ecological advantages, such as improving yields on neighboring cropland. Wetlands and floodplains have the capacity to act as "sinks" and retain water within the watershed in ways that impact the flood patterns on surrounding fields. There may also be changes in producer input allocation and cultivation strategies that lead to yield gains.

This paper documents the effects and externalities of the easement programs on agricultural production. It adds to the literature on the relationship between agricultural systems and climate change. I provide evidence that these conservation policies allow farmers to adapt in ways that have a concrete and meaningful impact on agricultural resilience. This paper also complements the cost-benefit conservation literature that quantifies the impact of conserved land habitats. My paper provides an economic estimate of some of the non-market values that wetlands and floodplains provide. In a back-of-the-envelope cost-benefit analysis, I find that doubling wetland easement land share is cost effective and yield benefits exceed US$7 billion.

My work contributes to the literature on adaptation to heightened agronomic yield risk. Burke and Emerick (2016) find evidence suggesting that long-run adaptation has been limited and insignificant. However, more recent work by Mérel and Gammans (2021) suggests that panel models may not be reflective of climate adaptation in the long term and alternate specifications do find evidence of long-run climate adaptation for crop yields. Other researchers take a different approach and instead focus on the effects of specific adaptation measures; there is evidence that various adaptation practices can be effective at increasing resiliency. Producers can manage risk through insurance (Annan and Schlenker 2015); technology (Goodwin and Piggott 2020); planting date adjustments (Kucharik 2008; Zipper, Qiu, and Kucharik 2016); cultivar selection (Sloat et al. 2020; Hagerty 2021); irrigation (Hornbeck and Keskin 2014); and conservation practices (Schulte et al. 2017; Fleckenstein et al. 2020). My work adds to this literature by shedding light on the ex-post effects of easements as well as the implications of conservation programs in a world with higher temperatures and more frequent, extreme weather events.

There is a burgeoning literature focused on comparing the costs and benefits of conservation efforts. The cost-benefit papers seek to identify optimal parcels and best targeting strategies to meet desired conservation goals (Heimlich 1994; Wu, Zilberman, and Babcock 2001; Costello and Polasky 2004; Newburn, Berck, and Merenlender 2006; Gelso, Fox, and Peterson 2008; Fleming, Lichtenberg, and Newburn 2018). Others quantify benefits by estimating how additional wetland and floodplain acreage impact property damages from flooding (Watson et al. 2016; Gourevitch et al. 2020; Taylor and Druckenmiller 2022). There are some smaller field-level/regional studies as well as anecdotal evidence of the ecosystem benefits of the NRCS easement programs (NRCS 2011; Mushet and Roth 2020). Yet there remains a gap in understanding the effects of these easement programs on agricultural outcomes at a broader level. I contribute the first work at a national scale over the entire duration of the program life span.

Another vein of the conservation literature examines the relationship between prices and easement quantity and quality. Many studies measure the impact of easements on land sales prices (Brown 1976; Shoemaker 1989; Nickerson and Lynch 2001; Shultz and Taff 2004; Kousky and Walls 2014; Lawley and Towe 2014). These works consistently find that the land discount on eased land adequately captures the forgone agricultural profits. A complementary literature uses auction modeling techniques to estimate the reservation value of retiring land from agricultural production (Kirwan, Lubowski, and Roberts 2005; Ferraro 2008; Brown et al. 2011; Narloch, Pascual, and Drucker 2013; Hellerstein, Higgins, and Roberts 2015; Boxall et al. 2017). These studies look at how easements impact prices; Parker and Thurman (2018) quantify how tax incentives (price benefits) influence easement growth and conservation land quality. My paper looks beyond the easement quantity-price relations and reveals program externalities including yield spillovers and changes in cultivation choices.

Quantifying the effect of easements on agricultural systems has implications for climate change adaptation policy, land value estimates, and conservation cost-benefit analyses. My results offer insights into how easements offer a strategy to remove marginal land from production, improve crop yields, and decrease risk in the face of a changing climate. The remainder of the paper proceeds as follows: Section 2.2 includes a discussion on program background, the relationship between climate and agriculture, and the role of insurance. Section 2.3 lays out the theoretical framework. Sections 2.4 and 2.5 present the data and empirical models respectively. Section 2.6 covers empirical results and discusses their implications. Section 2.7 summarizes the main findings. The results tables are in the appendix, and the coefficient plots are in an online appendix.[1]

2.2   Background

2.2.1   What Is an Easement?

The NRCS floodplain easement and wetland restoration programs allow agricultural producers to retire frequently flooded land from agricultural use. Producers apply to the program and, if selected, receive a lump-sum payment to forgo the right to crop on that field. The easement contract grants the NRCS surface rights and the right to restore the land. The landowners retain ownership and pay property taxes on the land. Landowners are also granted the rights to control public access, quiet enjoyment, and recreational use such as hunting and fishing. There is also a possibility of authorizing compatible use activities (CUA) such as cropping certain commodities, timber harvest, grazing or periodic haying when consistent with long-term enhancement of the easement functions and values. Easements often occur on lower-yielding land that is costly to manage and at higher risk of losses. Easements remove marginal land from production by increasing the opportunity cost of operating in high-risk areas for producers.

The NRCS strives to maximize the environmental benefits of easements. The NRCS states that the main purpose of the wetland restoration program is to "achieve the greatest wetland function and values, along with optimum wildlife habitat, on every acre enrolled in the program" (NRCS 2021c). The NRCS goal of the floodplain easement program is to "restore, protect, maintain and enhance the functions of floodplains while conserving their natural values such as serving fish and wildlife habitat, improving water quality, retaining flood water, and recharging groundwater" (NRCS 2021a). The NRCS pays for the majority of the restoration and carries it out themselves.

---

1. See http://www.nber.org/data-appendix/c14694/appendix.pdf

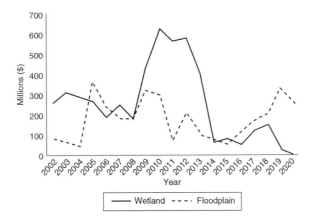

Fig. 2.2   **NRCS wetland and floodplain program funding over time**

The restoration process includes removing structures that impede water flow, removing or breaking tiling, building topographic features such as ridges and swales, and planting native vegetation. Wetland and floodplain easements retain water within a watershed and impact flood patterns in the area.

The NRCS pays the landowner for the right to restore the land. After ranking and selecting the optimal parcels, the NRCS offers the producer compensation that the producer can choose to accept or not. Easement compensation is based on the lowest of three values: fair market appraisal, geographic area rate cap, and a voluntary offer by a landowner. Most often, compensation is based on the geographic area rate cap (GARC), which stems from a market survey of cropland in the area. Landowners rarely posit a voluntary offer. Based on interviews with policy directors, it is most often the case that farmers who are not selected continue to crop on the land. Easement programs directly reduce insurance spending on future crop losses, since farmers would have continued farming otherwise.

The amount of easement projects selected depends on the individual state budget for each program. The easement programs are funded federally, but each state NRCS department oversees implementation. The wetland restoration program receives regular funding from Farm Bill appropriation. Funding for floodplain easements is provided by a congressional act, often after large-scale flooding in the US. Figure 2.2 demonstrates trends in funding from 2002 to 2020. Wetland and floodplain funding experience a sharp increase in 2009 and for a few years afterwards; the American Recovery and Reinvestment Act included stimulus spending for agricultural programs to counter the 2008 recession.

These easement programs date back to the early 1990s. Almost half of the natural wetlands in the US had been drained and filled for agricultural and development purposes by 1984 (NRCS 2021b). To slow the destruction of wetlands, Congress added wetland and conservation protection to the 1985 and 1990 Farm Bill. In 1985, the Swampbuster provision prohibited farmers from draining wetlands while participating in USDA programs and receiving any type of aid. This offered some of the first protections to wetlands. Then in 1990, the first wetland restoration program was authorized as an option for farmers to retire land that had been drained and to conserve eligible wetlands. Wetland restoration led to a reversal of wetland losses and often led to net increases in wetland coverage.

Easement programs have gone through multiple names and iterations. NRCS wetland restorations have been offered under the Emergency Wetlands Reserve Program, Wetlands Reserve Program, and Agricultural Conservation Easement Program for Wetlands Reserve Easements. The Emergency WRP was established in 1993 and became today's floodplain easement program (Hebblethwaite and Somody 2008). The Emergency WRP Program was funded after receiving emergency appropriations following severe flooding in the Midwest in the 1990s. This study includes all these easement types.

Most of the basics underlying the floodplain and wetland programs remain the same, but wetland projects tend to require higher investment and more management. To be eligible for a wetland easement, land needs to be farmed wetland or converted wetland with the potential to be restored in a cost-effective manner; priorities are put on easements with high potential for protecting and enhancing the habitat. Ranking criteria include drainage conditions, portion of hydric acres, protection potential of certain species, adjacency to other conserved areas and wetlands, and water quality improvement estimates. Wetland easements can be permanent, 30-year easements, 30-year contracts, or 10-year cost-share agreements. The most common type of wetland restoration are permanent.

The floodplain easement process varies slightly from the wetland easement process. In order to be eligible for a floodplain easement, the proposed acreage must be in a floodplain that has been damaged by a flood once in the calendar year or flooded at least twice in the past decade. Land that is in danger of being adversely impacted by a dam breach is also eligible. Other parcels may also be eligible if they enhance the floodplain system, improve erosion control, or promote easement management. Ranking criteria include flooding history, proximity to other protected land or public access points, adjacency to existing easements, acreage of proposed easement in the flood zone and associated flood hazard, percentage of acreage in different land use classes, estimated restoration costs, other parties' contribution of the cost, and existence of rare species within a certain buffer. All floodplain easements are permanent.

Wetlands and floodplains—both natural and man-made—are associated with many ecological and hydrological benefits that have been studied by economists, ecologists, hydrologists, and conservationists. Floodplains and wetlands have the potential to serve as flood protection by storing water and acting as natural buffers in the event of extreme flooding. Wetlands reduce damage from floods by lowering flood heights and reducing the water's destructive potential (Gleason, Laubhan, and Euliss 2008). Restored floodplains and wetlands are also associated with improved water quality, ground water reservoir replacement vital for irrigation systems, carbon sequestration, reduced greenhouse gases, and wildlife habitat (Bostian and Herlihy 2014; Roley et al. 2016; Sonnier et al. 2018; Speir, Tank, and Mahl 2020). There have been a few studies of NRCS wetland restoration projects, such as regional studies from the USDA's Conservation Effects Assessment Project (CEAP). These studies show that easements have been successful in supporting habitat and biodiversity, pollution management, surface water and floodwater containment, greenhouse gas emission management, and water sustainability (NRCS 2011, 2012; Hansen et al. 2015; Mushet and Roth 2020).

2.2.2   How Do Climate and Weather Impact Crop Production?

There is a large body of knowledge explaining how weather patterns and underlying climate impact crop production (Wing, De Cian, and Mistry 2021; Ortiz-Bobea 2021).

Extreme temperatures associated with climate change are projected to become more intense and frequent in upcoming years. Extreme heat exposure beyond a certain threshold reduces the quality and yields of agricultural crops (Schlenker and Roberts 2009). Heat stress adversely affects plant development, pollination, and reproductive processes (Hatfield and Prueger 2015). Extreme temperatures coupled with water scarcity—drought conditions—can also reduce productivity. Decreased soil moisture stunts crop growth and increases vulnerability to pests. Drought conditions are especially prevalent in the western half of the country.

While some areas are faced with worsening drought conditions, extreme precipitation is projected to be more frequent in other areas of the US, especially the central Midwest (Rosenzweig et al. 2002; Shirzaei et al. 2021). Excess precipitation coupled with higher temperatures are detrimental climate patterns for crop production (Eck et al. 2020). Excess spring moisture reduces yields by 1–3 percent yearly, but these losses can range up to 10 percent during extremely wet springs (Urban et al. 2015). Flooding impacts agriculture by delaying or preventing planting, damaging standing crops, and carrying away topsoil and nutrients.

When flooding occurs during planting season in the spring, farmers may be delayed or prevented from planting since their machines are unable to work on the inundated soil (Urban et al. 2015; Boyer, Park, and Yun 2022). Delayed planting increases production costs and risk by shortening the growing season as well as exposing crops to late-season freezes. In 2019, heavy precipitation led to a record 19 million acres prevented planted (English et al. 2021). Excess rain can also be harmful later in the season. If there is an abundance of water, flooding destroys crops by washing them away, decreasing oxygen intake and respiration, building up toxic compounds in the soil, inhibiting plant growth, and making plants prone to disease, insects, or mold (Hatfield et al. 2011). This type of water stress increases uncertainty and reduces profits. Extreme precipitation can also have more long-term impacts by reducing the soil quality over time by draining nutrients out of the soil or washing away the topsoil altogether.

Both heat and water stress can indirectly lead to losses by making crops prone to disease and insects (Deutsch et al. 2018; Jabran, Florentine, and Chauhan 2020). Higher temperatures and varying moisture levels have expanded the breeding ground of certain insects and changed their feeding habits: increased metabolisms lead to larger appetites and lower yields. Changing weather conditions have led to a wider range and distribution of pathogens that have increased the risk of plant diseases. There is large variation in top pest concerns dependent on crop type, geography, timing, and weather conditions (Savary et al. 2019).

Easements have the adaptation potential to improve agricultural resiliency, especially in the face of a changing climate. Escalating temperatures and extreme weather events make easing land a more lucrative option for producers. Easing marginal land that is at high risk of losses offsets climate-caused indemnities. Insurance premiums, subsidies, and indemnities are expected to increase (Tack, Coble, and Barnett 2018). Easements provide one potential pathway to reducing agricultural risk by reallocating land and improving the resiliency of land remaining in agricultural production.

2.2.3   What Is the Role of Insurance?

Crop insurance can be purchased to protect agricultural producers against the loss of crops from natural disasters such as excess heat, flooding, fire, drought, disease, insect damage, and destructive weather. Multiple peril crop insurance (MPCI) protects producers against lower than expected yields and revenues. MPCI is serviced by private sector insurance companies, which the USDA subsidizes, regulates, and re-insures. Glauber (2013) provides a thorough history of crop insurance. The government typically subsidizes 60 percent of a producer's premiums in addition to offering assistance after natural disasters (Congressional Budget Office 2019). There are more than 290 million acres insured in the US, which account for more than 80 percent of acres planted. In 2020, MPCI insured nearly US$110 billion in liability and cost taxpayers US$6.4 billion in premium subsidies and US$1.5 billion in delivery costs (Goodwin and Piggott 2020).

Producers can choose from a variety of policies and coverage options. Yield-based policies insure producers against crop-specific yield losses. Revenue-based policies protect against volatility in yields and prices, but are more expensive. Yield-based policies are the most accessible and have existed the longest. A producer pays a premium to the insurance company in order to purchase coverage on their commodities. Yield-based policies are based on the actual production history (APH) of a parcel and pay an indemnity for low yield states. The APH is an average of the past four to ten years of yields on a parcel and represents the expected yields of that parcel. The APH is used to determine the liability. The liability represents the expected value of a commodity and the maximum value that is insured by a policy. In the event of a loss, the indemnity payment is determined by taking the difference between the liability and the actual value of production.

The liability and any potential indemnity values depends on the coverage level selected by a producer. Coverage levels vary from 50–85 percent in 5 percent increments. A minimal amount of acreage in the US is covered at the 50 percent level.[2] A majority of producers choose to pay a premium and purchase additional coverage, called buy-up coverage. A producer is able to choose the percentage of the commodity value to insure. The coverage level can be thought of as a deductible. For example, a policy with an 80 percent coverage level insures against yield losses greater than 20 percent of the liability but does not provide indemnities for losses that total less than 20 percent of the liability.

To set insurance rates and premiums, the Risk Management Agency (RMA) uses a loss cost ratio (LCR) approach.[3] The RMA uses historical data on individual producers and calculates LCRs for each year and each producer. They do this by dividing a producer's indemnities by their liabilities. Then the RMA averages the LCRs across the the county level and over time. This resulting county-level average LCR is the base rate the RMA charges producers for coverage in that area.[4] The LCR represents the yield risk of a commodity in that county. The RMA sets the premium rate equal to the rate of expected losses over the total value of commodities. The loss ratio (LR) is the proportion of indemnities to the premiums paid by a producer. The loss ratio represents the actuarial fairness of the insurance policy. When the indemnities equate the premium paid (and the loss ratio is equal to one), expected losses are equal to the payment of the coverage for that specified risk.

Most previous work primarily links climate to crop yields. This is something that is done in this paper as well, but I believe that limiting the analysis to this approach has shortfalls. Looking strictly at yields does not capture

---

2. On the low end of coverage, there exists a specific policy called catastrophic crop insurance (CAT). CAT reimburses farmers for severe crop losses exceeding 50 percent of average historical yields at a payment rate of 55 percent of the established commodity price. No premium is required for this type of coverage except for an administrative fee—which has increased from $60 to $655 per crop per county in the past 20 years.

3. The history and details of how rates and premiums are devised are laid out in detail in the Federal Crop Insurance Primer (Congressional Research Service 2021) and other academic papers (Schnapp et al. 2000; Woodard, Sherrick, and Schnitkey 2011).

4. There are also other adjustments made for the base rate. Usually, the RMA also applies a spatially smoothing procedure, caps and cups rate changes, and a state excess load.

whether production is becoming more or less risky. This may underestimate the impact of climate and any potential adaptation measures on yield sensitivity. For this reason, I also estimate the effect of easements on the loss cost ratio and loss ratio. Some researchers have used the variance of yields but this measure is deficient, since the distribution of yields is ever evolving and changes in this coefficient are hard to interpret. Using the loss cost ratio and loss ratio has been gaining popularity because these measures capture the risks of individual producers. For example, Perry, Yu, and Tack (2020) use the loss cost ratio when estimating how warming impacts the agricultural risk of corn and soybeans. Goodwin and Piggott (2020) use the loss cost ratio and loss ratio when analyzing how seed innovations impact agricultural risk and insurance rate-making behavior.

It is also interesting to consider the role that insurance may have on the easement decision-making process. A common concern with insurance products is the moral hazard that they introduce. There are a number of studies that evaluate the moral hazard implications of subsidized multiperil crop insurance in agriculture (Horowitz and Lichtenberg 1993; Smith and Goodwin 1996; Coble et al. 1997; Glauber 2004; Kim and Kim 2018; Yu and Sumner 2018; Yu and Hendricks 2020; Wu, Goodwin, and Coble 2020). Moral hazard occurs since producers act in ways that are more risky, as they do not take on the full cost of the risks. In the easements context, insurance presents an additional hurdle to retiring agricultural land that would perhaps be better suited for easement. Not only does insurance impact the decision to ease a field but once a producer eases some land, the insurance decisions for surrouding land may change as well. If a farmer takes their most risky land out of production, they may be more willing to take on additional risks in other ways. The farmer could change the coverage levels on their remaining agricultural fields. Other potential risk-altering behavior could include changes in cultivation decisions, changes in acres planted, or changes in fertilizer, pesticide, and herbicide application.

2.3   Theoretical Model

I develop a theoretical model to draw intuition about why, when, and where easements are implemented and at what price. I consider the decision-making process for the farmer and the conservation agent. The farmer chooses the share of land to enroll in an easement program in order to maximize profits. The conservation agent chooses which land to ease and implicitly sets the price of easements. The conservation agent maximizes the environmental benefits of the land. I add to the framework by considering the role of insurance. This is a one-period model that does not consider leaving the land fallow or the option value of waiting to ease. For a more comprehensive theoretical framework on the easement decision-making process that considers dynamics, see Miao et al. (2016).

I start by considering L field parcels that are denoted by l_i. Each field is the same size and $i = 1, 2, \ldots, L$. Each field differs in its agricultural yields (y_i), costs of planting (c_i), and environmental benefits (b_i). I assume there is one commodity type that can be produced and the price of the commodity p is determined by the market.

2.3.1 Farmer's Problem

The farmer aims to increase profits by making land use decisions that will maximize income. The farmer with L land parcels determines what to do with each field l_i. The farmer can put field l_i into agricultural production ($a_i = 1$) or enroll the land into the easement program ($e_i = 1$).

For each field in agricultural production, the farmer makes a profit based on the commodity price (p), yield (y_i), and cost (c_i), where $\pi_i = py_i - c_i$. The yield can be high or low depending on whether an extreme weather event occurs. The probability of a disaster occurring on a field is f_i. If there is no disaster, yields are y_i. If there is a disaster such as a flood or drought, yields are $\delta_i y_i$ where $\delta_i \in (0, 1)$. The producer insures their fields against the risk of a disaster by paying a premium that is included in the cost function, c_i. The producer pays the cost of insurance in the event that there is a flood or not. The insurance company covers α of the expected yield value, and the coverage level is the same for each field $\alpha \in (0.5, 0.9)$. When a disaster does occur, the producer receives an indemnity payment: $m_i = p(\alpha y_i - \delta_i y_i)$. The indemnity payment is the commodity price multiplied by the difference between the covered yields in the non-disaster state and the yields in the disaster state. The expected agricultural profits on field i for the producer is the weighted sum of the income in the non-disaster and disaster state.

$$(1) \qquad \mathbb{E}(\pi_i) = \underbrace{(1 - f_i)(py_i - c_i)a_i}_{\text{profit in non-disaster state}} + \underbrace{f_i(p\delta_i y_i + m_i - c_i)a_i}_{\text{profit in disaster state}}$$

When a field is eased, the farmer receives a payment of r_i for retiring the land from agricultural production. The farmer is subject to their land constraint, $a_i + e_i \leq 1$ and non-negativity constraints, $a_i \geq 0, e_i \geq 0$. The farmer chooses a_i and e_i for each l_i to maximize profits. To solve the farmer's problem, I set up a Lagrangean and take first-order conditions.

$$(2) \qquad \max_{a_i, e_i} \sum_i^L (1 - f_i)(py_i - c_i)a_i + f_i(p\delta_i y_i + m_i - c_i)a_i + \sum_i^L r_i e_i$$

$$s.t. \ \forall i : a_i + e_i \leq 1, \ a_i \geq 0, e_i \geq 0$$

$$\mathcal{L} = \sum_i^L (1 - f_i)(py_i - c_i)a_i + f_i(p\delta_i y_i + m_i - c_i)a_i + \sum_i^I r_i e_i$$

$$+ \sum_i^L \mu_i(1 - a_i - e_i) + \sum_i^L \theta_i e_i + \sum_i^L \sigma_i a_i$$

$$[a_i] : (1 - f_i)(py_i - c_i) + f_i(p\delta y_i + m_i - c_i) - \mu_i - \sigma_i = 0$$

$$[e_i] : r_i - \mu_i - \theta_i = 0$$

$$[\mu_i] : 1 - a_i - e_i = 0$$

$$[\theta_i] : \theta_i e_i = 0$$

$$[\sigma_i] : \sigma_i a_i = 0$$

At the solution, the Kuhn-Tucker conditions show that the first-order conditions are satisfied (1), the original constraints hold (2), the Lagrange multipliers are non-negative (3), and complementary slackness holds (4).

1. $(1 - f_i)(py_i - c_i) + f_i(p\delta_i y_i + m_i - c_i) = \mu_i + \sigma_i, r_i = \mu_i + \theta_i$

2. $a_i + e_i = 1, a_i \geq 0, e_i \geq 0$

3. $\mu_i \geq 0, \theta_i \geq 0, \sigma_i \geq 0$

4. $\mu_i(1 - a_i - e_i) = 0, \theta_i e_i = 0, \sigma_i a_i = 0$

I use the complementary slackness conditions to explicitly define the optimal e_i and a_i. The farmer will ease field i when the retirement payment is greater than or equal to the expected agricultural profits of a field. When the retirement payment is less than the agricultural profits, the farmer will put that entire field toward agricultural production. This model also informs us of the qualities of land that are more likely to be eased. Land with lower yields, higher risk of flooding, lower flood-damage yields, higher costs of planting, and higher environmental benefits are more likely to be put under easement.

$$e_i^* = \begin{cases} 1 & \text{if } (1 - f_i)(py_i - c_i) + f_i(p\delta_i y_i + m_i - c_i) \leq r_i \\ 0 & \text{if } (1 - f_i)(py_i - c_i) + f_i(p\delta_i y_i + m_i - c_i) > r_i \end{cases}$$

$$a_i^* = \begin{cases} 1 & \text{if } (1 - f_i)(py_i - c_i) + f_i(p\delta_i y_i + m_i - c_i) > r_i \\ 0 & \text{if } (1 - f_i)(py_i - c_i) + f_i(p\delta_i y_i + m_i - c_i) \leq r_i \end{cases}$$

2.3.2   Conservation Agent's Problem

Babcock et al. (1996) compare different targeting strategies for conservation policy makers: maximizing the benefit-to-cost ratio, maximizing total benefits, and minimizing total costs. I use their model as a baseline when considering the conservation agent's problem.

The conservation agent is trying to maximize environmental benefits subject to their budget constraint. These benefits are idiosyncratic to a field. The conservation agent chooses which fields to enroll e_i while simultaneously choosing the price to offer a farmer to retire that field r_i. It is most often the case that the easement payment is equal to the geographical area rate cap. This can be interpreted as the average land value in a county. In my model, the agent sets the price equal to the average land value of the L fields. I call this price $\bar{r}$. The conservation agent uses the average expected agricultural profits for all L fields to determine $\bar{r} = 1 / L \sum_i^L (1 - f_i)(py_i - c_i) + f_i(p\delta y_i + m_i - c_i)$. The conservation agent is also subject to total budget T. I assume that the budget is positive $T > 0$ and that the conservation agent cannot exceed the budget $\sum_i^L \bar{r}e_i \leq T$. I also include the condition that the easement cannot be larger then the field itself $e_i \leq 1$. I can write out the conservation agent's objective function as a constrained maximization problem.

(3) $$\max_{e_i} \sum_i^L b_i e_i \; st. \sum_i^L \bar{r}e_i \leq T, \forall i : 0 \leq e_i \leq 1$$

To solve for the optimal e_i for the conservation agent, I set up another Lagrangean. I ignore the non-negativity constraint since it is not optimal for the conservation agent to have zero easements.

$$\mathcal{L} = \sum_i^L b_i e_i + \lambda \left(T - \sum_i^L \bar{r}e_i \right) + \sum_i^L \omega_i (1 - e_i)$$

$$[e_i] : b_i - \lambda\bar{r} - \omega_i = 0$$

$$[\lambda] = T - \sum_i^L \bar{r}e_i = 0$$

$$[\omega_i] : 1 - e_i = 0$$

Again, I write out the Kuhn-Tucker conditions that hold when the agent is at the optimal solution.

$$1. \; b_i - \lambda\bar{r} - \omega_i = 0$$

$$2. \; \sum_i^L \bar{r}e_i \leq T, e_i \leq 1$$

$$3. \; \lambda \geq 0, \omega_i \geq 0$$

$$4. \; \lambda \left(T - \sum_i^L \bar{r}e_i \right) = 0, \omega_i (1 - e_i) = 0$$

I use the Kuhn-Tucker conditions to derive the explicit solution of the conservation agent. The conservation agent will ease field i when the benefit to cost ratio of that field exceeds the shadow price. The shadow price λ represents the marginal benefit of relaxing the budget constraint, or the associated change in environmental benefits when the budget is increased by one unit. As long as the ratio of field easement benefits over the cost of acquisition exceeds the shadow value, the conservation agent will ease the parcel. The conservation agent will enroll the fields with the highest benefit-cost ratio first and will continue to enroll the most beneficial fields until the budget T is depleted.

$$e_i^\# = \begin{cases} 1 \text{ if } \dfrac{b_i}{\bar{r}} \geq \lambda \\[2ex] 0 \text{ if } \dfrac{b_i}{\bar{r}} < \lambda \end{cases}$$

2.3.3   Solving for Equilibrium

I combine the solutions of the farmer and conservation agent to find the equilibrium. The farmer will not ease a field unless the easement payment from the conservation agent exceeds the expected agricultural profits. When the conservation agent sets the price equal to average expected profits of all the land, the fields that are lower in agricultural profits are the ones that farmers will ease. Mathematically, this means that $e_i = 1$ if $(1 - f_i)(py_i - c_i) + f_i(p\delta y_i + m_i - c_i) \leq \bar{r}$. The conservation agent eases land when the environmental benefits over the shadow price are greater than the easement payment price: $e_i = 1$ if $\bar{r} \leq b_i / \lambda$. Land will be eased when both these conditions are met. A field will be eased when the benefit to cost ratio exceed the shadow price. Otherwise, the land will stay in agricultural production.

$$e_i^* = \begin{cases} l_i \text{ if } \dfrac{b_i}{(1 - f_i)(py_i - c_i) + f_i(p\delta_i y_i + m_i - c_i)} \geq \lambda^* \\[3ex] 0 \text{ if } \dfrac{b_i}{(1 - f_i)(py_i - c_i) + f_i(p\delta_i y_i + m_i - c_i)} < \lambda^* \end{cases}$$

$$e_i^* = \begin{cases} l_i \text{ if } \dfrac{b_i}{(1 - f_i)(py_i - c_i) + f_i(p\alpha y_i - c_i)} \geq \lambda^* \\[3ex] 0 \text{ if } \dfrac{b_i}{(1 - f_i)(py_i - c_i) + f_i(p\alpha y_i - c_i)} < \lambda^* \end{cases}$$

2.3.4   Comparative Statics and Hypotheses

This model predicts that fields with ample environmental benefits and low agricultural productivity are the most likely to be eased. The fields with high benefit to cost ratios will be eased. If the price of the commodity increases, then fields are less likely to be eased since the opportunity cost is higher. If the cost of production increases—for example, if insurance premiums increase—then more fields would go into the easement program.

I can also consider the impact of climate change. If there is frequent flooding or more frequent drought conditions, expected yields would be lower, making easing land more attractive. Or if damages from disasters were higher, easing fields would also be more likely to occur. If the expected agricultural profits of a field were higher due to increased insurance coverage, it would be less likely for land to go into easement. A field that may have been better off eased may remain in production because of the guaranteed income from the insurance coverage. This emphasizes some of the moral hazard issues that insurance introduces to the easement process. This also highlights that insurance and easements are substitute adaptation strategies, not complementary.

Consider the effect of easements on the overall land, total indemnities, and average yields. Increasing easements will decrease the acres in agricultural production. This is a mechanical result. If lower-yielding and high-risk land is eased as our model predicts, then average expected losses for the land will decrease. Decreasing acres in production will decrease indemnities paid out $\sum_i^L m_i$ and acres damaged $\sum_i^L \delta_i y_i a_i$. The average yields on the remaining land in production $\bar{y} = 1/L \sum_i^L y_i$ are expected to increase.

The hypotheses tested empirically are as follows:

I. Easements increase average yields.
II. Easements decrease indemnities.
III. Easements decrease acres in agricultural production.
IV. Easements decrease acres failed and prevented planted.

2.4 Data

I compile from a wide array of sources to build a comprehensive data set to address my research questions. Administrative data are collected from various branches at the USDA: NRCS, NASS, RMA, and FSA. The remote sensing Parameter-elevation Regressions on Independent Slopes Model (PRISM) data is the source of the weather and climate controls. Each observation is at the county-year level. The data spans from 1989–2020 and includes about 1,700 farming counties. The commodities of focus are corn, soybeans, and wheat. Key summary statistics for the main counties east of the 100th meridian are presented in table 2.1.

2.4.1 NRCS Easements

The NRCS has a detailed database with information on completed floodplain easements and wetland restorations. There are 1,613 completed floodplain easements and 17,751 completed wetland restorations as of 2020. On average, there are 615.9 acres of wetland easement and 56 acres of floodplain easement in a county over the sample period. The mean land share in a county of wetland easement is 0.00160 and 0.00012 for floodplain easements. I differentiate between the cropland and non-cropland easement acres in order to parse out the direct and indirect effects. I also integrate the data on the geographical area rate cap to estimate the approximate per acre easement cost. The geographical area rate cap is the rate that most often corresponds to the per acre purchasing cost of easements and averages around $3,126 per acre. I estimate the average floodplain easement cost for new

Table 2.1   **Summary statistics**

Variable	Mean	SD	Min	Max	N
NRCS					
Wetland Acres	615.9	2,234	0	46,608	53,241
Crop Wetland Acres	356.9	1,535	0	30,394	53,241
Floodplain Acres	55.95	488.0	0	12,651	53,241
Crop Floodplain Acres	20.38	180.1	0	6,250	53,241
Wetland Acres/County Acres	0.00160	0.00577	0	0.117	53,241
Crop Wetland Acres/County Acres	0.000939	0.00399	0	0.0787	53,241
Floodplain Acres/County Acres	0.000120	0.000837	0	0.0192	53,241
Crop Floodplain Acres/County Acres	4.89e-05	0.000385	0	0.0118	53,241
Geographical Area Rate Cap	3,162	9,953	0	792,500	72,424
Wetland Easement Cost Per Acre (est)	2,657	1,840	232	20,064	6,044
Floodplain Easement Cost Per Acre (est)	2,691	2,035	319	15,774	475
NASS					
Corn yield (bushel/acre)	122.2	38.55	0	246.7	50,343
Soybean yield (bushel/acre)	37.57	11.07	0.700	80.40	45,877
Wheat yield (bushel/acre)	48.90	14.76	0	109.7	34,867
Corn Planted Acres	46,580	56,077	50	397,000	50,358
Soybean Planted Acres	48,673	51,985	10	541,000	45,879
Wheat Planted	17,493	38,208	50	500,000	34,880
Corn Harvested Acres	43,223	54,505	20	394,000	50,325
Soybean Harvested Acres	47,981	51,595	10	539,000	45,877
Wheat Harvested Acres	15,057	33,367	30	480,000	34,842
Corn Failed Harvest Acres	3,387	5,771	0	124,500	50,324
Soybean Failed Harvest Acres	694.3	1,555	0	71,000	45,877
Wheat Failed Harvest Acres	2,455	8,092	0	253,000	34,842
RMA					
Corn Indemnity	845,110	3.427e+06	0	1.396e+08	53,241
Soybean Indemnity	388,800	1.120e+06	0	3.692e+07	53,241
Wheat Indemnity	143,686	790,740	0	3.550e+07	53,241
Corn Liability	1.248e+07	2.549e+07	0	2.860e+08	53,241
Soybean Liability	7.642e+06	1.378e+07	0	1.689e+08	53,241
Wheat Liability	1.314e+06	4.589e+06	0	1.234e+08	53,241
Corn Premium	1.040e+06	2.000e+06	0	3.207e+07	53,241
Soybean Premium	651,643	1.199e+06	0	2.240e+07	53,241
Wheat Premium	176,620	722,158	0	2.883e+07	53,241
Corn Loss Cost Ratio	0.0996	0.158	−5.73e-05	1.245	44,950
Soybean Loss Cost Ratio	0.0927	0.138	0	1.354	42,906
Wheat Loss Cost Ratio	0.123	0.180	0	1.366	37,859
Corn Loss Ratio	0.887	1.386	−0.00101	20.03	44,950
Soybean Loss Ratio	0.769	1.052	0	15.41	42,906
Wheat Loss Ratio	1.050	1.692	0	34.45	37,859

Table 2.1   **(cont.)**

Variable	Mean	SD	Min	Max	N
FSA					
Corn Planted Acres	49,547	58,567	0	378,953	18,303
Soybean Planted Acres	48,889	53,528	0	536,339	18,303
Wheat Planted Acres	11,599	32,113	0	374,145	18,303
Corn Prevented Planted Acres	1,619	7,820	0	260,914	18,303
Soybean Prevented Planted Acres	723.0	3,083	0	89,229	18,303
Wheat Prevented Planted Acres	590.7	3,095	0	122,702	18,303
Corn Failed Acres	119.1	644.6	0	22,474	18,303
Soybean Failed Acres	34.98	334.9	0	19,759	18,303
Wheat Failed Acres	116.2	774.9	0	42,701	18,303
PRISM					
Max. Temperature (C)	26.18	3.060	17.74	36.88	53,241
Min. Temperature (C)	13.97	3.152	5.495	23.67	53,241
Average Temperature (C)	20.08	3.063	11.85	29.79	53,241
Precipitation (total mm)	623.4	168.0	75.77	1,697	53,241
Moderate degree days (hrs)	3,508	229.4	2,198	4,170	53,241
Extreme degree days (hrs)	463.1	356.5	0	2,194	53,241

easements in the sample period to be $2,691 while wetland easement costs are a little higher at $2,657.

Figure 2.3 depicts the cumulative acres enrolled in wetland and floodplain easements over time. Wetland enrollment increases slowly at first and then spikes in the late 1990s and early 2000s. The growth rate plateaus until the passage of the American Recovery and Reinsurance Act in 2008, which provides the NRCS with additional funding. Wetland enrollment increases for a few years after ARRA before flattening again. Floodplain easement enrollment is milder. Floodplain easements are funded though congressional acts that are infrequent. Funding for floodplains spikes after severe agricultural flooding events such as in the late 1990s and 2008. This additional funding corresponds to high floodplain enrollment. The NRCS Easement data record dates of importance such as application date, agreement start date, enrollment date, closing date, recording date, and restoration completion dates. Each step in the process is defined in detail in table 2.2. Whether a producer can crop on the land or insure the land with the USDA during that time is also noted. Producers are encouraged to crop on the land until the NRCS is ready to actively restore the land. A floodplain takes an average of 2.8 years to go from application to restoration completion. The wetland restoration process is more intensive and takes 4.1 years on average to complete. Figure 2.4 shows a breakdown of each step's duration. There is a large range in terms of how long it takes to finish the easement process—there are cases in which it takes less than a year and others that take closer to nine years.

I focus my analysis on the closing date, the day the easement contract becomes official. The conservation agent has approval to purchase the ease-

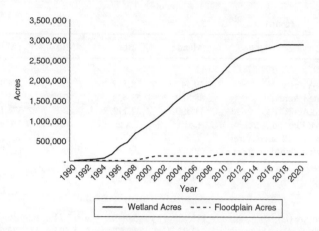

Fig. 2.3 **Cumulative acres in wetland and floodplain easement program**

Table 2.2   **Steps in the easement process**

Step	Description	Cropable?	Insurable?
Application	Application received by NRCS from producer.	Yes.	Yes
Agreement Start	Parcel selected and producer agrees to continue.	Yes.	Yes.
Enrollment	Parcel enrolls in program.	Yes.	Yes.
Closing	Attorneys sign off. Landowner receives payment.	CUA only.	No.
Recording	Transaction recorded in court.	CUA only.	No
Restoration Complete	Parcel restored.	No.	No.

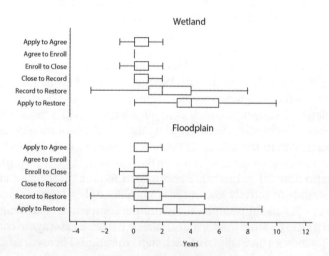

Fig. 2.4   **Duration in each easement step by program type**

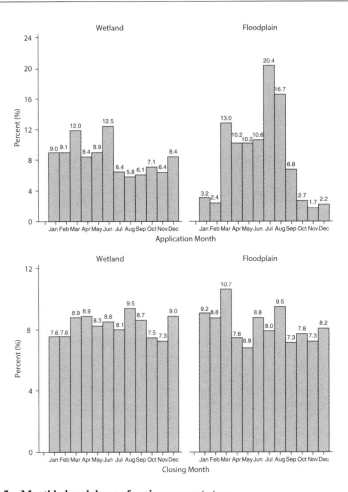

Fig. 2.5   Monthly breakdown of main easement steps

ment and the landowner is paid. After this date, farmers can no longer receive benefits on that field or insure the eased field. Notably, the farmer may still be able to plant on the field with a compatible use authorization until the restoration is complete, although they bear the full risk of production during that period. It is not until the restoration is complete that producers are prohibited from cropping on the easement. I therefore expect to observe direct and indirect effects of the easement decision beginning at the contract closing. Although, it is also possible that the indirect effects may increase after the restoration completion date.

I investigate the timing of the easement process in order to better understand when changes in agricultural production and risk may occur. I plot the distribution of the key steps in the easement process by month of occurrence in figure 2.5. Application timing is likely to be endogenous, as the decision to

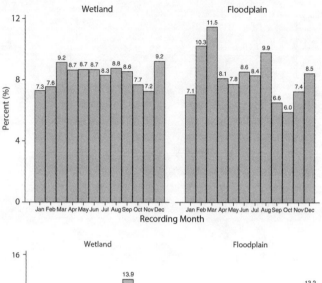

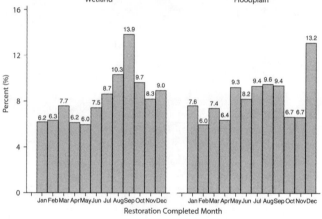

Fig. 2.5   (cont.)

apply to an easement program may be driven by agricultural losses. Wetland applications are more frequent in heavy precipitation months, March and June. Over 36 percent of floodplain applications are received in July and August, after producers have realized their yields. The uptick in applications is likely driven by farmers retiring marginal cropland after facing losses. The work completed by the NRCS, closings and recordings, is relatively evenly distributed across the year. There are seasonal patterns in restoration completion, since restorations require planting native flora. Wetlands tend to be finished by the end of summer around September. Floodplain restorations most commonly take place in late summer and December. The closing date seems to be reasonably exogenous and the best predictor of when easement effects are expected to occur.

2.4.2   National Agricultural Statistics Survey

Data on most agricultural outcomes stem from the USDA National Agricultural Statistics Survey (NASS). NASS includes yearly estimates of county-level yields. Figures 2.6, 2.7, and 2.8 map the average yields for corn, soybeans, and wheat, respectively. Corn production is centralized in

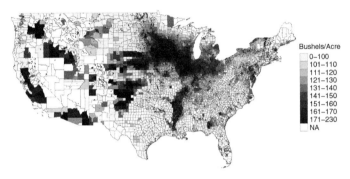

Fig. 2.6   Map of corn average yields and easements from 1989–2020

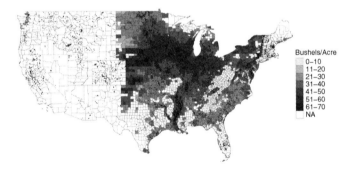

Fig. 2.7   Map of soybean average yields and easements from 1989–2020

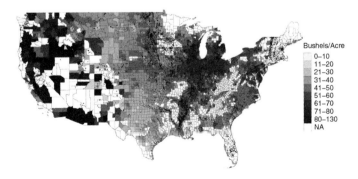

Fig. 2.8   Map of wheat average yields and easements from 1989–2020

the "corn belt" states: Nebraska, Iowa, Illinois, and Indiana. The mean corn yield during these three decades is 122 bushels per acre. Soybean production is more focused in the eastern half of the US. Soybean yields average 38 bushels per acre. Wheat production occurs in the Midwest of the US, but the highest yielding wheat counties are in the western states. Wheat yields average around 49 bushels per acre. The yield maps are overlaid with the easement locations to see the correlation between where production occurs and where easements take place.

NASS also provides statistics on acres planted and harvested since 1989. I create a measure of acres failed to harvest by subtracting the acres planted by acres harvested. On average, a county plants 46,000–48,000 acres of corn and soybean in a year. Wheat acreage is much lower at 17,000 acres per year in a county. Most of the acreage is harvested and the proportion of acres failed to harvest is low; usually a couple thousand acres are failed to harvest.

2.4.3   RMA Cause of Loss and Summary of Business

The Cause of Loss (COL) data set from the RMA (Risk Management Agency) provides valuable information on monthly indemnities for each county from 1989 to 2020. I aggregate each type of loss to the county-year level. Figure 2.9 compares the magnitude of losses by cause of loss. The biggest cause of loss is drought, with indemnities totaling over $35 billion. The second biggest cause of loss is excess moisture, with indemnities close to $30 billion. I expect easements to mitigate losses related to excess water and flooding. However, I also consider the overall loss cost ratio and other losses as well, since crops that face water stress are more liable to damages caused by disease, insects, and wildlife.

The Cause of Loss data are merged with the Summary of Business (SOB) data file. The SOB data record the acres planted, liabilities, premiums, subsidies, coverage levels, and chosen policies. I calculate the loss cost ratio by divid-

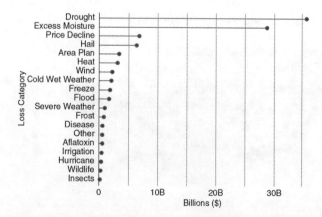

Fig. 2.9   Indemnity totals for corn, soybeans, and wheat by category from 1989 to 2020

ing the COL indemnity by the SOB liabilities. I do the same for the loss ratio by dividing the indemnities by the total premium amount. To create a balanced panel, I assume that reported indemnities are zero for county-years with no reported losses. I focus on the subset of counties that face indemnities (counties that have non-zero indemnities in that year). Figures 2.10, 2.11, and 2.12

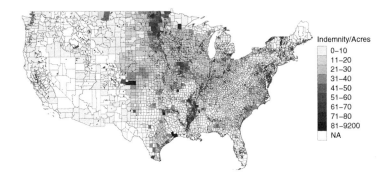

Fig. 2.10   Map of average indemnities per acre planted for corn and easements from 1989 to 2020

Fig. 2.11   Map of average indemnities per acre planted for soybeans and easements from 1989 to 2020

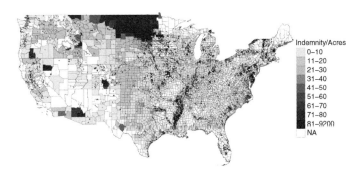

Fig. 2.12   Map of average indemnities per acre planted for wheat and easements from 1989 to 2020

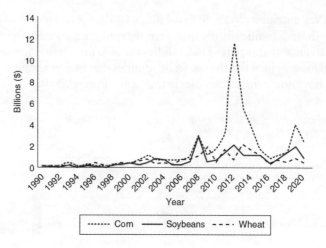

Fig. 2.13   Indemnities by commodity over time

show the extent of indemnities per acre planted for corn, soybeans, and wheat. There are corn indemnities scattered throughout the country, but there are high losses in the Dakotas and along the coasts. Soybean indemnities follow a similar spatial pattern but have lower average indemnities per acre compared to corn. Wheat indemnities are the highest of the three crops, especially in the northern edge of the United States. Easements seem to be concentrated in counties with high losses.

To explore when indemnities occur, I graph changes in total indemnities for corn, soybeans, and wheat in figure 2.13. Indemnities are relatively low and stable in the first decade of my sample. Spikes in losses become more frequent in the mid-2000s. It is important to note that acres enrolled and liability totals change drastically during this period. But some of these increases are also due to the changing climate. Losses are notable in 2008 (US$6 billion for wheat and corn combined) and 2012 (US$12 billion for corn), two years remembered for their extreme weather events. Extreme flooding throughout the Midwest in 2008 is associated with increased easement funding through ARRA. Record-breaking heat and limited precipitation led to a severe drought in 2012 in two-thirds of the US. It is expected that these billion-dollar weather disasters will increase in frequency.

2.4.4   FSA Crop Acreage

Producers who participate in FSA programs are required to self-report on acreage outcomes each year to the FSA. Records include the sum of planted acres, volunteer acres, failed acres, prevented acres, and net planted acres. These reports are used to calculate losses for various disaster assistance programs. Observations are aggregated to the county level for each year and are publicly available. Unlike the other USDA data, the FSA only spans from 2009 to 2020.

I consider how easements impact acres prevented planted and failed. Prevented planting is the inability to plant the intended crop acreage with proper equipment by the final planting date for a specific crop type. Failed acreage is acreage that is planted with the intent to harvest but is unable to be brought to harvest. The average number of acres that are prevented from planting in a county is 1,619, 723, and 590 for corn, soybeans, and wheat. Failure is less common with an average of 119, 35, and 116 acres failed for corn, soybeans, and wheat. I use these data to test whether easements reduce agricultural risk by reducing acres prevented planted during planting season or if easements reduce risk later on in the season by reducing acres failed.

2.4.5   PRISM Weather and Climate

Following the approach of Schlenker and Roberts (2009) and Ortiz-Bobea (2021), I control for weather variables in my models using PRISM data. I filter pixels that are classified as cropland or pastureland by the USGS National Land Cover Data Base. I aggregate monthly weather data over the growing season (April to September) to create a yearly panel. My precipitation measure represents the total millimeters of precipitation that a county receives in a growing season. I also include a squared precipitation term, since precipitation has a nonlinear effect on the agricultural outcomes of interest (Schlenker and Roberts 2009). Instead of focusing on average temperatures, I include the exposure of varying temperature levels by binning the hours spent at each Celsius degree. Similar to Annan and Schlenker (2015), my model includes moderate temperature exposure (total exposure from 10°C to 29°C) and extreme temperature exposure (total exposure at or above 30°C).

2.5   **Empirical Model**

2.5.1   Panel Model with Two-Way Fixed Effects

The main specification in this paper uses a panel model with two-way fixed effects (TWFE) to estimate how easements impact agricultural outcomes. My equation takes the form

$$Y_{it} = \beta_1 Wetland_{it} + \beta_2 Floodplain_{it} + \Gamma X_{it} + \alpha_i + \delta_t + \varepsilon_{it}.$$

The outcome variable of interest, Y, is the yield, loss cost ratio, and loss ratio for county i in year t. The crops of interest in this study are corn, soybeans, and wheat. When studying potential mechanisms, Y takes the value of acres planted, prevented planted, failed, and failed to harvest. I cluster standard errors at the state level, since state NRCS departments administer these programs. I take the inverse hyperbolic sine (IHS) of all the outcome variables except for the risk ratios. I prefer this transformation, as there are many zero-valued observations in the data and coefficients can then be interpreted as percent changes. I also apply a mean transformation to correct the magnitudes of the coefficients so I can interpret them as elasticities (Bellemare and Wichman 2020). When both the outcome (y) and treatment (x) are IHS, the elasticity equals $\left(b * \bar{x} * \sqrt{(\bar{y}^2 + 1)}\right) / \left(\bar{y} * \sqrt{(\bar{x}^2 + 1)}\right)$, where b is the coefficient after regressing IHS(y) on IHS(x), $\bar{x}$ is the mean of x, and $\bar{y}$ is the mean of y. When the treatment x is IHS but the outcome y is not, the semi-elasticity is $\left(b * \bar{x} * \sqrt{(\bar{y}^2 + 1)}\right) / \bar{y}$. The standard errors for the elasticities are then calculated using the delta method.

I include county-level fixed effects to account for observed and unobserved county factors that are time invariant. This allows me to use within county variation to reduce the threat of omitted variable bias. I also include year fixed effects. These control for both observable and unobservable factors changing across time that are consistent across counties. My identification strategy relies on the underlying assumption that conditional on the county and year, treatment is exogenous. To reduce the threat of omitted variable bias, I include relevant controls in my model. I account for planting-relevant variables that are common in the literature, such as precipitation, precipitation squared, moderate degree days, and extreme degree days.

The main treatment variables, *Floodplain* and *Wetland*, represent the floodplain and wetland eased land as a proportion of a county's total land area. The IHS of the treatment variables is used for ease of interpretation as there are many counties with zero easement acreage. The main source of identifying variation stems from variation across time and space in the closing of easement acres. The coefficient of interest β_1 measures the elasticity response of the chosen agricultural outcomes to a 100 percent increase in land share of wetland easement. The coefficient β_2 represents the elasticity response to a 100 percent increase in land share of floodplain easement. For ease of legibility, instead of measuring the response to a 1 percent increase in easement land share, I consider a "doubling" of land share in wetland and floodplain easement, or a 100 percent increase. Since land in easement is such a small percentage of acreage on average, a 100 percent increase in easement land share for a county is reasonable.

I use the closing year in my preferred specification, since this date is the most reasonably exogenous and the point in time that is associated with reduced risk. This is also the point at which a producer can no longer insure the parcel. The application date is heavily influenced by recent flooding and previous indemnities. This means that the treatment and outcome variables are co-determined. However, once a farmer decides to apply and enroll into the program, the rest of the process is in the hands of the NRCS. Meetings with NRCS directors and agents have shed light on the fact that the NRCS steps including closing, court recording, and restoration completion are somewhat random. Many potential hurdles may delay the process. It is often the case that various legal issues delay the closing and restoration process. For example, a previous utility contract may be unearthed and an agreement must be worked out between the different parties. Alternatively, sometimes there is trouble with accessing the parcel of land for the NRCS because of legalities with railroads and private roadways. There are many legal documents and processes that take a quasi-random amount of time to complete. For these reasons, I believe the timing of closing is reasonably exogenous.

The sample is restricted to counties that are east of the 100th meridian except for when I look at region heterogeneity. I include a county in the sample when that commodity is planted at least once during my time horizon. I use the NASS acreage and FSA acreage variable to create these sample groups.[5] So, for example, counties that plant corn at least once during the 30 years are included in the corn sample. Counties that never plant soybeans are omitted from the soybean sample. When calculating the mean of the treatment and outcome variables for the elasticity transformation, I use means specific to each commodity sub-sample. The mean of easement land share varies depending on the commodity sub-sample.

2.5.2   Limitations and Trends in TWFE Models

It is worth noting the limitations and current updates regarding panel models with two-way fixed effects. The two-way fixed effect strategy can also be interpreted as a difference-in-differences (DID) setup but with a staggered, continuous treatment variable. There has been a lot of recent work in the DID setting: decomposing the treatment effects, discerning how they are weighted, and understanding the underlying assumptions (Goodman-Bacon 2021; De Chaisemartin and D'haultfoeuille 2020; Callaway and Sant'Anna 2021). Alternative estimators have been specified to create the correct counterfactual groups and accurately weigh observations to find the average treatment effect in a variety of settings, especially in the canonical two-period DID setting. Currently, the literature is applying this logic to multi-period settings and cases when treatment is staggered and continuous (De Chaisemartin, D'haultfoeuille, and Guyonvarch 2019; Callaway et al. 2021). Callaway, Goodman-Bacon, and Sant'Anna (2021) propose a specification to correctly identify the causal effects of interest in a multi-period setting with variation in treatment timing and intensity as well as the needed parallel trend assumptions. The code for this alternate specification is still being developed. I use the traditional TWFE model here.

5. For NASS outcome variables, I use the NASS acreage commodity subsamples. For the FSA outcomes, I use the FSA acreage commodity sub-samples. There is not perfect overlap between the FSA and NASS groups. This is because the FSA sample is shorter and covers a shorter time span. But there are about 200 observations that belong in the FSA sample but are not in the NASS sample. I use the NASS sample of counties for the FSA outcomes and find similar results as a robustness check.

2.6   Results

This section reviews my findings from using a TWFE model. Tables 2A.1–2A.13 present regressions, and the online appendix presents coefficient plots to make the results easy to follow.[6]

2.6.1   How Do Easements Impact Crop Yields?

Table 2A.1 shows how wetland and floodplain easements impact corn, soybean, and wheat yields. As hypothesized by the theoretical model (hypothesis I), easements positively impact yields. For wetland easements, a 100 percent increase in land share of wetland easement is associated with 0.34 percent, 0.77 percent, and 0.46 percent increase in yields for corn, soybeans, and wheat. The estimates on floodplain easements are also positive but no longer statistically significant for corn and soybeans. There is evidence of significant increases in wheat yields of 0.13 percent after an increase in floodplain easement land share.

Table 2A.2 differentiates by the original land use of the easement. Eased land can be classified as cropland or non-cropland. Non-cropland is eased in order to connect eased cropland, improve drainage outcomes, and create more robust ecosystems. In 2020 in the main sample, there are 173,088 acres under floodplain easement of which 70,995 acres were originally cropland (41 percent). There are 2,133,094 wetland easement total acres closed and 1,198,473 acres were cropland (56 percent). Differentiating by the original land use uncovers the direct and indirect effect of easements. The estimates on cropland wetland and floodplain acres represent the direct and indirect effect of easements. The direct effect is the mechanical effect of taking land out of production and producers re-allocating their remaining resources. The indirect effect captures the effect of restoring land into a wetland and floodplain. The estimates on the non-cropland wetland and floodplain easements represent just the indirect effects of easements.

Table 2A.2 shows that doubling wetland crop acres has a positive, significant effect for soybeans and wheat. Doubling cropland in wetland easement increases corn yields by 0.14 percent, soybeans by 0.82 percent, and wheat by 0.33 percent. Doubling the land share of non-cropland into wetlands has a 0.22 percent, 0.29 percent, and 0.11 percent increase in yields for corn, soybeans and wheat; however, only the estimate for soybeans is significant. The results for floodplains differ in the fact that they are smaller in magnitude, and even negative at times. I believe the small magnitude is because of low variation and acreage in floodplain easement. Easing cropland into a floodplain has an insignificant effect for corn and wheat yields. Unexpectedly, doubling land share of cropland in floodplains decreases soybean yields by 0.06 percent. However, the indirect effect of floodplain easements is posi-

6. See http://www.nber.org/data-appendix/c14694/appendix.pdf

tive and significant for all three commodities. Doubling the share of land in non-cropland floodplain easements leads to a 0.14 percent, 0.06 percent, and 0.09 percent increase in corn, soybean, and wheat yields. This evidence lends support to hypothesis I that easements have an overall positive effect on agricultural production by increasing the average yields within a county. There seems to be different effects based on the easement type and the original use of the land.

Next, in table 2A.12, I explore the potential weather pathways by taking an approach similar to Annan and Schlenker (2015). The researchers look at how the portion of land that is insured impacts the effect of precipitation and degree days on crop yields. Similarly, I interact the share of wetland easements with moderate degree days, extreme degree days, precipitation, and precipitation squared. This allows me to see through which type of weather pathways easements impact crop yields. For corn, wetland easements reduce the effect of moderate degree days. Moderate degree days positively impact yields, so more land in easement will reduce the effect of moderate degree days. There is a similar story explaining the negative and significant interaction between wetland easement land share and precipitation. The interaction between wetland easements and precipitation squared is positive and significant for corn (although smaller than the interaction coefficient with just precipitation). This could emphasize that easements are effective at improving corn yields when precipitation is further from the optimal level and more extreme. For soybeans, I find that wetland easement land share mitigates the effect of extreme degree days on yields. Extreme degree days decrease soybean yields, and doubling wetland easements reduces this negative effect. Soybean fields are being taken out of production post-easement and yields are improving due to less damages from extreme degree days. For wheat, I do not find a significant effect of easements interacted with the weather pathways.

2.6.2   How Do Easements Impact Indemnities and Risk?

The next set of results evaluates how easements impact indemnities and agricultural risk (hypothesis II). I measure yield risk as the ratio of indemnities to liabilities as well as the ratio of indemnities to premiums paid. I do not take the inverse hyperbolic sine of the risk ratios so these results are interpreted as semi-elasticities. The subset of data here include only county observations that have a non-zero indemnity in that year.[7] Table 2A.3 shows how floodplain easements and wetland easements impact the loss cost ratio. Unlike with yields, I do not find a strong relationship between easement closing and reduced risk. I find no significant effects of wetland and floodplain easement land share on corn and wheat loss cost ratios. However, I do find that an increase in easement wetland acres reduces the loss cost ratio for

7. If a county has zero indemnities in a year, the loss cost ratio and loss ratio are undefined.

soybeans. Increasing wetland easements by 100 percent decreases soybean losses by $2.26 per dollar of liability. There is some evidence showing that soybean production is less risky post easement implementation.

To try to understand the types of agricultural losses that may be prevented by easements, I calculate the loss cost ratio for different subsets of indemnity types. Specifically, I create a separate loss cost ratio for excess moisture, flooding, drought, heat, disease, and insect losses. Table 2A.4 explores how the loss cost ratio for these different climate-related indemnities changes after an increase in easements. Even though disease and insects are not directly related to weather, research shows the changing climate has exacerbated pest and pathogen issues. Moreover, crops that experience extreme weather stress are more susceptible to disease and insect losses.

I find evidence that wetland easements significantly reduce indemnity losses from excess moisture, heat, disease, and insects. For soybeans, increasing wetland easements decreases losses from excess moisture by $3.59, from heat by $6.07, and from disease by $11.23 per dollar of liability. Doubling wetland easements significantly reduces insect losses by $8.50 per dollar of liability for corn; the coefficient for soybeans is almost identical but insignificant. These findings suggest that wetland easements could be used to improve agricultural resiliency, especially for soybean crops. Considering that climate change research predicts worsening excess moisture, heat, and disease conditions, easements provide a potential solution to mitigate costly crop losses.

I also find some evidence of increased drought risk associated with higher easement land share. Increasing wetland easements by 100 percent is associated with $3.80 more drought losses per dollar of wheat liability. Increasing floodplain easements by 100 percent leads to $0.46 and $0.32 more drought indemnities per dollar of corn and soybean liability, respectively. I posit that easements change the water patterns within a watershed and this may leave less water on remaining agricultural fields. This could increase the risk of drought for some fields.

To investigate how producer risk is impacted by easements, I regress the loss ratio on wetland and floodplain acres in table 2A.5. All the estimates are insignificant and noisy.

2.6.3   What Are the Potential Mechanisms?

This section explores the potential mechanisms through which easements may be impacting agricultural production. I look at how an increase in wetland and floodplain easement acres impact acres planted, harvested, failed, and prevented planted.

I start by looking at how planting behavior changes and examining how easement land share impacts acres planted. My model predicts that acres planted will decrease after an increase in easement land share (hypothesis III). Table 2A.6 uses NASS data on acreage planted that span the entire panel period. The estimates for floodplain easements are small and insignificant. I find that increasing wetland land share by 100 percent decreases acres planted of soybeans by 2 percent and acres planted of wheat by 1 percent. This is consistent with my hypothesis since easements take land out of production. Surprisingly, doubling wetland easement acres is associated with a 3 percent increase in corn acreage.

Table 2A.7 uses FSA acreage data, which have a shorter panel of data from 2009 to 2020, as a robustness check. The findings for wetland easements are similar but often smaller in magnitude and less significant. The results for floodplains are again insignificant and close to zero. Notably, doubling wetland easement land share leads to a −17 percent change in wheat acreage planted. Easements were focused on the wheat-growing regions from 2010 to 2020 and that led to a sizeable reduction in wheat acreage planted. The results in Table 2A.8 show that doubling easement land share impacts acres harvested. The acres harvested findings are almost identical to the acres planted results.

To test hypothesis IV, I estimate how easements affect acres failed to harvest, acres failed, and acres prevented planted. In table 2A.9a, I find that easements are associated with positive and negative changes in acres failed to harvest. Doubling wetland acres leads to a 3.17 percent increase in corn acres failed to harvest. This finding is consistent with the slippage narrative. More corn is planted, and this leads to higher corn harvest failure. Wetland easement land share has a negative but insignificant effect on soybean and wheat acres failed to harvest. Doubling floodplain acres has no significant effect on corn acres failed to harvest, increases soybean acres failed to harvest by 0.98 percent, and decreases wheat acres to harvest by −1.31 percent.

Table 2A.9b shows the results from the same regression as in table 2A.9a, but conditional on counties experiencing non-zero acres failed to harvest. The patterns are similar to the findings in table 2.9a. Increasing wetland easements land share increases corn acres failed to harvest by 1.66 percent. But, now doubling wetland easement land share significantly reduces acres failed to harvest for both soybeans and wheat by approximately 1 percent. Doubling floodplain easement still has no significant effect on corn, but continues to increase soybeans acres failed to harvest by less than 1 percent and decrease wheat acres failed to harvest by 0.83 percent.

In table 2A.10, I estimate how easement land share impacts acres failed using FSA data. Increasing wetland easements in a county by 100 percent decreases soybean acres failed by 10 percent and wheat acres failed by 21 percent. Increasing floodplain acres also decreases corn acres failed by about 5 percent while increasing wheat acres failed by 5 percent. There is some support for the hypothesis that easements reduces acres failed in some contexts, but also some contradictory findings. It is unclear from these results whether easements are associated with a reduction in acres failed.

Since easements are most likely to reduce losses from excess water and floods, I look at how easement land share impacts acres prevented planted using FSA data from 2009 to 2020. For wetland easements, increasing land share by 100 percent decreases acres prevented planted of soybeans and increases acres prevented planted of wheat, but these estimates are insignificant. Again, unexpectedly, doubling wetland easement land share increases acres prevented planted for corn by 43 percent. This deepens the implications, of the slippage effect. It seems that more land is being put toward corn production post-easement and this may be leading to higher corn losses. Floodplain easements are associated with reductions in acres prevented planted. Doubling land share in floodplain easement reduces prevented planted acreage for corn by 14 percent and soybeans by 9 percent. It seems that floodplain easements are successful at reducing the risks associated with prevented planting.

My results show that increasing wetland easement land share increases corn acreage planted, corn acreage failed to harvest, and corn acreage prevented planted. These findings run counter to my hypotheses, but evidence of a similar spillover effect, called slippage, has been associated with other conservation programs. Wu (2000) finds that a 100-acre increase in the Conservation Reserve Program (CRP) leads to the conversion of 20 acres from non-cropland into cropland, which offsets the ecosystem benefits of the program by 9–14 percent. The slippage effect is driven by increased output prices from the reduced supply as well as substitution effects in which producers begin producing on lower quality land. However, Roberts and Bucholtz (2005) replicate Wu's findings and do not find conclusive evidence of a slippage effect. More recently, Fleming (2014) uses satellite data and finds evidence of a mild slippage effect: an additional 100 acres of land enrolled in CRP leads to the conversion of 4 acres to cropland. Uchida (2014) uses Census of Agriculture panel data and also finds robust evidence of farm-level slippage effects of about 14 percent. Lichtenberg and Smith-Ramírez (2011) find evidence that land is reallocated to crop production when there are increases in participation in cost-sharing conservation programs. Pfaff and Robalino (2017) provide an overview of the slippage literature and a deeper discussion of the mechanisms behind it.

There are a few potential explanations for why the wetland easement program is associated with producers switching their production toward corn. This could be a case of input reallocation (Pfaff and Robalino 2017). Easements free up inputs such as labor and capital and producers look for new ways to raise profits on their lands. Producers often rotate a field between corn and soybeans to diversify their commodities, improve production, renew the soil nutrients, reduce erosion, balance the pest and weed communities, and decrease the need for fertilizers and herbicides. It is likely that the producers retired their riskiest fields and have less land to plant on post-easement. Because of these changes in their production choice set, it could be the case that farmers try to maximize profits by planting a more

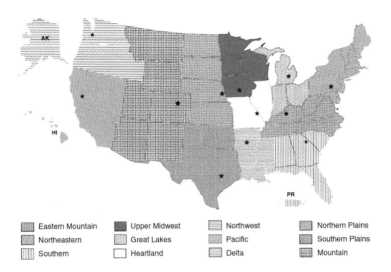

Eastern Mountain	Upper Midwest	Northwest	Northern Plains
Northeastern	Great Lakes	Pacific	Southern Plains
Southern	Heartland	Delta	Mountain

Fig. 2.14   Map of NASS regions from USDA

profitable commodity on their remaining fields. There is also some evidence suggesting that corn is more resilient against excess moisture and flooding. Producers that put land into easement may be taking other adaptive steps by producing more corn over soybeans. This mechanism is referred to as *learning* in Pfaff and Robalino (2017); conservation programs may encourage producers and their neighbors to engage in new practices. Continually cropping corn is more profitable but also more risky. Continuous corn cropping is also associated with some negative environmental externalities that could be counterproductive to easement goals.

2.6.4   How Do Easements Impact Each Region?

Finally, I explore heterogeneity in easement effects by region in table 2A.13. I look at how wetland easement and floodplain easement land share impact corn, soybean, and wheat yields by region. The NASS has 12 regional offices that are responsible for the statistical work of an area. These are often grouped by similarities in production. A map in figure 2.14 shows which are included in each region. This regional analysis deepens our understanding of which areas of the US are driving the results. The Southern region and Northwest region of the US actually experience decreased soybean and wheat yields after increased easement land share. However, most regions find positive or insignificant impacts of easement land share on yields. The Heartland, Northern Plains, and Southern Plains seem to be the most impacted by easement land share and see the biggest increases in yields. It may be interesting to further explore these states that see significant easement effects. This regional analysis may point policy makers toward which regions derive the highest agricultural benefits from easements.

2.6.5   How Do Benefits and Costs Compare?

I use the yield estimates from table 2A.1 to conduct a simple cost-benefit analysis of wetland easements. I calculate the ratio of yield benefits to the cost of easing land (analogous to the solution for the conservation agent's problem laid out in my theoretical model). This analysis does not include all ecosystem benefits associated with easements, nor does it include the administrative, maintenance, or enforcement costs of the program. However, the back-of-the-envelope accounting exercise is beneficial in understanding the potential scope of wetland easement effects as well as their cost-effectiveness.

Table 2.3 depicts the benefits and costs of doubling a county's land share in wetland easement by 100 percent and then aggregates to all counties in the main sample. The benefit equals the change in average revenue for a county from corn, soybeans, and wheat crops. Average revenue is the product of the average price per bushel, the average yield, and average acres planted of each commodity (averaged over 1989–2020). The change in revenue post-

Table 2.3   Cost benefit analysis of wetland easements

	Corn	Soybeans	Wheat	Total
Benefit				
Average price per bushel	3.36	8.18	4.50	
Average yield (bushels per acre)	122.20	37.57	48.90	
Average acres planted	46,580	48,673	17,493	
Average county revenue	19,125,375	14,958,313	3,849,335	
Estimate of yield increase after 100% increase in wetland easements	0.00338	0.00766	0.00457	
Increase in revenue after 100% increase in wetland easements	64,644	114,581	17,591	196,816
Present value of increased revenue at 5% interest rate	1,292,875	2,291,614	351,829	3,936,318
Number of counties in sample	1,871	1,762	1,716	
Total increase in revenue	2,419,969,825	4,037,823,054	603,738,885	7,060,531,764
Cost				
Average county wetland easement acres				616
Average per acre cost of wetland easement				2,657
Cost of 100% increase in wetland easements in a county				1,636,446
Number of counties in sample				1,871
Total cost				3,061,791,027
Benefit to cost ratio				
One county				2.41
All counties in sample				2.31

easement is the product of the average revenue and the yield estimates. Next, I find the present value (in perpetuity) by dividing by a 5 percent interest rate. The present value of benefits for one county is $3,936,318. After aggregating for all the counties in the sample, the present value of yield benefits is over US$7 billion.

To increase wetland acres by 100 percent, it would cost on average $1,636,446 for one county and $3 billion for all the counties in the main sample. To find the cost of doubling wetland easements, I find the product of the average acres in wetland easement and the average cost of purchasing easement acreage. The overall benefit to cost ratio is 2.41 for a county and 2.31 overall, which means the program is cost-effective. With an interest rate of 3 percent the benefit cost ratio becomes 4.01 (3.84). An interest rate of 10 percent brings the benefit cost ratio down to 1.20 (1.15). There is currently a small share of land in wetland easements, and these estimated benefits may represent the point that the highest marginal benefits are achieved. In the future, if there is more land in easement and less in cropland, the marginal benefit would be expected to diminish.

I find that my cost-benefit estimates are within range of other cost-benefit analyses estimates completed on the Conservation Reserve Programs and Wetland Restoration Programs. Miao et al. (2016) finds that CRP is not cost-effective when comparing the environmental benefits index to the contract rental rates. In the Indian Creek watershed of Iowa, Johnson et al. (2016) calculate a cost-benefit ratio of 1.3–4.8 after estimating the water quality, flood mitigation, air quality, and climate benefits. In a CEAP study, Hansen (2007) calculates a cost-benefit ratio of 0.70–0.85 for CRP lands when considering the soil erosion and wildlife habitat benefits. In a report on wetland ecosystems, Hansen et al. (2015) summarizes some cost-benefit analyses of wetland restorations. They predict average easement costs are between $160 and $6,100 per acre. Duck hunting benefits range from $0–$143 per acre, while reduced greenhouse gas emission benefits range from $0–$129 per acre. The review on flood mitigation benefits had high variance in reported benefits and did not lead to any conclusive estimates of avoided damages on croplands. My estimates of yield benefits may prove useful in future program cost-benefit analyses by capturing an additional ecosystem benefit of wetlands.

2.7 Conclusion

This paper presents novel evidence that wetland and floodplain easements increase crop yields for corn, soybeans, and wheat. I parse out the direct and indirect effect of easements by distinguishing by original land use. As expected, easing cropland directly improves yields by removing lower yielding land from production and reallocating inputs and labor toward surrounding cropland. Importantly, I also find that easing non-cropland increases some crop yields indirectly by improving production on non-eased fields. I study the mechanisms through which easements impact agricultural production. I find some evidence suggesting that easements reduce losses due to excess moisture, heat, disease, and insects. Easements increase yields by mitigating the effect of extreme precipitation and extreme degree days. There are mixed findings regarding how easements affect acres prevented and failed. Unexpectedly, I also find evidence of a slippage effect in which producers switch to more corn production following easement closing. This slippage effect may actually increase agricultural risk and be associated with increased drought risk and more corn acreage prevented planting. This study is a step toward a better understanding of the NRCS easement programs. Accounting for these spillover effects may be important when considering future field selection into the program. Quantifying these benefits may impact how policy makers fund future conservation efforts to adapt to a changing climate.

Appendix

Table 2A.1   **Effect of easements on crop yields (bushels/acre)**

	Corn yield (1)	Soybean yield (2)	Wheat yield (3)
100% Wetland Easement Acres	0.338***	0.766***	0.457***
	(0.108)	(0.191)	(0.113)
100% Floodplain Easement Acres	0.113	0.012	0.126***
	(0.087)	(0.056)	(0.047)
Observations	50,261	45,836	34,818
Number of Counties	1,871	1,762	1,716
R-squared	0.449	0.486	0.315
County FE	Yes	Yes	Yes
Year FE	Yes	Yes	Yes
Controls	Yes	Yes	Yes

Note: Estimates are transformed to interpret results as elasticities. Delta method used to calculate standard errors. *** $p < 0.01$; ** $p < 0.05$; * $p < 0.10$.

Table 2A.2   **Effect of cropland/non-cropland easements on crop yields (bushels/acre)**

	Corn Yield (1)	Soybean Yield (2)	Wheat Yield (3)
100% Crop Wetland Easement Acres	0.140	0.824***	0.333***
	(0.113)	(0.226)	(0.098)
100% Non-crop Wetland Easement Acres	0.217*	0.287**	0.115
	(0.117)	(0.117)	(0.111)
100% Crop Floodplain Easement Acres	−0.047	−0.061**	0.024
	(0.032)	(0.025)	(0.052)
100% Non-crop Floodplain Easement Acres	0.141**	0.060**	0.087**
	(0.062)	(0.027)	(0.035)
Observations	50,261	45,836	34,818
Number of Counties	1,871	1,762	1,716
R-squared	0.450	0.486	0.315
County FE	Yes	Yes	Yes
Year FE	Yes	Yes	Yes
Controls	Yes	Yes	Yes

Note: Estimates are transformed to interpret results as elasticities. Delta method used to calculate standard errors. *** $p < 0.01$; ** $p < 0.05$; * $p < 0.10$.

Table 2A.3   **Effect of easements on loss cost ratio**

	Corn Loss Cost Ratio (1)	Soybean Loss Cost Ratio (2)	Wheat Loss Cost Ratio (3)
100% Wetland Easement Acres	−0.421	−2.264**	0.337
	(1.155)	(1.023)	(0.708)
100% Floodplain Easement Acres	0.003	−0.001	−0.182
	(0.211)	(0.141)	(0.208)
Observations	41,905	42,869	37,830
R-squared	0.202	0.172	0.131
Number of Counties	1,775	1,664	1,603
County FE	Yes	Yes	Yes
Year FE	Yes	Yes	Yes
Controls	Yes	Yes	Yes

Note: Estimates are transformed to interpret results as semi-elasticities. Delta method used to calculate standard errors. *** $p < 0.01$; ** $p < 0.05$; * $p < 0.10$.

Table 2A.4   **Effect of easements on loss cost ratios with different indemnity causes**

	Corn Loss Cost Ratio (1)	Soybean Loss Cost Ratio (2)	Wheat Loss Cost Ratio (3)
Excess Moisture			
100% Wetland Easement Acres	0.978	–3.594*	0.203
	(2.282)	(1.843)	(1.501)
100% Floodplain Easement Acres	–0.092	–0.023	–0.306
	(0.504)	(0.209)	(0.329)
R-squared	0.149	0.137	0.168
Flood			
100% Wetland Easement Acres	–6.415	1.359	–12.098
	(5.681)	(3.376)	(8.468)
100% Floodplain Easement Acres	–0.081	–0.929	–1.008
	(1.308)	(1.158)	(2.091)
R-squared	0.034	0.054	0.018
Drought			
100% Wetland Easement Acres	–0.188	–1.594	3.797*
	(0.751)	(1.084)	(2.029)
100% Floodplain Easement Acres	0.458**	0.317*	–0.082
	(0.221)	(0.175)	(0.303)
R-squared	0.300	0.291	0.063
Heat			
100% Wetland Easement Acres	–2.397	–6.070***	–3.056
	(1.941)	(2.222)	(2.041)
100% Floodplain Easement Acres	–0.724	0.418	0.159
	(0.692)	(0.431)	(1.140)
R-squared	0.087	0.055	0.010
Disease			
100% Wetland Easement Acres	–6.822	–11.228***	1.488
	(8.920)	(4.116)	(2.649)
100% Floodplain Easement Acres	0.500	–1.409	–1.501
	(1.633)	(1.601)	(2.226)
R-squared	0.009	0.003	0.082
Insects			
100% Wetland Easement Acres	–8.502*	–8.512	14.251
	(4.555)	(12.054)	(9.961)
100% Floodplain Easement Acres	–0.504	0.244	3.312
	(1.321)	(1.419)	(2.556)
R-squared	0.005	0.004	0.019
Observations	44,905	42,869	37,830
Number of Counties	1,775	1,664	1,603
County FE	Yes	Yes	Yes
Year FE	Yes	Yes	Yes
Controls	Yes	Yes	Yes

Note: Estimates are transformed to interpret results as semi-elasticities. Delta method used to calculate standard errors. *** $p < 0.01$; ** $p < 0.05$; * $p < 0.10$.

Table 2A.5   **Effect of easements on loss ratio**

	Corn Loss Ratio (1)	Soybean Loss Ratio (2)	Wheat Loss Ratio (3)
100% Wetland Easement Acres	–0.152	–0.258	0.520
	(1.101)	(0.648)	(0.902)
100% Floodplain Easement Acres	0.059	–0.267	–0.117
	(0.205)	(0.188)	(0.235)
Observations	44,905	42,869	37,830
Number of Counties	1,775	1,664	1,603
R-squared	0.225	0.194	0.150
County FE	Yes	Yes	Yes
Year FE	Yes	Yes	Yes
Controls	Yes	Yes	Yes

Note: Estimates are transformed to interpret results as semi-elasticities. Delta method used to calculate standard errors. *** $p < 0.01$; ** $p < 0.05$; * $p < 0.10$.

Table 2A.6   **Effect of easements on acres planted (NASS)**

	Corn Acres Planted (1)	Soybean Acres Planted (2)	Wheat Acres Planted (3)
100% Wetland Easement Acres	3.164***	–2.076***	–1.117*
	(0.872)	(0.506)	(0.679)
100% Floodplain Easement Acres	0.127	0.425	–0.295
	(0.216)	(0.364)	(0.277)
Observations	50,276	45,838	34,831
Number of Counties	1,871	1,762	1,717
R-squared	0.092	0.152	0.236
County FE	Yes	Yes	Yes
Year FE	Yes	Yes	Yes
Controls	Yes	Yes	Yes

Note: Estimates are transformed to interpret results as elasticities. Delta method used to calculate standard errors. *** $p < 0.01$; ** $p < 0.05$; * $p < 0.10$.

Table 2A.7   **Effect of easements on acres planted (FSA)**

	Corn Acres Planted (1)	Soybean Acres Planted (2)	Wheat Acres Planted (3)
100% Wetland Easement Acres	0.683	−3.260*	−16.777**
	(1.475)	(1.667)	(7.327)
100% Floodplain Easement Acres	0.246	−0.107	−0.526
	(0.431)	(0.650)	(1.797)
Observations	18,243	17,799	17,156
Number of Counties	1,785	1,706	1,653
R-squared	0.041	0.039	0.174
County FE	Yes	Yes	Yes
Year FE	Yes	Yes	Yes
Controls	Yes	Yes	Yes

Note: Estimates are transformed to interpret results as elasticities. Delta method used to calculate standard errors. *** $p < 0.01$; ** $p < 0.05$; * $p < 0.10$.

Table 2A.8   **Effect of easements on acres harvested (NASS)**

	Corn Acres Harvested (1)	Soybean Acres Harvested (2)	Wheat Acres Harvested (3)
100% Wetland Easement Acres	3.138***	−2.109***	−0.976
	(0.870)	(0.520)	(0.755)
100% Floodplain Easement Acres	0.140	0.417	−0.205
	(0.249)	(0.371)	(0.267)
Observations	50,243	45,836	34,793
Number of Counties	1,871	1,762	1,713
R-squared	0.095	0.158	0.237
County FE	Yes	Yes	Yes
Year FE	Yes	Yes	Yes
Controls	Yes	Yes	Yes

Note: Estimates are transformed to interpret results as elasticities. Delta method used to calculate standard errors. *** $p < 0.01$; ** $p < 0.05$; * $p < 0.10$.

Table 2A.9a   **Effect of easements on acres failed to harvest (NASS)**

	Corn Acres Failed to Harvest (1)	Soybean Acres Failed to Harvest (2)	Wheat Acres Failed to Harvest (3)
100% Wetland Easement Acres	3.166**	−1.204	−1.404
	(1.440)	(0.782)	(0.880)
100% Floodplain Easement Acres	0.119	0.979**	−1.305**
	(0.216)	(0.404)	(0.547)
Observations	50,242	45,836	34,793
Number of Counties	1,871	1,762	1,713
R-squared	0.046	0.053	0.054
County FE	Yes	Yes	Yes
Year FE	Yes	Yes	Yes
Controls	Yes	Yes	Yes

Note: Estimates are transformed to interpret results as elasticities. Delta method used to calculate standard errors. *** $p < 0.01$; ** $p < 0.05$; * $p < 0.10$.

Table 2A.9b   **Effect of easements on non-zero acres failed to harvest (NASS)**

	Corn Acres Failed to Harvest (1)	Soybean Acres Failed to Harvest (2)	Wheat Acres Failed to Harvest (3)
100% Wetland Easement Acres	1.656*	−1.186***	−1.380*
	(0.915)	(0.357)	(0.716)
100% Floodplain Easement Acres	0.055	0.662**	−0.826***
	(0.163)	(0.264)	(0.261)
Observations	49,112	42,006	33,050
Number of Counties	1,865	1,745	1,695
R-squared	0.075	0.116	0.095
County FE	Yes	Yes	Yes
Year FE	Yes	Yes	Yes
Controls	Yes	Yes	Yes

Note: Estimates are transformed to interpret results as elasticities. Delta method used to calculate standard errors. *** $p < 0.01$; ** $p < 0.05$; * $p < 0.10$.

Table 2A.10   **Effect of easements on acres failed (FSA)**

	Corn Acres Failed (1)	Soybean Acres Failed (2)	Wheat Acres Failed (3)
100% Wetland Easement Acres	3.159	−10.445**	−21.045***
	(7.581)	(5.073)	(4.738)
100% Floodplain Easement Acres	−5.770*	2.309	5.447**
	(3.069)	(2.299)	(2.724)
Observations	18,243	17,799	17,156
Number of Counties	1,785	1,706	1,653
R-squared	0.074	0.026	0.066
County FE	Yes	Yes	Yes
Year FE	Yes	Yes	Yes
Controls	Yes	Yes	Yes

Note: Estimates are transformed to interpret results as elasticities. Delta method used to calculate standard errors. *** $p < 0.01$; ** $p < 0.05$; * $p < 0.10$.

Table 2A.11   **Effect of easements on acres prevented planted (FSA)**

	Corn Acres Prevented Planted (1)	Soybean Acres Prevented Planted (2)	Wheat Acres Prevented Planted (3)
100% Wetland Easement Acres	43.446***	−5.716	4.373
	(13.855)	(6.529)	(9.977)
100% Floodplain Easement Acres	−14.014***	−8.936**	2.904
	(3.387)	(4.418)	(4.885)
Observations	18,243	17,799	17,156
Number of Counties	1,785	1,706	1,653
R-squared	0.278	0.282	0.154
County FE	Yes	Yes	Yes
Year FE	Yes	Yes	Yes
Controls	Yes	Yes	Yes

Note: Estimates are transformed to interpret results as elasticities. Delta method used to calculate standard errors. *** $p < 0.01$; ** $p < 0.05$; * $p < 0.10$.

Table 2A.12   **Effect of easements on yields through weather pathways**

	Corn Yield (1)	Soybean yield (2)	Wheat Yield (3)
100% Wetland	5.4413**	–0.6018	0.7384
	(2.4398)	(1.9986)	(3.4051)
Moderate degree days	0.0066***	0.0061***	0.0029
	(0.0013)	(0.0017)	(0.0024)
× 100% Wetland	–0.0206*	–0.0002	–0.0019
	(0.0118)	(0.0087)	(0.0152)
Extreme degree days	–0.0145***	–0.0135***	–0.0006
	(0.0023)	(0.0025)	(0.0033)
× 100% Wetland	0.0035	0.0391***	0.0060
	(0.0110)	(0.0091)	(0.0145)
Precipitation (100mm)	0.0901***	0.1108***	0.0407**
	(0.0174)	(0.0180)	(0.0156)
× 100% Wetland	–0.5553***	0.0188	–0.0749
	(0.1941)	(0.1908)	(0.3379)
Precipitation squared (100mm)	–0.0066***	–0.0071***	–0.0051***
	(0.0012)	(0.0011)	(0.0012)
× 100% Wetland	0.0316**	–0.0019	0.0069
	(0.0124)	(0.0117)	(0.0261)
Observations	50,261	45,836	34,818
Number of Counties	1,871	1,762	1,716
R-squared	0.451	0.489	0.314
County FE	Yes	Yes	Yes
Year FE	Yes	Yes	Yes

Note: Wetland easement estimates and interactions are transformed to interpret results as elasticities. Delta method used to calculate standard errors. $***\ p < 0.01$; $**\ p < 0.05$; $*\ p < 0.10$.

Table 2A.13   **Effect of easements on crop yields by NASS region**

	Corn yield	Soybean yield	Wheat yield
Northeastern			
100% Wetland	−0.037	−0.003	−0.162
	(0.119)	(0.060)	(0.145)
100% Floodplain	−0.026	−0.013	−0.093***
	(0.026)	(0.016)	(0.015)
N	4,157	2,945	2,600
Eastern Mountain			
100% Wetland	0.038	0.058*	0.144**
	(0.046)	(0.034)	(0.061)
100% Floodplain	−0.046	0.041	0.113***
	(0.049)	(0.033)	(0.036)
N	9,362	8,069	6,890
Southern			
100%	−0.379	−0.288**	−0.059
	(0.328)	(0.117)	(0.154)
100% Floodplain	−0.096	0.005	−0.562***
	(0.064)	(0.007)	(0.030)
N	4,566	3,691	2,997
Great Lakes			
100% Wetland	−0.166	−0.047	0.113
	(0.132)	(0.104)	(0.190)
100% Floodplain	0.026	0.104***	0.040
	(0.032)	(0.018)	(0.052)
N	7,228	6,694	5,943
Upper Midwest			
100% Wetland	0.219	0.029	0.064
	(0.197)	(0.158)	(0.285)
100% Floodplain	0.092	0.033	0.164
	(0.089)	(0.041)	(0.206)
N	7,528	7,339	2,215
Heartland			
100% Wetland	0.378**	0.406**	0.301
	(0.147)	(0.169)	(0.309)
100% Floodplain	0.050	0.065**	−0.001
	(0.045)	(0.033)	(0.049)
N	5,791	5,787	5,023
Delta			
100% Wetland	0.034	1.839***	0.692
	(0.259)	(0.399)	(0.489)
100% Floodplain	−0.084	0.010	−0.063
	(0.084)	(0.151)	(0.101)
N	2,845	3,488	2,128
Northern Plains			
100% Wetland	0.969***	0.291*	1.367***
	(0.191)	(0.152)	(0.294)
100% Floodplain	0.395***	0.079	0.232***
	(0.111)	(0.066)	(0.083)
N	8,820	7,536	7,567

	Corn yield	Soybean yield	Wheat yield
Southern Plains			
100% Wetland	−0.003	−0.237	1.037***
	(0.070)	(0.294)	(0.316)
100% Floodplain	0.031***	0.077***	0.003
	(0.005)	(0.009)	(0.029)
N	3,764	2,012	6,229
Mountain			
100% Wetland	1.076**	–	0.240
	(0.504)		(0.218)
100% Floodplain	0.319***	–	−0.033
	(0.079)		(0.057)
N	1,742	–	2,845
Northwest			
100% Wetland	−35.659**	–	0.369
	(15.949)		(0.399)
100% Floodplain	−2.974***	–	0.077
	(0.900)		(0.059)
N	721	–	2,088
Pacific			
100% Wetland	7.064*	–	0.021
	(4.053)		(0.954)
100% Floodplain	1.184	–	0.681**
	(0.854)		(0.286)
N	517	–	724

Note: Estimates are transformed to interpret results as elasticities. Mean of each region is used. Delta method used to calculate standard errors. *** $p < 0.01$; ** $p < 0.05$; * $p < 0.10$.

References

Annan, F., and W. Schlenker. 2015. "Federal Crop Insurance and the Disincentive to Adapt to Extreme Heat." *American Economic Review* 105 (5): 262–66.

Babcock, B., P. Lakshminarayan, J. Wu, and D. Zilberman. 1996. "The Economics of a Public Fund for Environmental Amenities: A Study of CRP Contracts." *American Journal of Agricultural Economics* 78: 961–71.

Bellemare, M. F., and C. J. Wichman. 2020. "Elasticities and the Inverse Hyperbolic Sine Transformation." *Oxford Bulletin of Economics and Statistics* 82 (1): 50–61.

Bostian, M. B., and A. T. Herlihy. 2014. "Valuing Tradeoffs between Agricultural Production and Wetland Condition in the U.S. Mid-Atlantic Region." *Ecological Economics* 105: 284–91.

Boxall, P. C., O. Perger, K. Packman, and M. Weber. 2017. "An Experimental Examination of Target Based Conservation Auctions." *Land Use Policy* 63: 592–600.

Boyer, C. N., E. Park, and S. D. Yun. 2022. "Corn and Soybean Prevented Planting Acres Response to Weather." *Applied Economic Perspectives and Policy*. February.

Brown, L. K., E. Troutt, C. Edwards, B. Gray, and W. Hu. 2011. "A Uniform Price Auction for Conservation Easements in the Canadian Prairies." *Environmental and Resource Economics* 50 (1): 49–60.

Brown, Ralph J. 1976. "A study of the impact of the wetlands easement program on agricultural land values." *Land Economics* 52 (4): 509–517.

Burke, M., and K. Emerick. 2016. "Adaptation to Climate Change: Evidence from US Agriculture." *American Economic Journal: Economic Policy* 8 (3): 106–40.

Callaway, Brantly, Andrew Goodman-Bacon, and Pedro HC Sant'Anna. "Difference-in-differences with a continuous treatment." arXiv preprint arXiv:2107.02637 (2021).

Callaway, B., and P. H. Sant'Anna. 2021. Difference-in-Differences with multiple time periods. *Journal of Econometrics* 225 (2): 200–230.

Coble, K. H., T. O. Knight, R. D. Pope, and J. R. Williams. 1997. "An Expected-Indemnity Approach to the Measurement of Moral Hazard in Crop Insurance." *American Journal of Agricultural Economics* 79 (1): 216–26.

Congressional Budget Office (CBO). 2019. "CBO's May 2019 Baseline for Farm Programs." pp. 55.

Congressional Research Service. 2021. "Federal Crop Insurance: A Primer." Technical report.

Costello, C., and S. Polasky. 2004. "Dynamic Reserve Site Selection." *Resource and Energy Economics* 26 (2): 157–74.

De Chaisemartin, C., and X. D'haultfoeuille. 2020. "Two-Way Fixed Effects Estimators with Heterogeneous Treatment Effects." *American Economic Review* 110 (9): 2964–2996.

De Chaisemartin, C., X. D'haultfoeuille, and Y. Guyonvarch. 2019. "Fuzzy Differences-in-Differences with Stata." *The Stata Journal* 1–22.

Deschênes, O., and M. Greenstone. 2012. "The Economic Impacts of Climate Change: Evidence from Agricultural Output and Random Fluctuations in Weather: Reply." *American Economic Review* 102 (7): 3761–3773.

Deutsch, C. A., J. J. Tewksbury, M. Tigchelaar, D. S. Battisti, S. C. Merrill, R. B. Huey, and R. L. Naylor. 2018. "Increase in Crop Losses to Insect Pests in a Warming Climate." *Science* 361: 916–19.

Eck, M. A., A. R. Murray, A. R. Ward, and C. E. Konrad. 2020. "Influence of Growing Season Temperature and Precipitation Anomalies on Crop Yield in the Southeastern United States." *Agricultural and Forest Meteorology* 291: 108053.

English, B. C., A. S. Smith, J. R. Menard, D. W. Hughes, and M. Gunderson. 2021. "Estimated
Economic Impacts of the 2019 Midwest Floods." *Economics of Disasters and Climate Change* 5: 431–48.

Ferraro, P. J. 2008. "Asymmetric Information and Contract Design for Payments for Environmental Services." *Ecological Economics* 65 (4): 810–21.

Fleckenstein, M., A. Lythgoe, J. Lu, N. Thompson, O. Doering, S. Harden, J. M. Getson, and L. Prokopy. 2020. "Crop insurance: A Barrier to Conservation Adoption? *Journal of Environmental Management* 276: 111223.

Fleming, D. A. 2014. "Slippage Effects of Land-Based Policies: Evaluating the Conservation Reserve Program Using Satellite Imagery." *Papers in Regional Science* 93 (S1): S167–S178.

Fleming, P., E. Lichtenberg, and D. A. Newburn. 2018. "Evaluating Impacts of Agricultural Cost Sharing on Water Quality: Additionality, Crowding In, and Slippage." *Journal of Environmental Economics and Management* 92: 1–19.

Gelso, B. R., J. A. Fox, and J. M. Peterson. 2008. "Farmers' Perceived Costs of Wetlands: Effects of Wetland Size, Hydration, and Dispersion." *American Journal of Agricultural Economics* 90 (1): 172–85.

Glauber, J. W. 2004. "Crop Insurance Reconsidered Published." *American Journal of Agricultural Economics* 86 (5): 1179–195.

Glauber, J. W. 2013. "The Growth of the Federal Crop Insurance Program, 1990–2011." *American Journal of Agricultural Economics* 95 (2): 482–88.

Gleason, R. A., M. K. Laubhan, and N. H. Euliss. 2008. "Ecosystem Services Derived from Wetland Conservation Practices in the United States Prairie Pothole Region with an Emphasis on the U.S. Department of Agriculture Conservation Reserve and Wetlands Reserve Programs." U.S. Geological Professional Paper 1745, 58.

Goodman-Bacon, A. 2021. "Difference-in-Differences with Variation in Treatment Timing." *Journal of Econometrics* 225 (2): 254–77.

Goodwin, B. K., and N. E. Piggott. 2020. "Has Technology Increased Agricultural Yield Risk? Evidence from the Crop Insurance Biotech Endorsement." *American Journal of Agricultural Economics* 102 (5): 1578–1597.

Gourevitch, J. D., N. K. Singh, J. Minot, K. B. Raub, D. M. Rizzo, B. C. Wemple, and T. H. Ricketts. 2020. "Spatial Targeting of Floodplain Restoration to Equitably Mitigate Flood Risk." *Global Environmental Change* 61.

Hagerty, N. 2022. "Adaptation to Surface Water Scarcity in Irrigated Agriculture." Working Paper.

Hansen, L. 2007. "Conservation Reserve Program: Environmental Benefits Update." *Agricultural and Resource Economics Review* 36 (2): 267–80.

Hansen, L. R., D. Hellerstein, M. Ribaudo, J. Williamson, D. Nulph, C. Loesch, and W. Crumpton. 2015. "Targeting Investment to Cost Effectively Restore and Protect Wetland Ecosystems: Some Economic Insights." Technical report, U.S. Department of Agriculture, Economic Research Service.

Hatfield, Jerry L., Kenneth J. Boote, Bruce A. Kimball, L. H. Ziska, Roberto C. Izaurralde, Donald Ort, Allison M. Thomson, and D. Wolfe. 2011. "Climate impacts on agriculture: implications for crop production." *Agronomy Journal* 103 (2): 351–370.

Hatfield, J. L., and J. H. Prueger. 2015. "Temperature Extremes: Effect on Plant Growth and Development." *Weather and Climate Extremes* 10: 4–10.

Hebblethwaite, J. F., and C. N. Somody. 2008. "Progress in Best Management Practices." Chapter 32 in *The Triazine Herbicides. Amsterdam*: Elsevier.

Heimlich, R. E. 1994. "Costs of an Agricultural Wetland Reserve." *Land Economics* 70 (2): 234–46.

Hellerstein, D., N. Higgins, and M. Roberts. 2015. "Options for Improving Conservation Programs: Insights from Auction Theory and Economic Experiments." *The Conservation Reserve Program: Issues and Considerations* 181; 79–139.

Hornbeck, R., and P. Keskin. 2014. "The Historically Evolving Impact of the Ogallala Aquifer: Agricultural Adaptation to Groundwater and Drought." *American Economic Journal: Applied Economics* 6 (1): 190–219.

Horowitz, J. K., and E. Lichtenberg. 1993. "Insurance, Moral Hazard, and Chemical Use in Agriculture." *American Journal of Agricultural Economics* 75 (4): 926–35.

IPCC. 2012. "Managing the Risks of Extreme Events and Disasters to Advance Climate Change Adaptation: Special Report of the Intergovernmental Panel on Climate Change." Technical report. Cambridge: Cambridge University Press.

Jabran, K., S. Florentine, and B. S. Chauhan. (2020). "Impacts of Climate Change on Weeds, Insect Pests, Plant Diseases and Crop Yields: Synthesis." In: *Crop Protection Under Changing Climate*, edited by K. Jabran, S. Florentine, and B. Chauhan. New York: Springer. https://doi.org/10.1007/978-3-030-46111-9_8.

Johnson, K. A., B. J. Dalzell, M. Donahue, J. Gourevitch, D. L. Johnson, G. S. Karlovits, B. Keeler, and J. T. Smith. 2016. "Conservation Reserve Program (CRP) Lands Provide Ecosystem Service Benefits That Exceed Land Rental Payment Costs." *Ecosystem Services* 18: 175–85.

Kim, T., and M.-K. Kim. 2018. "Ex-post Moral Hazard in Prevented Planting." *Agricultural Economics* 49 (6): 671–80.

Kirwan, B., R. Lubowski, and M. Roberts. 2005. "How Cost-Effective Are Land Retirement Auctions? Estimating the Difference between Payments and Willingness to Accept in the Conservation Reserve Program." *American Journal of Agricultural Economics* 87 (5): 1239–1247.

Kousky, C., and M. Walls. 2014. "Floodplain Conservation as a Flood Mitigation Strategy: Examining Costs and Benefits." *Ecological Economics* 104: 119–28.

Kucharik, C. J. 2008. "Contribution of Planting Date Trends to Increased Maize Yields in the Central United States." *Agronomy Journal* 100 (2): 328–36.

Lawley, C., and C. Towe. 2014. "Capitalized Costs of Habitat Conservation Easements." *American Journal of Agricultural Economics* 96 (3): 657–72.

Lichtenberg, E., and R. Smith-Ramírez. 2011. "Slippage in Conservation Cost Sharing." *American Journal of Agricultural Economics* 93 (1): 113–29.

Mérel, P., and M. Gammans. 2021. "Climate Econometrics: Can the Panel Approach Account for Long-Run Adaptation?" *American Journal of Agricultural Economics* 103 (4): 1207–1238.

Miao, R., H. Feng, D. A. Hennessy, and X. Du. 2016. "Assessing Cost-Effectiveness of the Conservation Reserve Program (CRP) and Interactions between the CRP and Crop Insurance." *Land Economics* 92 (4): 593–617.

Miao, R., D. A. Hennessy, and H. Feng. 2016. "Grassland Easement Evaluation and Acquisition: an Integrated Framework." *Agricultural and Applied Economics Association Annual Meeting*, 51.

Mushet, D. M., and C. L. Roth. 2020. "Modeling the Supporting Ecosystem Services of Depressional Wetlands in Agricultural Landscapes." *Wetlands* 40 (5): 1061–1069.

Narloch, U., U. Pascual, and A. G. Drucker. 2013. "How to Achieve Fairness in Payments for Ecosystem Services? Insights from Agrobiodiversity Conservation Auctions." *Land Use Policy* 35: 107–18.

Newburn, D. A., P. Berck, and A. M. Merenlender. 2006. "Habitat and Open Space at Risk of Land-Use Conversion: Targeting Strategies for Land Conservation." *American Journal of Agricultural Economics* 88 (1): 28–42.

Nickerson, C. J., and L. Lynch. 2001. "The Effect of Farmland Preservation Programs on Farmland Prices." *American Journal of Agricultural Economics* 83 (2): 341–51.

NOAA. 2020. "NOAA National Centers for Environmental Information, State of the Climate: National Climate Report for Annual 2019." Technical report.

NOAA National Centers for Environmental Information (NCEI). 2023. "U.S. Billion-Dollar Weather and Climate Disasters." Technical report. https://www .ncei.noaa.gov/access/billions/, DOI: 10.25921/stkw-7w73.

NRCS. 2011. "Conservation of Wetlands in Agricultural Landscapes of the United States: Summary of the CEAP-Wetlands Literature Synthesis." Technical report.

NRCS 2012. "Restoring America's Wetlands: A Private Lands Conservation Success Story." Technical report. Washington, DC.

NRCS. 2021a. EWP Floodplain Easement Program-Floodplain Easement Option EWPP-FPE. www.nrcs.usda.gov/wps/portal/nrcs/detail/national/programs/land scape/ewpp/?cid=nrcs143 008216.

NRCS. 2021b. Wetlands. www.nrcs.usda.gov/wps/portal/nrcs/main/national/water /wetlands/?cid=stelprdb1043554.

NRCS. 2021c. Wetlands Reserve Program. www.nrcs.usda.gov/wps/portal/nrcs /detail/null/?cid=nrcs143 008419.

Ortiz-Bobea, Ariel. 2021. "The empirical analysis of climate change impacts and adaptation in agriculture." In *Handbook of Agricultural Economics*, volume 5, 3981–4073. Amsterdam: Elsevier.

Parker, D. P., and W. N. Thurman. 2018. "Tax Incentives and the Price of Conservation." *Journal of the Association of Environmental and Resource Economists* 5 (2): 331–69.

Perry, E. D., J. Yu, and J. Tack. 2020. "Using Insurance Data to Quantify the Multidimensional Impacts of Warming Temperatures on Yield Risk." *Nature Communications* 11 (1): 1–9.

Pfaff, A., and J. Robalino. 2017. "Spillovers from Conservation Programs." *Annual Review of Resource Economics* 9: 299–315.

Roberts, M. J., and S. Bucholtz. 2005. "Slippage Effects of the Conservation Reserve Program or Spurious Correlation? A Comment." *American Journal of Agricultural Economics* 87 (1): 244–50.

Roley, S. S., J. L. Tank, J. C. Tyndall, and J. D. Witter. 2016. "How Cost-Effective Are Cover Crops, Wetlands, and Two-Stage Ditches for Nitrogen Removal in the Mississippi River Basin? *Water Resources and Economics* 15:43–56.

Rosenzweig, C., J. Elliott, D. Deryng, A. C. Ruane, C. Müller, A. Arneth, K. J. Boote, C. Folberth, M. Glotter, N. Khabarov, K. Neumann, F. Piontek, T. A. M. Pugh, E. Schmid, E. Stehfest, H. Yang, and J. W. Jones. 2014. "Assessing Agricultural Risks of Climate Change in the 21st Century in a Global Gridded Crop Model Intercomparison." *Proceedings of the National Academy of Sciences* 111 (9): 3268–3273.

Rosenzweig, C., F. N. Tubiello, R. Goldberg, E. Mills, and J. Bloomfield. 2002. "Increased Crop Damage in the US from Excess Precipitation under Climate Change." *Global Environmental Change* 12: 197–202.

Savary, S., L. Willocquet, S. J. Pethybridge, P. Esker, N. McRoberts, and A. Nelson. 2019. "The Global Burden of Pathogens and Pests on Major Food Crops." *Nature Ecology & Evolution* 3: 430–39.

Schlenker, W., and M. J. Roberts. 2009. "Nonlinear Temperature Effects Indicate Severe Damages to U.S. Crop Yields under Climate Change." *Proceedings of the National Academy of Sciences* 106 (37): 15594–15598.

Schnapp, F., J. Driscoll, T. Zacharias, and R. Josephson. 2000. "Ratemaking Consideration for Multiple Peril Crop Insurance." Report prepared for USDA/Risk Management Agency.

Schulte, L. A., J. Niemi, M. J. Helmers, M. Liebman, J. G. Arbuckle, D. E. James, R. K. Kolka, M. E. O'Neal, M. D. Tomer, J. C. Tyndall, H. Asbjornsen, P. Drobney, J. Neal, G. V. Ryswyk, and C. Witte. 2017. "Prairie Strips Improve Biodiversity and the Delivery of Multiple Ecosystem Services from Corn-Soybean Croplands." *Proceedings of the National Academy of Sciences* 114 (50): 11247–11252

Seifert, C. A., M. J. Roberts, and D. B. Lobell. 2017. "Continuous Corn and Soybean Yield Penalties across Hundreds of Thousands of Fields." *Agronomy Journal* 109 (2): 541–48.

Shirzaei, M., M. Khoshmanesh, C. Ojha, S. Werth, H. Kerner, G. Carlson, S. F. Sherpa, G. Zhai, and J. C. Lee. 2021. "Persistent Impact of Spring Floods on Crop Loss in U.S. Midwest." *Weather and Climate Extremes* 34: 100392.

Shoemaker, R. 1989. "Agricultural Land Values and Rents under the Conservation Reserve Program." *Land Economics* 65 (2): 131–37.

Shultz, S. D., and S. J. Taff. 2004. "Implicit Prices of Wetland Easements in Areas of Production Agriculture." *Land Economics* 80 (4): 501–12.

Sloat, L. L., S. J. Davis, J. S. Gerber, F. C. Moore, D. K. Ray, P. C. West, and N. D. Mueller. 2020. "Climate Adaptation by Crop Migration." *Nature Communications* 11 (1243): 1–9.

Smith, V. H., and B. K. Goodwin. 1996. "Crop Insurance, Moral Hazard, and Agricultural Chemical Use." *American Journal of Agricultural Economics* 78 (2): 428–38.

Sonnier, G., P. J. Bohlen, H. M. Swain, S. L. Orzell, E. L. Bridges, and E. H. Boughton. 2018. "Assessing the Success of Hydrological Restoration in Two Conservation Easements within Central Florida Ranchland." *PLoS ONE* 13 (7): e0199333.

Speir, S. L., J. L. Tank, and U. H. Mahl. 2020. "Quantifying Denitrification following Floodplain Restoration via the Two-Stage Ditch in an Agricultural Watershed." *Ecological Engineering* 155: 105945.

Tack, J., K. Coble, and B. Barnett. 2018. "Warming Temperatures Will Likely Induce Higher Premium Rates and Government Outlays for the U.S. Crop Insurance Program." *Agricultural Economics* 49 (5): 635–47.

Taylor, C. A., and H. Druckenmiller. 2022. "Wetlands, Flooding, and the Clean Water Act." *American Economic Review* 112 (4): 1334–1363.

Uchida, S. 2014. "Indirect Land Use Effects of Conservation: Disaggregate Slippage in the US Conservation Reserve Program." Working paper. University of Maryland.

Urban, D. W., M. J. Roberts, W. Schlenker, and D. B. Lobell. 2015. "The Effects of Extremely Wet Planting Conditions on Maize and Soybean Yields." *Climatic Change* 130 (2): 247–60.

USDA 2021. NRCS Conservation Programs. www.nrcs.usda.gov/Internet/NRCS RCA/reports/cp nat.html.

Watson, K. B., T. Ricketts, G. Galford, S. Polasky, and J. O'Niel-Dunne. 2016. "Quantifying Flood Mitigation Services: The Economic Value of Otter Creek Wetlands and Floodplains to Middlebury, VT." *Ecological Economics* 130: 16–24.

Wing, I. S., E. De Cian, and M. N. Mistry. 2021. "Global Vulnerability of Crop Yields to Climate Change." *Journal of Environmental Economics and Management* 109: 102462.

Woodard, J. D., B. J. Sherrick, and G. D. Schnitkey. 2011. "Actuarial Impacts of Loss Cost Ratio Ratemaking in U.S. Crop Insurance Programs." *Journal of Agricultural and Resource Economics* 36 (1): 211–28.

Wu, J. 2000. "Slippage Effects of the Conservation Reserve Program." *Journal of Environmental Economics and Management* 82 (November): 979–92.

Wu, J., D. Zilberman, and B. A. Babcock. 2001. "Environmental and Distributional Impacts of Conservation Targeting Strategies." *Journal of Environmental Economics and Management* 41 (3): 333–50.

Wu, S., B. K. Goodwin, and K. Coble. 2020. "Moral Hazard and Subsidized Crop Insurance." *Agricultural Economics* 51 (1): 131–42.

Yu, J., and N. P. Hendricks. 2020. "Input Use Decisions with Greater Information on Crop Conditions: Implications for Insurance Moral Hazard and the Environment." *American Journal of Agricultural Economics* 102 (3): 826–45.

Yu, J., and D. A. Sumner 2018. "Effects of Subsidized Crop Insurance on Crop Choices." *Agricultural Economics* 49 (4): 533–45.

Zipper, S. C., J. Qiu, and C. J. Kucharik. 2016. "Drought Effects on US Maize and Soybean Production: Spatiotemporal Patterns and Historical Changes." *Environmental Research Letters* 11: 094021.

The Cost-Effectiveness of Irrigation Canal Lining and Piping in the Western United States

R. Aaron Hrozencik, Nicholas A. Potter, and Steven Wallander

3.1 Introduction

Water resources are vital in meeting the caloric and health needs of a growing world population (Molden 2007). The expansion of irrigated agriculture in the past century has significantly increased the productivity of agriculture (Edwards and Smith 2018; Njuki and Bravo-Ureta 2019). However, global climate change is expected to increase water scarcity, threatening global food security (Hanjra and Qureshi 2010; Mancosu et al. 2015; Dinar, Tieu, and Huynh 2019; Siirila-Woodburn et al. 2021). Researchers and policy makers have heralded water conservation efforts as a means to mitigate the economic consequences of water scarcity (Gobarah et al. 2015). A growing economics literature has analyzed the efficacy of differing water conservation efforts in addressing water scarcity issues (Pfeiffer and Lin 2014; Gobarah et al. 2015; Koech and Langat 2018). This study contributes to this literature by examining the conservation potential of investments in irrigation infrastructure, specifically the lining and piping of water conveyance infrastructure to reduce water lost during transport. Water conveyance lining and piping have received relatively limited attention in the water conservation literature. However, these investments are receiving renewed

R. Aaron Hrozencik and Nicholas A. Potter are Research Agricultural Economists, and Steven Wallander is an Economist in the Conservation and Environment Branch of the Resource and Rural Economics Division of the USDA-Economic Research Service.

The findings and conclusions in this manuscript are those of the authors and should not be construed to represent any official USDA or US government determination or policy. All authors contributed equally to the writing of the manuscript, as such authors share co-first authorship. For acknowledgments, sources of research support, and disclosure of the or authors' material financial relationships, if any, please see https://www.nber.org/books-and -chapters/american-agriculture-water-resources-and-climate-change/cost-effectiveness -irrigation-canal-lining-and-piping-western-united-states.

attention from policy makers interested in promoting climate resilience in the agricultural sector (Fischer and Willis, 2020).

Farmers and policy makers have a suite of options at their disposal to address the growing scarcity of water resources. These options range from on-farm water management strategies such as irrigation scheduling (Wang et al. 2021), conservation tillage practices (Huang et al. 2021), cover cropping (Novara et al. 2021), and altered crop rotations (Williams, Wuest, and Long 2014); to the adoption of efficiency enhancing irrigation technologies such as drip irrigation systems (Van der Kooij et al. 2013) and low pressure center pivot irrigation systems (Pfeiffer and Lin 2014). There also exist opportunities to conserve water before it reaches irrigated farms and ranches, including managing forests to increase snowpack and streamflow (Gleason et al. 2021), covering water storage reservoirs to reduce evaporative losses (Lehmann, Aminzadeh, and Or 2019), and improving water delivery infrastructure to diminish the amount of water lost during conveyance (Plusquellec 2019). This paper focuses on the water conservation potential of investments in off-farm water conveyance infrastructure, specifically the lining and piping of canals.

Globally, many surface water-dependent agricultural production systems rely on conveyance infrastructure to deliver water from natural bodies of water to arable land. However, transporting water can result in conveyance losses as some water is lost to seepage or evaporation during transport.[1] In many cases water lost during conveyance imposes an economically significant cost on the irrigated agricultural sector. The economic cost of conveyance losses may grow as global climate change continues to increase water scarcity, particularly in snowpack dependent production systems (Reidmiller et al. 2019; Evan and Eisenman 2021; Siirila-Woodburn et al. 2021). Despite the current and potential future costs of conveyance losses, literature rigorously examining the costs and benefits of conveyance loss mitigating investments remains limited (Plusquellec 2019). A recent survey of 230 studies on water conservation investments only included 10 studies that estimated the conservation potential of canal lining or piping (Pérez-Blanco, Hrast-Essenfelder, and Perry 2020).

The sector structure and the related data sources are one reason for the limited focus on canal lining and piping. Typically irrigation with off-farm surface water involves three levels of decision making: (1) the farmer who is irrigating; (2) a local water delivery organization that manages conveyance infrastructure such as ditches, canals, and turnouts; and (3) a large water capture and storage project (often managed by a federal or state agency) that

---

1. In a broadly defined hydrologic system, conveyance losses are not an actual loss of water. Water seepage from main and lateral canals is stored in aquifers while evaporated water returns to the land in the form of precipitation. The water is lost in the sense that it is not immediately available for its intended use.

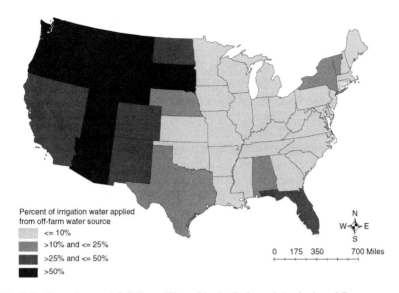

Fig. 3.1   Prevalence of Off-Farm Water Use by Irrigated Agricultural Sector

Note: Data for Connecticut and Rhode Island are suppressed due to disclosure concerns. Off-farm surface water is surface water from off-farm water suppliers, such as the US Bureau of Reclamation; irrigation districts; mutual, private, cooperative, or neighborhood ditches; commercial companies; or community water systems. It includes reclaimed water from off-farm livestock facilities, municipal, industrial, and other reclaimed water sources (USDA-NASS 2019).

Source: USDA-NASS, 2018 Irrigation and Water Management Survey

supplies water to the local delivery organization.[2] A significant amount of research and data collection has been focused on either the farm-level decision making or the large state and federal water projects. This study focuses on the decisions of irrigation delivery organizations using data from USDA's 2019 Survey of Irrigation Organizations, the first nationally representative data set of irrigation organizations collected in over forty years (Wallander, Hrozencik, and Aillery 2022). Irrigation water delivery organizations (e.g., irrigation districts, acequias, ditch companies, etc.) are important institutions in the western US, where the majority (see figure 3.1) of surface water-fed irrigated agriculture relies on off-farm water deliveries (USDA-NASS

2. Off-farm surface water refers to "water from off-farm water suppliers, such as the US Bureau of Reclamation; irrigation districts; mutual, private, cooperative, or neighborhood ditches; commercial companies; or community water systems. It includes reclaimed water from off-farm livestock facilities, municipal, industrial, and other reclaimed water sources" (USDA-NASS 2019). Meanwhile, on-farm surface water refers to "water from a surface source not controlled by a water supply organization. It includes sources such as streams, drainage ditches, lakes, ponds, reservoirs, and on-farm livestock lagoons on or adjacent to the operated land" (USDA-NASS 2019).

2019; Hrozencik 2021).[3] These organizations own and operate much of the infrastructure where conveyance losses occur. In 2019, more than 15 percent of all water brought into irrigation water delivery organization systems was lost during conveyance (USDA-NASS 2020).[4]

Investments in water conveyance infrastructure can diminish conveyance losses and help achieve water conservation objectives. Specifically, upgrading previously unlined (earthen) conveyance canals to lined canals or piped infrastructure can reduce seepage losses and, in the case of piping, can also reduce evaporation losses. However, most of the irrigation canals managed by irrigation organizations in the western US are unlined, and organizations cite the cost of upgrading canals as the primary barrier to investing in lining or piping (Hrozencik, Wallander, and Aillery 2021). The US Department of Agriculture's Natural Resources Conservation Service (USDA-NRCS) reports that lining one mile of a relatively small canal costs between $30 thousand and $228 thousand, depending on canal size and lining material (USDA-NRCS 2020a). Costs may be significantly higher for larger irrigation canals. For example, lining sections of the All-American canal, which is among the largest canals in the US, cost more than $1.8 million per quarter mile (CNRA 2009). Piping irrigation infrastructure is even more costly. Recent irrigation infrastructure piping projects funded by USDA-NRCS report per mile piping costs between $0.6 million and $3.2 million per mile. However, piped irrigation infrastructure requires less maintenance and lasts longer than most lined canals (Newton and Perle 2006).

When considering the potential for canal lining and piping, uncertainty about the benefits, i.e., the expected reductions in conveyance losses, is as important as costs. Despite the potential importance of the benefits of improved conveyance infrastructure, there are no standard estimates to inform public or private investment decisions. A number of studies from the engineering literature leverage analytical equations, simulation modeling, and flow measurements to estimate conveyance losses as a function of canal characteristics, e.g., soil type, lining, size, flow rate, etc. (see Taylor 2016 for an extensive review of the conveyance loss/seepage engineering literature). However, attempting to extrapolate from these studies to the full population of irrigation delivery organizations raises concerns of external validity. There are several reasons that by focusing on places in time and space where infrastructure investments reap the largest conservation

---

3. The prevalence of off-farm surface water use in the western US is related to the unique legal institutions defining water rights within the region. Notably, the doctrine of prior appropriation divorces riparian land ownership from the process of water right allocations and instead assigns water rights based on beneficial use (Haar and Gordon 1958). Allocating water based on beneficial use incentivizes water users to collectively invest in the infrastructure necessary to convey water from natural rivers and streams to arable land.

4. Conveyance losses of 15 percent fall within the range of losses reported in the hydrological and agricultural engineering literature (Todd 1970; Mohammadi, Rizi, and Abbasi 2019; Karimi Avargani et al. 2020).

benefits, these studies potentially overstate the water conservation impacts of canal lining for the average irrigation organization. First, the locations selected for engineering simulations may not reflect average conditions for the universe of conveyance infrastructure. In addition, given the high cost of lining canals, many of these studies occur in locations where conveyance losses were particularly large before infrastructure improvements (Baumgarten 2019). As such, these infrastructure improvement projects yield larger reductions in conveyance losses than might be expected from lining or piping an average unimproved canal. Finally, some of these studies do not account for the potentially rapid degradation of canal performance over time, which can be particularly acute for canals lined with rigid materials like concrete (Plusquellec 2019).

To the extent that the economics literature has treated the relationship between water conveyance infrastructure and losses, the research has primarily used theoretical modeling to understand how water lost during conveyance could influence the optimal allocation of scarce water resources (Tolley and Hastings 1960; Chakravorty and Roumasset 1991; Chakravorty, Hochman, and Zilberman 1995; Umetsu and Chakravorty 1998). Umetsu and Chakravorty (1998) stand out in this literature by explicitly modeling irrigation system investment decisions. They model investment as a function of canal seepage and return flows demonstrating how the benefits of diminished conveyance losses vary based on the availability of water losses for future use. Ward (2010) provides a comprehensive overview of the economic incentives and policy mechanisms determining irrigation infrastructure investments. This study extends this literature by providing the first econometric estimates of the average expected water conservation benefits of canal lining and piping by irrigation organizations in the western US. Our empirical approach estimates the impact of canal lining or piping on conveyance losses while conditioning on other factors—such as climate, region, and vegetation along canals (i.e., phreatophytes)—that are also drivers of conveyance losses. The theoretical model for this paper illustrates how an organization's decision to line or pipe conveyance infrastructure is likely to be driven by expected losses with and without irrigation, which suggests that in an econometric estimation the share of miles lined or piped could be endogenous with respect to observed conveyance losses. To test for bias due to such potential endogeneity, this study implements an instrumental variable control function approach. Our results suggest that on average, increasing the share of conveyance that is piped by 1 percentage point decreases conveyance losses by between 0.1 and 0.17 percentage points. We also find that lining canals reduces conveyance losses, however the magnitude of this effect is relatively smaller, ranging from 0.07 to 0.06 percentage points. We leverage these empirical estimates to develop water conservation supply curves for lining and piping canals. Results indicate that strategic investments in the piping and lining of canals can increase aggregate water availability by between 0.3 percent and 1.75 percent for a cost less than $20,000 per acre-foot conserved.

3.2 Theoretical Model

This research is focused on estimating the expected change in conveyance losses as a function of canal lining and piping. Conveyances losses are calculated as the percentage of water brought into a delivery system (w_{in}) that is not either directly delivered to agricultural users (w_{ag}) or otherwise discharged from the system (w_{other}). We are interested in estimating conveyance losses as a function of the percentage of an organization's conveyance infrastructure that is lined (γ_{lined}) or piped (γ_{piped}) and a vector of other variables, such as climate and organization characteristics (X_{CL}).

$$(1) \qquad CL \equiv 100 * \left[\frac{w_{in} - w_{ag} - w_{other}}{w_{in}} \right] = f(\gamma_{lined}, \gamma_{piped}, X_{CL}).$$

Conveyance losses are bounded ($CL \in [0,1]$) and hypothesized to be convex in lining and piping investment ($CL'(\gamma_{lined}) < 0$; $CL''(\gamma_{lined}) > 0$; $CL'(\gamma_{piped}) < 0$; $CL''(\gamma_{piped}) > 0$).

To illustrate how estimation of this equation might be impacted by endogeneity, we develop a simple theoretical model exploring organization decision making around canal lining and piping. The model posits an organization with a single output—water delivered to farms and ranches (w_{ag})—and assumes that this value is fixed due to limited water supply, conveyance size, rights to water, or other restrictions. The idea that irrigation water at the farm gate is a fixed quantity is found in other models of on-farm crop choice and irrigation investment decisions (Moore, Gollehon, and Carey 1994). We also assume that other water outflows (e.g., deliveries to residential customers, releases for downstream users, or environmental flow requirements) are also fixed by contract or water right obligations to constituents. These assumptions restrict a delivery organization's choice to the amount of water brought into the organization's system (w_{in}^*) and investment in lining and piping, captured as the percent of canals lined (γ_{lined}^*), and the percent of canals piped (γ_{piped}^*).

The organization chooses water inflows and conveyance lining and piping to minimize costs, reflecting the fact that most irrigation delivery organizations are either irrigation districts, which function more like regulated utilities, or ditch companies ("mutuals" or acequias), which function as cooperatives and do not operate as profit maximizing firms. Organizations face a marginal cost of water brought into the system (p_w) that is a composite price reflecting the marginal cost of water acquisition (such as through a contract with a federal or state water project), the cost of moving water (primarily the energy costs of operating pumping), and other input costs such as labor inputs that vary with w_{in}. Since lining and piping canals is a long-run capital investment decision, the marginal costs of canal improvement is expressed as an annualized cost (ac_{lined} and ac_{piped}). These variables enter into the following cost minimization problem.

$$\min_{P_w, \gamma_{lined}, \gamma_{piped}} Cost_{wag} = p_w * w_{in} + ac_{lined} * \gamma_{lined} + ac_{piped} * \gamma_{piped}$$

$$s.t.$$

(2)
$$(1 - f(\gamma_{lined}, \gamma_{piped}, X_{CL})) * w_{in} = w_{ag} + w_{other}$$

$$\gamma^*_{lined} \geq 0$$

$$\gamma^*_{lined} \leq 100$$

$$\gamma^*_{piped} \geq 0$$

$$\gamma^*_{piped} \leq 100$$

The first constraint captures the water budget in which the net water inflows, subject to conveyance losses, must equal the fixed water outflows. The last four constraints capture the possibility that organizations may face corner solutions in their lining and piping investment decisions. The Lagrangian form of this optimization problem incorporates shadow prices for all of these constraints.

(3)
$$\min_{P_w, \gamma_{lined}, \gamma_{piped}} \mathcal{L} = (p_w * w_{in} + ac_{lined} * \gamma_{lined} + ac_{piped} * \gamma_{piped})$$

$$+ \lambda_1((1 - f(\gamma_{lined}, \gamma_{piped}, X_{CL})) * w_{in} = w_{ag} + w_{other})$$

$$+ \lambda_2(\gamma^*_{lined} \geq 0) + \lambda_3(\gamma^*_{lined} \leq 100) + \lambda_4(\gamma^*_{piped} \geq 0)$$

$$+ \lambda_5(\gamma^*_{piped} \leq 100).$$

The first-order conditions for an interior solution, where λ_2 to λ_5 all equal zero, are:

$$\frac{\partial \mathcal{L}}{\partial w_{in}} = p_w - \lambda_1(1 - f(\gamma_{lined}, \gamma_{piped}, X_{CL})) = 0$$

$$\frac{\partial \mathcal{L}}{\partial \gamma_{lined}} = ac_{lined} + \lambda_1(w^*_{in}) \frac{\partial f(\gamma_{lined}, \gamma_{piped}, X_{CL})}{\partial \gamma_{lined}} = 0$$

$$\frac{\partial \mathcal{L}}{\partial \gamma_{piped}} = ac_{piped} + \lambda_1(w^*_{in}) \frac{\partial f(\gamma_{lined}, \gamma_{piped}, X_{CL})}{\partial \gamma_{piped}} = 0$$

To solve for the approximation of the optimal input decisions, we use a first-order Taylor series expansion around a baseline state of canal lining (γ^0_{lined} and γ^0_{piped}), which implies a baseline state of conveyance loss (CL^0).

(4) $\gamma^*_{lined} = \gamma^0_{lined} + \left(1 - \dfrac{p_w}{\lambda_1}\right)(f'(\gamma^*_{lined} \,|\, X_{CL}) - f'(\gamma_0 \,|\, X_{CL}))^{-1}.$

Substituting for λ_1 from the second of the first-order conditions gives:

(5) $\gamma^*_{lined} = \gamma^0_{lined} + \left(1 + \dfrac{p_w}{ac_{lined}}\right)((w_{in})f'(\gamma^*_{lined} \,|\, \gamma^0_{piped}, X_{CL}))$

$(f'(\gamma^*_{lined} \,|\, \gamma^0_{piped}, X_{CL}) - f'(\gamma^0_{lined} \,|\, \gamma^0_{piped}, X_{CL}))^{-1}.$

The implication of this model for the research question in this paper is that the optimal investment in canal lining or piping depends upon the price of bringing water into these systems and the cost of lining or piping as well as the expected change in conveyance losses. This raises the possibility that in an econometric estimation of the conveyance loss function, the percentage of the canal system lined or piped is potentially endogenous. The empirical model tests for such endogeneity.

3.3   Empirical Model

To understand how variation in water conveyance infrastructure influences conveyance loss, we estimate the following econometric model

(6)   *Conveyance Losses$_i$ = $G(\beta_0 + \beta_1 *$ Conveyance Lined$_i$*

$+ \beta_2 *$ Conveyance Piped$_i$ $+ \gamma X_i) + \varepsilon_i$,

where the dependent variable, Conveyance Losses$_i$, represents for the ith organization the fraction of total water diverted lost during conveyance and $G(\cdot)$ is the logistic function (Papke and Wooldridge 1996). Conveyance Lined$_i$ and Conveyance Piped$_i$ describe the fraction of the ith organization's total conveyance infrastructure that is lined and piped, respectively. The fraction of conveyance lined and conveyance piped are potentially interdependent and endogenous factors affecting conveyance loss. The associated parameters, β_1 and β_2, capture how changes in the lining and piping of an organization's conveyance infrastructure influence conveyance losses. The econometric model also includes an intercept term, β_0, and a matrix of other explanatory variables, X_i (e.g., state fixed effects, irrigable acres, the density of the organizations conveyance system, water scarcity indicators, water use reporting requirements, climate, etc.), with associated vector of estimated parameters, γ. Finally, ε_i is an idiosyncratic error term.

We model conveyance losses as a nonlinear function of an organization's conveyance infrastructure, differentiating between lined and piped infrastructure. Obtaining unbiased estimates of the model's parameters of interest, β_1 and β_2, is potentially complicated by endogeneity between conveyance losses and conveyance lining and piping decisions. Organizations with relatively large conveyance losses may have larger incentives to invest in the efficiency of conveyance infrastructure by lining main and lateral canals or installing piped conveyance. Under this scenario, causation runs bilaterally between the conveyance infrastructure characteristics and conveyance losses resulting in a downward bias in the estimates of β_1 and β_2. We take an instrumental variable (IV) approach to address this potential endogeneity. Because of nonlinearity in both the first and second stage, we employ a control function model to estimate effects.

However, the choices of how much conveyance to line and to pipe are interdependent, since lined conveyance cannot be piped and vice versa. The fraction of conveyance lined (Conveyance Lined$_i$), the fraction of conveyance piped (Conveyance Piped$_i$), and the fraction of conveyance that is neither lined nor piped (Unimproved$_i$) must sum to one. We address the fractional and interdependent nature of the endogenous covariates in the first stage of our model with the use of a fractional multinomial model following the methods outlined in Papke and Wooldridge (1996), with unlined being the reference case, specifically

(7) $Conveyance\{Unimproved_i, Lined_i, Piped_i\} = G(\lambda Z_i + \gamma X_i) + \varepsilon_i.$

This approach instruments for endogenously determined conveyance characteristics while recognizing the interdependence of lining, piping, and unlined/piped conveyance infrastructure.

Recognizing the bounded nature of conveyance losses expressed as a fraction of total diversions, we use a control function estimation to address the potential endogeneity of Conveyance Lined$_i$ and Conveyance Piped$_i$ in our second stage model. The control function approach is preferred over other methods (e.g., two stage least squares) when the second stage is nonlinear as control function methods allow for more straightforward hypothesis testing for model selection and covariate exogeneity (Wooldridge 2015). We estimate a fractional response model that takes the form

(8) $Conveyance\ Losses_i = G(\beta_0 + \beta_1 * Conveyance\ Lined_i$

$+ \beta_2 * Conveyance\ Piped_i + \gamma * W_i$

$+ \phi_1 v_{Lined} + \phi_2 v_{Piped}) + \varepsilon_i,$

where v_{Lined} and v_{Piped} are the residuals for Conveyance Lined and Conveyance Piped from the estimation of the first stage model represented by equation 5.

Valid instruments must be adequate predictors of the endogenous explanatory variables, Conveyance Lined$_i$ and Conveyance Piped$_i$, and meet the exclusion restriction—that is, only affect the dependent variable, Conveyance Losses$_i$, indirectly through the endogenous explanatory variable. We use a suite of organization-level characteristics as instruments for the potentially endogenous lining and piping variables. Our instruments leverage information on reasons for not lining canals, the importance of municipal water deliveries in organization operations, and the role that constituents have in organization decision making. Specifically, we instrument for potentially endogenous conveyance infrastructure characteristics with the following variables: (1) a dummy variable indicating whether an organization cited expense as a reason for not lining their conveyance infrastructure; (2) a variable capturing the share of water delivered to the municipal sector; and (3) a dummy variable indicating whether constituents have input in organization management decisions through direct voting or representatives on an elected or appointed board. Not improving conveyance infrastructure due to expense represents exogenous local material and construction costs. These costs have no impact on conveyance losses outside of their effect on infrastructure improvement decisions provided that local construction and material markets are not dominated by organization infrastructure projects. The share of water delivered to the municipal sector reflects the benefits of conserved water as organizations can sell additional water supplies to municipal customers. These benefits presumably do not influence conveyance losses except through lining and piping as they primarily indicate the extent of residential development within the organization's service area, which is likely orthogonal to geographical and geological characteristics impacting conveyance losses. Finally, constituent input likely increases the likelihood that organization decision making aligns with constituent priorities related to water supply reliability and conveyance losses. As such, this input affects infrastructure investment but is otherwise unrelated to conveyance losses provided that organizations do not adjust their governance structures in response to losses.

3.4   Data

To estimate the econometric model outlined in equation 4 we leverage novel data collected in the 2019 Survey of Irrigation Organizations (SIO). The 2019 SIO was the first nationally representative data collection effort focused on water delivery organizations since the 1978 Census of Irrigation Organizations. SIO data were collected by the US Department of Agriculture's National Agricultural Statistics Service (USDA-NASS) during the spring of 2020. SIO data were collected using a mailed paper questionnaire with web and telephone interviewing instruments also available for survey enumeration. The reported survey response rate was 44 percent (USDA-NASS 2020). SIO data represent the operations of the organizations delivering water directly to farms or directly influencing some aspect of on-farm groundwater use in the 24 states where these types of irrigation organizations are most common.

We focus our analysis on a subset of the survey responses collected in the 2019 SIO. Specifically, we use survey responses from 673 organizations that indicate delivering water to farms and ranches in 2019 and respond to the relevant sections of the survey instrument.[5] Here we describe the data used to estimate our empirical models, beginning with an in-depth discussion of the primary variables of interest, conveyance loss and conveyance infrastructure, and concluding with information about the remaining exogenous covariates and instrumental variables.

3.4.1 Conveyance Loss Data

Survey respondents were asked to report their conveyance losses at two points in the survey. First, the survey asked for all of the inflows and outflows from irrigation systems in terms of total acre-feet, specifying conveyance losses as part of outflows. Second, participants were asked to report conveyance loss as a percent of diversions.[6]

Nationally representative totals of reported inflows and outflows were summarized by USDA (USDA-NASS 2020). In 2019, irrigation water delivery organizations brought 70.1 million acre-feet of water into their delivery systems and had 10.7 million acre-feet of conveyance losses. This indicated a national conveyance loss of 15.3 percent. Notably, total outflows, including the conveyance losses, were only 67.3 million acre-feet, suggesting that about 4 percent of total inflows were either held back as storage within the irrigation systems, or outflows such as conveyance losses were underreported.

Due to the potential for underreporting conveyance losses in volumetric terms, this study relies primarily on self-reported conveyance loss in percentage terms. Missing or 0 percent conveyance losses are imputed from volumetric conveyance losses for those observations that reported volumetric

---

5. The 2019 Survey of Irrigation Organizations (SIO) collected 1,360 survey responses from irrigation water delivery and groundwater management organizations. An observation weighting methodology was utilized to account for survey non-response; more information on this weighting methodology can be found in USDA-NASS SIO data publication (USDA-NASS 2020). We use the following criteria to select the observation used in our empirical analysis. In parentheses after each inclusion criteria are the number of remaining observations that meet that criteria as well as all other preceding criteria. (1) Water delivery organization (1,262); (2) Delivered water to farms in 2019 (1,254); (3) Provided information on conveyance infrastructure (857); (4) Report less than one mile of conveyance infrastructure per irrigable acreage (845); (5) Report non-zero conveyance losses or have 100 percent of total conveyance piped and report zero conveyance losses (692); (6) Are located within a state with at least 5 observations (674); and (7) Report correct FIPS number for state where organization is located (673).

6. 484 respondents provided information on conveyance losses in terms of volume of water lost and the share of water lost. For these organizations, comparing the reported share of water lost to the reported volume of water lost (converted to a share of total water diverted) results in a correlation coefficient of 0.734. Approximately 71 percent of organizations that disclose conveyance losses in percentage and volumetric terms report values within 10 percentage points, when converting the volumetric conveyance loss data to a share using total water diverted.

Table 3.1   **Summary statistics for outcomes**

Statistic	Mean	St. Dev.
Conveyance Loss (share)	0.1486	0.1424
Conveyance Lined (share)	0.0986	0.2464
Conveyance Piped (share)	0.3064	0.4141

conveyance loss. Of the 673 respondents in our sample, 598 report positive conveyance losses as a percent of diversions or in acre-feet. Of these, 530 report positive conveyance losses in percentage terms. Observations that report zero conveyance loss in both measures are only included if they report that 100 percent of their conveyance infrastructure is piped.[7] About 15 percent of the "zero" conveyance loss organizations report that 100 percent of their infrastructure is piped.

Percent conveyances losses are skewed toward lower values. Over 70 percent of organizations report a conveyance loss below 20 percent. Table 3.1 presents conveyance loss summary statistics. The average conveyance loss is 14.9 percent. This is slightly smaller than the volume-based national estimate cited above, but is consistent with the possibility that the national number includes underreporting due to some of the "zero" conveyance loss organizations that are excluded from our sample.

3.4.2   Conveyance Infrastructure Data

Included as part of the data on the infrastructure operated by irrigation organizations, the survey asked respondents to report on the miles of canals and pipes used for delivering water to farms and ranches. Organizations reported the total number of miles of main and lateral canals, and for each of those they separately reported on miles that are unlined (i.e., unimproved), lined, or piped. Those six categories were summed to calculate the total miles of conveyance infrastructure for each organization. Based on that total, the shares for lined and piped miles were calculated.

About one-fourth of the survey respondents who reported delivering water did not provide any detail on conveyance infrastructure. Whether this was because the organizations did not keep records on miles of infrastructure or face unusual ownership arrangements in which they neither own nor operate the conveyance infrastructure is not clear from the survey questions. The observations are excluded from the study.

Figure 3.2 demonstrates the heterogeneity of organization-level conveyance infrastructure characteristics by plotting the percent of organizations and acreage served by an organization falling within differing, mutually exclusive categories of conveyance characteristics. Approximately 43 per-

---

7. Piping can conceivably reduce conveyance losses to zero (Newton and Perle 2006).

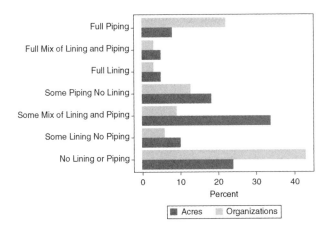

Fig. 3.2   Categories of lining and piping shares by percentage of organizations and acreage
Source: USDA-NASS, 2019 Survey of Irrigation Organizations.

cent of organizations have no lined or piped conveyance infrastructure. These organizations account for 23 percent of acreage served, indicating that organizations with no lined or piped canals are generally small in terms of acreage and service area. Meanwhile, 9 percent of organizations delivering water to 34 percent of acreage have some mix of lined and piped infrastructure, suggesting that larger organizations are more likely to invest in both lined and piped conveyance. Finally, 22 percent of organizations serving 8 percent of acreage have a fully piped conveyance system. Organizations with a fully lined or a full mix of lined and piped conveyance are relatively less common, each accounting for about 3 percent of organizations and 2 percent and 5 percent of acreage, respectively.

3.4.3   Exogenous Covariates and Instrumental Variables

Table 3.2 presents summary statistics for the exogenous covariates and instrumental variables used to estimate equation 4. Note that these summary statistics represent a sample of the full set of organizations surveyed in the SIO. As such, reported statistics may differ from those reported by USDA-NASS (USDA-NASS 2020).

The exogenous covariates included in our empirical model of conveyance losses consist of the following: "Irrigable Acres," "Conveyance Density," "Sufficient Water in 2019," "Required to Report Use," "Phreatophyte Problems," "July Mean Daily Temperature," "Water Stress," and "Drought Risk." "Irrigable Acres" refers to the amount of land that could have received water from the organization in 2019, which could be larger than the amount of land irrigated using water delivered by the organization. Since organizations that serve larger areas move water over greater distances through larger

Table 3.2   **Summary statistics of covariates**

Statistic	Mean	St. Dev.
Exogenous Covariates		
Irrigable Acres (000s)	11.1041	31.9556
Conveyance Density (mi/acre)	0.0214	0.0592
Sufficient Water in 2019 (0/1)	0.2734	0.4460
Required to Report Use (0/1)	0.5468	0.4982
Phreatophyte Problems (0/1)	0.5290	0.4995
July Mean Daily Temperature (°C)	20.3693	3.1288
Water Stress	0.8602	1.3503
Drought Risk	2.7153	0.2147
Unlined due to:		
GW Recharge (0/1)	0.2036	0.4029
Min. Seepage (0/1)	0.1516	0.3589
Other (0/1)	0.0951	0.2936
Instruments		
Unlined due to Expense (0/1)	0.5587	0.4969
Municipal Deliveries (share)	0.0574	0.1511
Can Vote (0/1)	0.9287	0.2576

systems, the expectation is that greater irrigable acres will be associated with higher conveyance losses. "Conveyance Density" records the total conveyance infrastructure per irrigable acres and is measured in miles per acre. The expectation is that higher conveyance density will be associated with higher conveyance losses. "Sufficient Water in 2019" is a dummy variable indicating whether an organization cited sufficient water supplies in 2019 as a reason for not engaging in water marketing. Since 2019 was an above average precipitation year in most areas of the western US, the expectation is that a positive response will indicate that an organization had average or above average quantities of water moving through their system. If conveyance losses increase with percent utilization of conveyance capacity, then a positive response would be expected to be associated with higher conveyance losses. "Required to Report Use" is a dummy variable indicating whether the organization is required to report water use for irrigation to users/shareholders, water project managers of state or federal suppliers, or any other regulatory authority. If reporting requirements lead to more efficient management, then the expectation is that this would be associated with lower conveyance losses. "Phreatophyte Problems" is a dummy variable expressing whether the organization reported having issues with vegetation (e.g., salt cedar, willow, etc.) along ditches and canals. Since such vegetation is directly responsible for conveyance losses, the expectation is that a positive response will be associated with higher conveyance losses. "Unlined Due to:(x)" are a suite of dummy variables signaling whether the organization reported that the following were reasons for leaving conveyance unlined: unlined canals provide groundwater recharge ("GW Recharge"), water loss is minimal due to soils and geology ("Min. Seepage"), and "Other" to account reasons not listed.

"July Mean Daily Temperature" measures the average July daily temperature for the county where the irrigation organization is primarily located, using 30-year normal temperature data reported by PRISM (PRISM 2021). The expectation is that higher average temperatures will be associated with higher conveyance losses as greater volumes of water are lost to evaporation (Wang et al. 2013). "Water Stress" is an index variable based HUC-8 level output from the Water Supply Stress Index Model (WaSSI) (Sun et al. 2015). HUC-8 level historical water stress output from WaSSI is mapped to county cropland geospatial data to derive a county-level measure of historical water stress, which is then matched to the organizations within that county. In cases where multiple HUC-8s occur within a county, the county-level measure is an area weighted average. "Drought Risk" is the standard deviation of July Palmer Modified Drought Index (PDMI) data calculated using weather station data spanning the past century (Mo and Chelliah 2006; Wallander et al. 2013). Weather station point data are spatially interpolated to the county level, which is then matched to organizations within the county. Including climatic and water stress related variables aims to control for the role that expectations related to water scarcity and drought have in determining conveyance losses.

We instrument for potentially endogenous "Conveyance Lined" and "Conveyance Piped" with a set of three variables representing information on reasons for not lining canals, the importance of municipal water deliveries in organization operations, and the role that constituents have in organization decision making. Specifically, "Unlined Due to Expense" is a dummy variable indicating whether an organization cited expense as a reason for not improving conveyance infrastructure. "Municipal Deliveries" is the share of total organization water outflows delivered to municipal customers. "Can Vote" is a dummy variable representing whether constituents have input in organization management decisions through direct voting or representatives on an elected or appointed board.

3.5   Results

Table 3.3 presents results estimating the empirical model outlined in equation 4 using the data described in section 3.4. Column 1 presents results from a linear version of our primary econometric model. Column 2 displays estimation results from the nonlinear, fractional response logistic model presented in equation 4 but does not instrument for potentially endogenous conveyance lining or piping decisions. Column 3 also presents fractional response logistic model results but instruments for potential endogeneity in the conveyance infrastructure covariates using a control function approach.

Table 3.3   Conveyance loss empirical model results

	Linear Uninstrumented	Logistic Uninstrumented	Logistic Control Function
Conveyance Lined (share)	−0.0684**	−0.0623*	0.0620
	(0.0243)	(0.0250)	(0.0569)
Conveyance Piped (share)	−0.1004***	−0.1328***	−0.1750***
	(0.0156)	(0.0192)	(0.0310)
Unlined due to:			
GW Recharge	0.0275	0.0247*	0.0239
	(0.0150)	(0.0119)	(0.0123)
Min. Seepage	−0.0212	−0.0170	−0.0228
	(0.0155)	(0.0138)	(0.0135)
Other	0.0130	0.0107	0.0108
	(0.0182)	(0.0139)	(0.0141)
Log Acres	0.0149***	0.0151***	0.0131***
	(0.0038)	(0.0034)	(0.0035)
Conveyance Density	0.1349	0.1211	0.1236
	(0.0905)	(0.0697)	(0.0700)
Sufficient Water in 2019	−0.0105	−0.0134	−0.0103
	(0.0112)	(0.0112)	(0.0112)
Required to Report Use	−0.0049	−0.0038	0.0014
	(0.0110)	(0.0107)	(0.0111)
Phreatophyte Problems	0.0420***	0.0443***	0.0328*
	(0.0126)	(0.0122)	(0.0140)
July Mean Daily			
Temperature (°C)	−0.0028	−0.0031	−0.0024
	(0.0022)	(0.0022)	(0.0023)
Water Stress	0.0051	0.0050	0.0044
	(0.0040)	(0.0037)	(0.0038)
Drought Risk	0.0738**	0.0820**	0.0822**
	(0.0257)	(0.0277)	(0.0276)
R^2	0.2575	0.2690	0.2698
Adj. R^2	0.2312	0.2419	0.2404
Num. obs.	673		
n		673	673
DF		649	647
Sigma		0.1255	0.1255

Note: All models include state fixed effects. Robust standard errors are shown in parenthesis. All models have 673 observations that include all irrigation organizations with some conveyance loss as well as those with 100 percent of conveyance piped and no conveyance loss.
*** $p < 0.001$; ** $p < 0.01$; * $p < 0.05$.

To facilitate result interpretation and comparison between the linear and nonlinear model results all nonlinear model results are presented as average marginal effects following methods outlined in Ramalho, Ramalho, and Murteira (2011).

Nearly all model specifications yield negative and statistically significant estimates of β_1 and β_2, the parameters of interest from equation 4. Parameter estimates of β_1 indicate that, for the average organization, increasing the amount of conveyance that is lined by 1 percentage point decreases conveyance losses by approximately 0.06 percentage points. However, this result is only statistically significant and negative for the specifications which do not instrument for potential endogeniety between conveyance lining/piping and losses. The control function IV specification yields a parameter estimate suggesting a positive relationship between lining and conveyance losses, however this result is not statistically significant. All parameter estimates of β_2 are negative and statistically significant, suggesting that for the average organization, increasing the share of conveyance that is piped by 1 percentage point decreases conveyance losses by between 0.1 and 0.175 percentage points.

The difference in the average marginal effect of piping versus lining conveyance infrastructure may be related to the relevant lifespan of each canal improvement option. Namely, piped canals generally have a longer lifespan than lined canals (Newton and Perle 2006). Lined conveyance infrastructure can degrade quickly without costly routine maintenance to address cracked lining materials, which can significantly increase conveyance losses (Plusquellec 2019). Given that the data used to estimate the effect of lining and piping on conveyance losses do not include information on the age of the improved infrastructure, our estimated effects relate to the efficiency of lined and piped canals of an average age within our data. These estimated effects may underestimate the conveyance loss mitigation potential of newly improved canals, particularly lined canals.

Parameter estimates for the suite of explanatory variables citing why organizations leave canals unimproved (Unlined Due to:[x]) follow intuition. Citing groundwater recharge (GW Recharge) as a reason for not improving conveyance is associated with higher conveyance losses as these losses contribute toward potential groundwater recharge objectives. Reporting minimal seepage (Min. Seepage) as a reason for not improving canals is negatively linked to conveyance losses as cited soil and geologic attributes diminish losses. Finally, the catch-all Other category for reasons for not lining canals is positively correlated with conveyance losses.

Other explanatory variable parameter estimates also generally follow intuition. Log transformed irrigable acres (Log Acres) increases conveyance losses and is statistically significant across all model specifications. Organizations with expansive service areas have larger conveyance losses as water deliveries must generally travel longer distances. This relationship holds even conditioning on the density of the organization's conveyance infrastructure (Conveyance Density), which also increases losses but is not statistically significant. Organizations that did not engage in water marketing due to sufficient water (Sufficient Water in 2019 = 1) are negatively associated conveyance losses but the relationship is not statistically significant.

Logistic model estimates indicate that water use reporting requirements are also generally associated with lower conveyance losses. However, these estimates are not statistically significant and the sign of the parameter estimate changes for the control function logistic model specification. Model results also indicate a positive and statistically significant relationship between conveyance losses and organization-level issues with phreatophytes. This relationship follows intuition as phreatophytes may be responsible for a portion of conveyance losses as root systems in and around conveyance infrastructure uptake water during transport.

Finally, the suite of parameters associated with variables capturing the effects of climate and water scarcity on conveyance losses yields somewhat surprising results. Increasing water stress and drought risk are both associated with increased conveyance losses, however this relationship is only statistically significant for drought risk. Higher conveyance losses in locations with higher incidence of drought potentially suggest that other climatic conditions that covary with drought (e.g., air temperature, solar radiation), which increase evaporative losses, may be driving the estimated relationship. Parameter estimates for July mean temperature run contrary to this reasoning as all three model specifications find a negative relationship between July temperatures and conveyance losses. However, this relationship is not statistically significant.

3.5.1   Statistical Tests and First-Stage Model Results

In table 3.4 we report relevant test statistics for the IV control function specification. We conduct a Wu-Hausman test of the null hypothesis that both the uninstrumented model (column 2 in table 3.3) and the instrumented model (column 3 in table 3.3) are consistent (Hausman 1978). We also conduct tests examining instrument strength for the IV control function specification. The standard F-Test for weak instruments (Stock, Wright, and Yogo 2002; Stock and Yogo 2005; Staiger and Stock 1997) does not apply in the nonlinear case, so we instead report the Wald statistic for the joint null hypothesis that in the first stage the coefficients of all instruments are not different from zero.

The Wu-Hausman test statistic fails to reject the null hypothesis that both models are consistent, suggesting that endogeneity is not a significant issue for the uninstrumented model. As such, our preferred specification, based on efficiency criteria, is the uninstrumented logistic specification (column 2

Table 3.4   **IV tests**

Test	Statistic	DF	Endog DF	p-value
Wald (Conveyance Lined)	96.9672	3		0.0000
Wald (Conveyance Piped)	260.0475	3		0.0000
Wu-Hausman	0.4221	1	649	0.5159

Table 3.5   **First-stage model results**

	Dependent Variables: Share of Conveyance	
	Lined	Piped
	(1)	(2)
	Instruments	
Unlined due to Expense	−2.8705***	−3.6111***
	(0.3107)	(0.2287)
Municipal Deliveries (share)	2.1602**	1.1836
	(0.7413)	(0.6202)
Can Vote	−0.2438	0.7623*
	(0.3889)	(0.3514)
	Exogenous Covariates	
Unlined due to:		
GW Recharge	−1.4177**	−2.1157***
	(0.4793)	(0.4142)
Min. Seepage	−0.7078	−2.0006***
	(0.3712)	(0.3653)
Other	−2.6238***	−3.0544***
	(0.4867)	(0.3247)
Log Acres	0.3368***	0.1345*
	(0.0709)	(0.0582)
Conveyance Density	5.3123*	5.3715***
	(2.3253)	(1.4192)
Sufficient Water in 2019	−0.1561	0.2684
	(0.2670)	(0.2006)
Required to Report Use	−0.0608	0.4145*
	(0.2734)	(0.1987)
Phreatophyte Problems	−0.0074	−1.0203***
	(0.2702)	(0.2039)
July Mean Daily Temperature (°C)	−0.0020	0.0810^*
	(0.0499)	(0.0361)
Water Stress	−0.0589	−0.1467
	(0.1135)	(0.0965)
Drought Risk	0.0811	0.2710
	(0.5588)	(0.3868)

in table 3.3). Finally, Wald test statistics suggest that instrumental variables explain a significant degree of variation in the share of conveyance lined and conveyance piped, suggesting that weak instruments are not a concern.

While the Wu-Hausman test statistic reveals a preference for the uninstrumented model specification, results from the first stage of the IV control function specification are useful in understanding factors that influence organization conveyance infrastructure characteristics. Table 3.5 presents these first-stage model results related to the IV control function specification. We begin with a discussion of the estimated relationship between instrumental variables and canal lining and piping and then briefly discuss how other exogenous covariates influence conveyance infrastructure.

The expense of improving conveyance infrastructure is negatively associated in a statistically significant manner with the share of organization conveyance that is lined and piped, indicating the importance of exogenous costs in determining conveyance improvement investments. Meanwhile, the share of water delivered to municipal customers is positively correlated with the share of conveyance that is lined and piped, however this relationship is only statistically significant for canal lining. This result demonstrates the importance of the opportunity cost of water lost in conveyance in determining organization conveyance investment decisions. Having a means to sell water conserved increases the share of lined and piped conveyance infrastructure. Finally, constituent ability to influence organization decision making yields mixed results with respect to lining and piping. Constituent voting decreases canal lining but increases canal piping, however this relationship is only statistically significant for the share of piped conveyance. This result potentially indicates a preference among constituents for canal piping compared to lining.

The remaining exogenous covariates also reveal informative relationships concerning the share of lined and piped conveyance. The suite of variables concerning reasons organizations do not improve conveyance canals all yield the expected negative relationship. Additionally, organizations with larger and more dense conveyance systems have larger shares of lined and piped canals. This relationship suggests the importance of capital constraints in determining conveyance characteristics as larger, potentially less capital-constrained organizations have a larger share of their canals lined and/or piped.

3.5.2 Conditional Marginal Effect of Lining and Piping Conveyance

The average marginal effects of canal lining and piping presented in table 3.3 belie important effect heterogeneity based on the current share of an organization's conveyance that is lined or piped. Namely, the marginal impact of increasing the share of conveyance that is piped by 1 percentage point may differ for an organization that has 50 percent of its conveyance piped compared to an organization that has none of its conveyance piped. We explore effect heterogeneity as a function of current conveyance in figure 3.3, which separately plots the conditional marginal effects for differing shares of lining and piping. Specifically, figure 3.3 calculates the marginal effect for the full range of observed shares of conveyance that is lined or piped using regression results from column (2) of table 3.3 and conditioning on the mean or mode of all covariates.[8] The left panel of figure 3.3 plots the conditional

8. To calculate conditional marginal effects we set all continuous covariates at their mean and all binary covariates at their mode. State-level effects are not included.

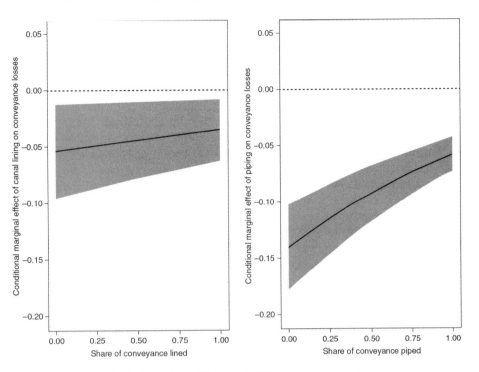

Fig. 3.3   Marginal effect of canal lining and piping on conveyance losses

Note: Marginal effects are calculated using methods outlined in Ramalho, Ramalho, and Murteira (2011). The shaded area represents the 95 percent confidence interval for the marginal effect estimated at a given level of the share of conveyance lined or piped. The marginal effects of lining and piping are calculated setting all continuous variables as their mean and all dummy variables as their mode except for state-level effects, which are set to zero.

marginal effect of lining and demonstrates that the effect of lining becomes marginally smaller across the [0,1] range. The right panel of figure 3.3 plots the conditional marginal effect of piping and indicates that the impact of piping also wanes across the [0,1] range. For example, increasing the share of conveyance piped for an organization with no piped infrastructure by 1 percentage point leads to an approximate 0.15 percentage point reduction in conveyance losses. Meanwhile, the same increase in piped conveyance for an organization with 75 percent of its conveyance piped yields approximately a 0.07 percentage point reduction in conveyance losses.

3.5.3   Simulation of Water Conservation Supply Curve

Based on the estimated conveyance loss function, we construct a simple supply curve for water conservation based on an assumed series of projects that would line or pipe 100 percent of an organization's unimproved canals (either unlined or unpiped). This exercise illustrates how a coordinated water conservation effort that begins with least cost conservation options would initially capture a fair amount of low-cost conservation but will rapidly progress to more expensive options. This exercise also provides a useful means to compare the relative cost-effectiveness of investments in lining versus piping canals.

We estimate the change in water availability due to investments in the lining and piping of conveyance infrastructure using results from the logistic model specification (see column 2 of table 3.3) to calculate, for each organization, the predicted change in conveyance losses if all unimproved infrastructure was lined or piped. To estimate this change in organization level conveyance losses we use a linear approximation of the conditional marginal effect functions (see figure 3.3), conditioning based on the organization-level observed covariate values. We integrate this function between each organization's current level of lining or piping and 100 percent lined or piped to find the total change in conveyance losses associated with fully lining or piping remaining unimproved infrastructure. For example, consider an organization that currently has 10 percent of its conveyance lined, 10 percent of its conveyance piped, and 80 percent of its conveyance is unimproved. To simulate how fully lining or piping the remaining unimproved canals affects conveyance losses, we estimate how losses change when lining or piping the remaining 80 percent of the organization's conveyance, taking into account how the marginal effect changes as a larger share of infrastructure is lined or piped. Reductions in conveyance losses are then aggregated across all organizations and converted into percentages of total water inflows to facilitate comparison with aggregate conveyance losses. Finally, we leverage estimated canal lining and piping costs to calculate, for each organization, the cost of fully lining/piping all unimproved canals. Specifically, we integrate cost estimates provided by the USDA's Natural Resources Conservation Service (NRCS) and organization-level data on the length of unimproved canals to calculate the cost associated with fully lining or piping (USDA-NRCS 2020a; USDA-NRCS 2020b). Many construction options exist when lining and piping canals. For example, canals may be lined with concrete or less expensive geomembranes. As such, we calculate lining and piping costs using three cost estimates ("Low," "Medium," and "High").[9]

The combination of estimated changes in conveyance losses and lining/ piping costs provides a marginal cost of conservation for each organization. Ordering these organization-level marginal costs yields supply curves for water conservation resulting from lining and piping, which are introduced

---

9. Low, Medium, and High canal lining costs are $30,000, $60,000, and $228,000, respectively, per mile of lined canals. These costs are drawn from an NRCS publication and correspond to minimum, mean, and maximum cost estimates. The Low, Medium, and High piping cost supply curves assume costs of $629,000, $1,512,000, and $3,239,000 per mile which correspond to the minimum, mean, and maximum per mile costs reported in recently funded PL-566 projects involving the piping of irrigation infrastructure (USDA-NRCS 2020b).

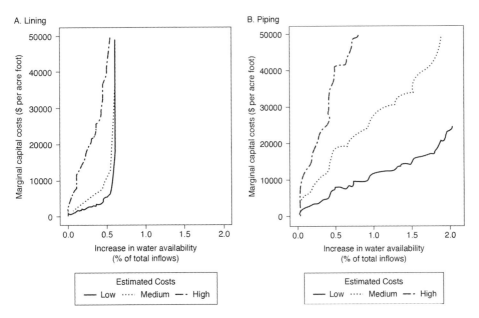

Fig. 3.4   Supply curve of water conservation through lining and piping conveyance infrastructure investments

Note: Panels A and B represent the water conservation supply curves for lining and piping, respectively. The Low, Medium, and High costs for lining canals refer to $30,000, $60,000, and $228,000 per mile of lined canals, respectively (USDA-NRCS 2020a). The Low, Medium, and High costs for piping canals refer to $629,000, $1,512,000, and $3,239,000 per mile of piped canal, respectively. Marginal capital costs represent private costs for lining and piping infrastructure which in some cases may differ from the total social costs of improving water conveyance infrastructure. For example, conveyance losses may be recharging an aquifer that supplies water for a wetland habitat. Lining or piping conveyance could potentially impose additional social costs if diminishing losses reduce water flows to the wetland and damage the habitat. The effects of lining and piping on water availability presented here relate to the average age of infrastructure in the data used to estimate our empirical model (see table 3.3). Newly lined and piped canals may yield larger increases in water availability than those estimated here.

in figure 3.4. Figure 3.4a demonstrates the water conservation potential of investments in canal lining. Our simulation exercise indicates that strategic investments in canal lining can increase total water availability by 0.3 percent to 0.6 percent, depending on the cost scenario. In the low cost scenario these increases are achieved for less than $20,000 per acre-foot conserved and correspond to between a 2 percent and 4 percent reduction in aggregate conveyance losses. Figure 3.4b presents the water conservation capacity of piping investments. Depending on the cost scenario, strategic investments in piping irrigation conveyance can yield between 0.3 percent to 1.75 percent increases in total water availability. In the lost cost scenario these increases are obtained for less than $20,000 per acre-foot conserved. These changes in water availability correspond to between a 2 percent and 12 percent decrease in total conveyance losses. As increases in water availability due to canal lining and piping occur annually, the price paid for this additional water is similar to an organization purchasing a water right. These costs are relatively similar to observed water market transactions in the western US, suggesting that lining or piping may be more cost effective than purchasing rights on the open market (Schwabe et al. 2020).

Comparing figure 3.4a and figure 3.4b demonstrates the differences in the relative cost efficiency of canal piping and lining. Namely, canal lining is relatively more cost effective than piping when aiming to achieve small (between 0.1 percent and 0.5 percent depending on the cost scenario) aggregate increases in water availability. For larger increases in water availability piping canals is more effective as the low price of lining projects is outweighed by their relatively smaller reduction in conveyance losses. Together these results suggest that a combination of investments in lining and piping may be optimal to achieve water conservation objectives. Finally, given that our empirical estimates of the impact of conveyance improvements correspond to effects for the average age of lined and piped infrastructure within our sample, it may be the case that newly lined or piped canals yield larger increases in available water, making initial investments in conveyance improvements more cost effective than calculated here.

3.6   Conclusion

This paper analyzes the relationship between water conveyance infrastructure attributes and conveyance losses to characterize the benefits of investments in irrigation infrastructure. This research builds on past work in the engineering literature by utilizing novel survey data describing the operations and infrastructure of irrigation water delivery organizations in the western US to empirically characterize the water conservation benefits of investments in conveyance infrastructure. Our results constitute a representative estimate of the impact of canal lining and piping on conveyance losses using a data set that provides external validity for policy-relevant simulations. We find that, for the average organization, increasing the share of their conveyance that is piped decreases conveyance losses by between 0.1 and 0.17 percentage points. We also find that lining canals generates reductions in conveyance losses, however these effects are smaller in magnitude ranging from 0.06 to 0.07 percentage points.

A simple simulation exercise focused on the costs and benefits of conveyance lining and piping demonstrates how investments in improved water conveyance infrastructure can provide cost-effective water conservation, initially at costs near that of procuring new supplies via market transaction (Schwabe et al. 2020). These simulations demonstrate that conveyance investments can increase total water availability by between 0.3 percent and 1.75 percent, which corresponds to between a 2 percent and 12 percent decrease in aggregate conveyance losses. For smaller increases in water availability lining canals is more cost effective than piping. However, for larger increases in water availability piping is more cost effective, indicating that a mix of both lining and piping investments is likely optimal to meet water conservation objectives.

Together our empirical and simulation modeling results provide important evidence informing the use of conveyance infrastructure improvements to conserve water. Growing water scarcity concerns throughout the western US and globally underscore the importance of understanding the costs and benefits of the range of policy mechanisms and investments available to enhance water availability (Hanjra and Qureshi 2010: Mancosu et al. 2015: Dinar, Tieu, and Huynh 2019; Siirila-Woodburn et al. 2021). Ample research explores the water conservation potential of farm-level practices and technology adoption (Van der Kooij et al. 2013; Pfeiffer and Lin 2014; Williams, Wuest, and Long 2014; Wang et al. 2021; Huang et al. 2021; Novara et al. 2021). Our research builds on this extensive literature by providing novel evidence regarding how investments in off-farm infrastructure can increase water availability, affording policy makers another tool to address water scarcity and support the irrigated agricultural sector.

The estimated water conservation potential of investments in conveyance infrastructure invite additional research questions which merit attention within the literature. For example, our empirical modeling does not specifically address the longevity of lined and piped canals which may be particularly important for lined canals which degrade relatively quickly (Plusquellec 2019). Additional research is needed characterizing how the dynamics of conveyance infrastructure longevity affect investment decisions. Finally, our simulation model focuses solely on the water conservation returns of the initial capital costs for installing improved conveyance infrastructure. However, there are potentially maintenance and operation costs which may influence organization investment decisions and water conservation outcomes. Additional research is needed to understand these dynamics and their impact on optimal public and private investment in conveyance infrastructure improvements.

References

Baumgarten, B. 2019. "Canal Lining Demonstration Project Year 25 Durability Report." U.S. Department of the Interior, Bureau of Reclamation, ST-2019-1743-01. Created December 1, 2021.

Chakravorty, U., E. Hochman, and D. Zilberman. 1995. "A Spatial Model of Optimal Water Conveyance." *Journal of Environmental Economics and Management* 29 (1): 25–41.

Chakravorty, U., and J. Roumasset. 1991. "Efficient Spatial Allocation of Irrigation Water." *American Journal of Agricultural Economics* 73 (1): 165–73.

CNRA. 2009. "California Natural Resources Agency (CNRA), Bound Accountability, All-American Canal Lining Project." GAO-06–314. Created December 1, 2021.

Dinar, A., A. Tieu, and H. Huynh. 2019. "Water Scarcity Impacts on Global Food Production." *Global Food Security* 23: 212–26.

Edwards, E. C., and S. M. Smith. 2018. "The Role of Irrigation in the Development of Agriculture in the United States." *The Journal of Economic History* 78 (4): 1103–41.

Evan, A., and I. Eisenman. 2021. "A Mechanism for Regional Variations in Snowpack Melt under Rising Temperature." *Nature Climate Change* 11 (4): 326–30.

Fischer, B., and B. Willis. 2020. "Western Piorities in the 2018 Farm Bill." In *Western Economics Forum*, volume 18, 11–16.

Gleason, K. E., J. B. Bradford, A. W. D'Amato, S. Fraver, B. J. Palik, and M. A. Battaglia. 2021. "Forest Density Intensifies the Importance of Snowpack to Growth in Water-Limited Pine Forests." *Ecological Applications* 31 (1): e02211.

Gobarah, M. E., M. Tawfik, A. Thalooth, and E. A. E. Housini. 2015. "Water Conservation Practices in Agriculture to Cope with Water Scarcity." *International Journal of Water Resources and Arid Environments* 4 (1): 20–29.

Haar, C. M., and B. Gordon. 1958. "Riparian Water Rights vs. A Prior Appropriation System: A Comparison." *Boston University Law Review* 38: 207.

Hanjra, M. A., and M. E. Qureshi. 2010. "Global Water Crisis and Future Food Security in an Era of Climate Change." *Food Policy* 35 (5): 365–77.

Hausman, J. A. 1978. "Specification Tests in Econometrics." *Econometrica: Journal of the Econometric Society 46 (6)*: 1251–1271.

Hrozencik, R. A. 2021. "Trends in US Irrigated Agriculture: Increasing Resilience under Water Supply Sscarcity." Available at SSRN 3996325.

Hrozencik, R. A., S. Wallander, and M. Aillery. 2021. "Irrigation Organizations: Water Storage and Delivery Infrastructure." U.S. Department of Agriculture, Economic Research Service Economic Brief No. 32.

Huang, Y., B. Tao, Z. Xiaochen, Y. Yang, L. Liang, L. Wang, P.-A. Jacinthe, H. Tian, and W. Ren. 2021. "Conservation Tillage Increases Corn and Soybean Water Productivity across the Ohio River Basin." *Agricultural Water Management* 254: 106962.

Karimi Avargani, H., S. M. Hashemy Shahdany, S. E. Hashemi Garmdareh, and A. Liaghat. 2020. "Determination of Water Losses through the Agricultural Water Conveyance, Distribution, and Delivery System, Case Study of Roodasht Irrigation District, Isfahan." *Water and Irrigation Management* 10 (1): 143–56.

Koech, R., and P. Langat. 2018. "Improving Irrigation Water Use Efficiency: A Review of Advances, Challenges and Opportunities in the Australian Context." *Water* 10 (12): 1771.

Lehmann, P., M. Aminzadeh, and D. Or. 2019. "Evaporation Suppression from Water Bodies Using Floating Covers: Laboratory Studies of Cover Type, Wind, and Radiation Effects." *Water Resources Research* 55 (6): 4839–4853.

Mancosu, N., R. L. Snyder, G. Kyriakakis, and D. Spano. 2015. "Water Scarcity and Future Challenges for Food Production." *Water* 7 (3): 975–92.

Mo, K. C., and M. Chelliah. 2006. "The Modified Palmer Drought Severity Index Based on the NCEP North American Regional Reanalysis." *Journal of Applied Meteorology and Climatology* 45 (10): 1362–1375.

Mohammadi, A., A. P. Rizi, and N. Abbasi. 2019. "Field Measurement and Analysis of Water Losses at the Main and Tertiary Levels of Irrigation Canals: Varamin Irrigation Scheme, Iran." *Global Ecology and Conservation* 18: e00646.

Molden, D. 2007. *Water for Food, Water for Life: A Comprehensive Assessment of Water Management in Agriculture.* London/Colombo: Earthscan/International Water Management Institute.

Moore, M. R., N. R. Gollehon, and M. B. Carey. 1994. "Multicrop Production Decisions in Western Irrigated Agriculture: The Role of Water Price." *American Journal of Agricultural Economics* 76 (4): 859–74.

Newton, D., and M. Perle. 2006. "Irrigation District Water Efficiency Cost Analysis and Prioritization." DWA final report. USBR.

Njuki, E., and B. E. Bravo-Ureta. 2019. "Examining Irrigation Productivity in US Agriculture Using a Single-Factor Approach." *Journal of Productivity Analysis* 51 (2): 125–36.

Novara, A., A. Cerda, E. Barone, and L. Gristina. 2021. "Cover Crop Management and Water Conservation in Vineyard and Olive Orchards." *Soil and Tillage Research* 208: 104896.

Papke, L. E., and J. M. Wooldridge. 1996. "Econometric Methods for Fractional Response Variables with an Application to 401(k) Plan Participation Rates." *Journal of Applied Econometrics* 11 (6): 619–32.

Pérez-Blanco, C. D., A. Hrast-Essenfelder, and C. Perry. 2020. "Irrigation Technology and Water Conservation: A Review of the Theory and Evidence." *Review of Environmental Economics and Policy.*

Pfeiffer, L., and C.-Y. C. Lin. 2014. "Does Efficient Irrigation Technology Lead to Reduced Groundwater Extraction? Empirical Evidence." *Journal of Environmental Economics and Management* 67 (2): 189–208.

Plusquellec, H. 2019. "Overestimation of Benefits of Canal Irrigation Projects: Decline of Performance Over Time Caused By Deterioration of Concrete Canal Lining." *Irrigation and Drainage* 68 (3): 383–88.

PRISM. 2021. Prism Climate Group, Oregon State University. http://prism.oregon state.edu. created August 1, 2021.

Ramalho, E. A., J. J. Ramalho, and J. M. Murteira. 2011. "Alternative Estimating and Testing Empirical Strategies for Fractional Regression Models." *Journal of Economic Surveys* 25 (1): 19–68.

Reidmiller, D., C. Avery, D. Easterling, K. Kunkel, K. Lewis, T. Maycock, and B. Stewart. 2019. "Fourth National Climate Assessment." *Volume II: Impacts, Risks, and Adaptation in the United States. U.S. Global Change Research Program.*

Schwabe, K., M. Nemati, C. Landry, and G. Zimmerman. 2020. "Water Markets in the Western United States: Trends and Opportunities." *Water* 12 (1): 233.

Siirila-Woodburn, E. R., A. M. Rhoades, B. J. Hatchett, L. S. Huning, J. Szinai, C. Tague, P. S. Nico, D. R. Feldman, A. D. Jones, W. D. Collins, et al. 2021. "A Low-to-No Snow Future and Its Impacts on Water Resources in the Western United States." *Nature Reviews Earth & Environment* 2 (11): 800–819.

Staiger, D., and J. H. Stock. 1997. "Instrumental Variables Regression with Weak Instruments." *Econometrica* 65 (3): 557.

Stock, J. H., J. H. Wright, and M. Yogo. 2002. "A Survey of Weak Instruments and Weak Identification in Generalized Method of Moments." *Journal of Business & Economic Statistics* 20 (4): 518–29.

Stock, J. H., and M. Yogo. 2005. "Testing for Weak Instruments in Linear IV Regression." In *Identification and Inference for Econometric Models: Essays in Honor of Thomas J. Rothenberg.* Cambridge, UK: Cambridge University Press.

Sun, S., G. Sun, P. Caldwell, S. McNulty, E. Cohen, J. Xiao, and Y. Zhang. 2015.

"Drought Impacts on Ecosystem Functions of the US National Forests and Grasslands: Part II Assessment Results and Management Implications." *Forest Ecology and Management* 353: 269–79.

Taylor, D. 2016. "Modelling Supply Channel Seepage and Analysing the Effectiveness Mitigation Options." PhD Dissertation. University of Southern Queensland Faculty of Health, Engineering and Sciences.

Todd, D K. *Water encyclopedia*. United States: N. p., 1970. Web.

Tolley, G. S., and V. Hastings. 1960. "Optimal Water Allocation: The North Platte River." *The Quarterly Journal of Economics* 74 (2): 279–95.

Umetsu, C., and U. Chakravorty. 1998. "Water Conveyance, Return Flows and Technology Choice." *Agricultural Economics* 19 (1–2): 181–91.

USDA-NASS. 2019. "Irrigation and Water Management Survey, 2018." National Agricultural Statistics Service (NASS), Agricultural Statistics Board, United States Department of Agriculture (USDA). Created December 1, 2021.

USDA-NASS. 2020. "Irrigation Organizations." National Agricultural Statistics Service (NASS), Agricultural Statistics Board, United States Department of Agriculture (USDA). Created December 1, 2021.

USDA-NRCS. 2020a. "Lining Cost Scenarios." Created December 1, 2021.

USDA-NRCS. 2020b. "Pl-566 Funded Projects." Created December 1, 2021.

Van der Kooij, S., M. Zwarteveen, H. Boesveld, and M. Kuper. 2013. "The Efficiency of Drip Irrigation Unpacked." *Agricultural Water Management* 123: 103–10.

Wallander, S., M. Aillery, D. Hellerstein, and M. Hand. 2013. The Role of Conservation Programs in Drought Risk Adaptation." *Economic Research Service ERR*, 148.

Wallander, S., R. A. Hrozencik, and M. Aillery. 2022. "Irrigation Organizations: Drought Planning and Response." U.S. Department of Agriculture, Economic Research Service Economic Brief No. 33.

Wang, C., J. Zhao, Y. Feng, M. Shang, X. Bo, Z. Gao, F. Chen, and Q. Chu. 2021. "Optimizing Tillage Method and Irrigation Schedule for Greenhouse Gas Mitigation, Yield Improvement, and Water Conservation in Wheat–Maize Cropping Systems." *Agricultural Water Management* 248: 106762.

Wang, W., S. Liu, T. Kobayashi, and M. Kitano. 2013. Evaporation from Irrigation Canals in the Middle Reaches of the Heihe River in the Northwest of China: A Preliminary Study.

Ward, F. A. 2010. "Financing Irrigation Water Management and Infrastructure: A Review." *International Journal of Water Resources Development* 26 (3): 321–49.

Williams, J., S. Wuest, and D. Long. 2014. "Soil and Water Conservation in the Pacific Northwest through No-Tillage and Intensified Crop Rotations." *Journal of Soil and Water Conservation* 69 (6): 495–504.

Wooldridge, J. M. 2015. "Control Function Methods in Applied Econometrics." *Journal of Human Resources* 50 (2): 420–45.

Center Pivot Irrigation Systems as a Form of Drought Risk Mitigation in Humid Regions

Daniel Cooley and Steven M. Smith

4.1 Introduction

Farmers confronting climate change can undertake a variety of adaptation strategies to cope. Broadly speaking they can hedge financially (i.e., crop insurance or income diversification) or adapt production processes, such as crop and seed variant choices, fertilizer and pesticide use, soil management, planting and harvesting adjustments, and irrigation (Smit and Kinner 2002). While all options may prove instrumental to a farmer's success and subsequently our aggregate food security, we focus on the role of new irrigation in this chapter. This adaptation is particularly important to understand because it has the potential to interact with other adaptation strategies (such as crop insurance or switching crops) and has deleterious externalities to others relying on the water resources. The latter concern is exacerbated by the fact that newly installed irrigation tends to be in more humid regions where fewer laws govern the use of water, and other non-irrigators such as power plants and municipal water supplies are likely to be affected (Fuchs et al.

Daniel Cooley is a PhD candidate in the Department of Economics and Business at the Colorado School of Mines.

Steven M. Smith is Associate Professor in the Departments of Economics and Business, and of Hydrologic Science and Engineering at the Colorado School of Mines.

For helpful comments and suggestions, we thank Jared Carbone, Ariel Dinar, Ben Gilbert, Zeynep Hansen, Gary Libecap, Haoluan Wang, and conference participants at the Midwestern Economic Association Annual Meeting (2022). We acknowledge funding support from NSF, DISES program (Grant #2108196), NSF/USDA INFEWS program (Grant #67020-30130), and USDA, AFRI program (Grant #67023-29421). The authors have no financial interests that are relevant to this research. All errors are our own. For acknowledgments, sources of research support, and disclosure of the authors' material financial relationships, if any, please see https://www.nber.org/books-and-chapters/american-agriculture-water-resources-and-climate-change/center-pivot-irrigation-systems-form-drought-risk-mitigation-humid-regions.

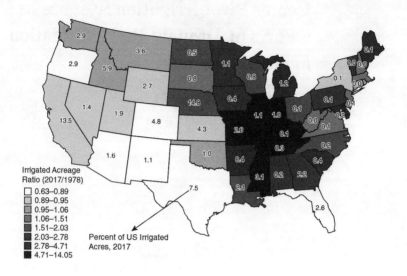

Fig. 4.1   Irrigated acres in the US, 1978–2017

Note: States with darker shading experienced a greater percent change in irrigated acres over the observed period. States with larger numbers contain a greater percentage of the nation's irrigated acreage as of 2017.

Source: Underlying data come from the USDA Census (Haines, Fishback, and Rhode 2018; USDA 2019).

2012). As farmers take up more irrigation in response to climate change, it is important to begin to understand some of the causes and effects of this adaption strategy in historically rainfed regions.

Irrigation has long served as an adaptation to increase crop yields and provide resilience during severe drought in arid regions (Troy, Kigpen, and Pal 2015; Zhang, Lin, and Sassenrath 2015; Tack, Barkley, and Hendricks 2017; Edwards and Smith 2018; Zaveri and Lobell 2019). The 17 arid states of the US, commonly delineated as those to the west of the 100th meridian, were quick to adapt water policy to spur irrigation in the 19th century (Leonard and Libecap 2019). This region attracted huge public investment to build large dams in the 20th century, and tapped groundwater resources with abandonment after the 1940s (Edwards and Smith 2018). In total, these efforts brought some 40 million acres under irrigation by 1978. Since then, however, these states have stagnated in irrigated acreage. Meanwhile, there is a persistent trend in irrigation uptake in humid states, as can be seen in figure 4.1. These humid states, on average, have more than tripled their irrigated acreage since 1978, and today, over one-third of irrigated acreage in the US is east of the 100th meridian. This chapter explores the benefits of irrigation investment in more humid regions in the context of climate change trends and existing crop insurance coverage. Specifically, we look at Illinois, where irrigated acreage has increased nearly fivefold since 1978, to consider how irrigation investment has emerged and its effect on cropping patterns, crop yields, and federal crop insurance payouts.

Economists have not paid significant attention to the irrigation potential of rainfed agriculture. This is an oversight given that freshwater availability will shape irrigation adaptation over the long run. Freshwater availability limitations in arid and historically irrigated regions (e.g., the western US) means climate change adaptation in those regions may not take the form of additional irrigation, but more humid regions, like the eastern US, can still expand irrigation (Elliot et al. 2014). In a global study, Rosa et al. (2020) estimated that irrigation could be expanded as an adaptation strategy with little negative effects on water resources on 35 percent of current rainfed crops. This, of course, will take costly investments to accomplish (Elliott et al. 2014), and similarly scaled projects to those in the mid-20th century in the US may be unreasonable to expect (Schlenker, Hanemman, and Fisher 2005). However, shorter distances from the more abundant streams in humid areas and technological advances have lowered the costs to tap underground water resources efficiently, reducing the need for large projects to garner economies of scale. Instead, individuals can now make the investment decision on their own.

Irrigation expansion in the US has often been by center pivot irrigation systems (CPIS) since their invention in 1948 by Frank Zybach (Anderson 2018). Today, some 57 percent of irrigated acreage in the US is fed by a sprinkler with higher percentages in humid regions (Hrozencik and Aillery 2021). In arid states, these self-propelled irrigation systems transformed the agricultural industry; farmers were able to adopt more water-intensive plants, sustain higher crop yields, and utilize more land as cropland (Evans 2001). With the line of aridity shifting east (Seager et al. 2018), these upsides may explain continued irrigation increases in the Great Plains region. But in the Midwest, where natural rainfall was already sufficient to feed the more profitable, more water-intensive crops like corn and soybeans, average rainfall has *increased*.

Despite the increasing average rainfall across the Midwest, the variability of precipitation extremes is also increasing (Ford, Chen, and Schoof 2021). Additionally, the length and frequency of dry periods during the summer is predicted to increase across the Midwest throughout the 21st century (Grady, Chen, and Ford 2021), and the intensity of droughts in humid regions may also be greater when they do occur (Trenberth et al. 2014). This is particularly important for corn and soybeans, major crops in the Midwest, as summertime drought is correlated with low corn and soybean yields (Mishra and Cherkauer 2010).

Negri, Gollehon, and Aillery (2005) found that the tails of weather distributions matter more than the means in predicting irrigation uptake, albeit based on cross-sectional variation. Temporally, farmers also tend to invest in irrigation shortly after experiencing a drier year (Smith and Edwards 2021). In Illinois, severe, statewide droughts occurred in 1988, 2005, and 2012 with the most recent severe drought prior to 1988 occurring in 1964 (State Climatologist Office for Illinois 2015). The 2012 drought caused a large decline in crop yields bringing statewide corn production down to 105 bushels per acre from 157 bushels per acre in 2011 (Illinois Department of Natural Resources 2013). Irrigated acreage in Illinois, overwhelmingly consisting of CPIS, has increased by 54 percent from 1997 to 2015 (Stubbs 2016), and 833 new CPIS were installed statewide in the two years following the 2012 drought (ISWS 2015).

We explore the trends in CPIS adoption and their benefits as a form of drought risk mitigation in Illinois given the upward trends in precipitation means and variation as well as irrigation. The question is interesting because CPIS are expensive and alternatives do exist. The capital costs to set up a new CPIS on 160 acres (irrigating about 128 acres) is upwards of $153,000 (Sherer 2018). Meanwhile, farmers have crop insurance to insulate themselves from droughts and other disasters, and Illinois farmers have been increasing their coverage, going from 87 percent and 85 percent for corn and soybeans respectively in 2016 to 96 percent and 93 percent in 2020 (USDA 2021). We consider whether CPIS uptake confers the benefits found in the West—crop switching, increased yields, resilience, and expanded cropland—and how crop insurance payouts are affected.

A critical part of this research is knowing where and when CPIS are installed. The agricultural census reported on this specific irrigation technology at the county level only in 1959 and 1969, and state records vary greatly. To circumvent this data shortage, we leverage a deep learning model to identify the locations of CPIS from satellite imagery. The model extends that of Cooley, Maxwell, and Smith (2021), which was utilized to identify CPIS over the Ogallala aquifer, a much more arid region. Our focus on Illinois among the Midwest states is largely because the Illinois State Water Survey conducted a survey in 2012 and 2014 in which CPIS were manually identified from aerial photography and well records for the entire state, providing a ground truth with which to train the model. Still, deploying the model in a humid region where the iconic "circles" of CPIS are less detectable limits performance. Accordingly, we run the model on drought years when the circles are most apparent and fill in the interim years via a linear trend. Results are robust to alternative assumptions for the non-drought years.

We aggregate 30 × 30 m resolution CPIS data to the county level and combine it with annual crop production from USDA's National Agricultural Statistics Service (NASS), and Illinois crop insurance data from the USDA's Risk Management Agency (RMA). In addition, we draw on county-level statistics from the USDA Agriculture Censuses. We also collect weather data from both NOAA and PRISM to address climate variation and various other hydrologic and topographic data.

We find that adoption of CPIS in Illinois is strongly correlated with the presence of alluvial aquifers (and not other groundwater or streams) and often spurred by experiencing relatively drier years. In addition, larger farms are more likely to adopt CPIS. Notably, soil suitability and topography offer little predictive value, but counties with slightly less valuable farmland prior to irrigation have adopted CPIS more extensively.

In terms of the effects of CPIS installation, we find some evidence that where CPIS are installed there is a shift from soybeans to corn or a shift in the frequency of corn in the crop rotation, which is particularly interesting given that corn has greater water use efficiency than soybeans (Dietzel et al. 2015). This is a significant result as irrigation improvements in other regions with a more arid or semiarid environment have resulted in increased average crop yield, a switch to thirstier crops, and even an increase in cropland (Pfeiffer and Lin 2014). Furthermore, there is no significant correlation between CPIS installation and average crop yield in Illinois. However, corn yield during drought years shows a positive correlation with CPIS presence at the county level, and the sum of money paid by the insurance to the insured, known as indemnity, is negatively correlated with CPIS presence at the county level during drought years for both corn and soybeans. Taken in combination, these results imply that the primary benefit of installing a CPIS in Illinois is drought risk mitigation despite the high rate of crop insurance coverage in Illinois. This result is particularly interesting in the light of previous research that suggests greater crop insurance coverage disincentivizes farmers from adapting to drought due to moral hazard (Annan and Schlenker 2015).

Broadly, our results contribute to the literature regarding agricultural adaptation to climate change. Most notably, we explore a novel setting that has been largely neglected in previous work more directly related to irrigation improvements (e.g., Koundouri, Nauges, and Tzouvelekas 2006; Baerenklau and Knapp 2007; Torkamani and Shajari 2008; Pfeiffer and Lin 2014; Christine et al. 2012). In the economics literature, irrigated agriculture has often been discarded when analyzing the effect of climate change (e.g., Schlenker and Roberts 2009; Burke and Emerick 2016) under the argument that it is poor proxy for non-irrigated areas as similar adaptations are not expected (Schlenker, Hanemman, and Fisher 2005). Yet, the East is increasing irrigation, although average effects on production are not well identified (Smith and Edwards 2021).

We also expand on the small literature regarding supplemental irrigation. The value of supplemental water reserves for irrigated areas goes back to work by Tsur (1990) and Tsur and Graham-Tomasi (1991) that identified and quantified the quasi-option value of groundwater reserves. More recent work has shown the supplemental water rights in the western US as an adaptation to reduced rain and higher temperatures (Bigelow and Zhang 2018) that adds real value to the farms (Brent 2017). Our efforts are distinct in that we consider new irrigation adoption solely in a primarily rainfed region.

Our work also builds on the substantial body of research concerning the effects of advancements in technology on agricultural production (e.g., Griliches 1957; Ruttan 1960; Just, Schmitz, and Zilberman 1979; Zilberman 1984; Lau and Yotopoulos 1989). Furthermore, this research focuses on the shift from unirrigated land to land irrigated by CPIS rather than marginal improvements to existing irrigation technologies (e.g., Pfeiffer and Lin 2014). Technological choice is important as CPIS have been associated with more resilience than other forms of irrigation (Cooley, Maxwell, and Smith 2021). The results also speak to the potential for moral hazard with crop insurance, where insured farmers strategically underinvest in yield enhancing activities during extreme weather events (e.g., Smith and Goodwin 1996; Annan and Schlenker 2015; Connor and Katchova 2020; Wang, Rejesus, and Aglasan 2021).

Finally, the deep learning model used to automatically identify CPIS for this study exemplifies the use of machine learning methods to extract and classify information from unstructured data in economics (e.g., Athey 2019; Storm, Baylis, and Heckelei 2020), and to the growing literature regarding automated CPIS identification by examining a humid region rather than the more arid regions where similar models have been deployed (e.g., Zhang et al. 2018; Deines et al. 2019; Saraiva et al. 2020; Valencia et al. 2020; Tang et al. 2021; Cooley, Maxwell, and Smith 2021).

4.2   Background and Theoretical Framework

4.2.1   Illinois Agriculture and Climate

At nearly $8,000 per acre as of 2021, Illinois's cropland is among the most valuable in the US. More valuable cropland is found in the Northeast, where non-productivity factors likely drive the land values higher, and in California (USDA 2021). Illinois produced over $1 billion worth of crops in 2017, just behind Iowa and, more distantly, California. Unlike California, where specialty crops are prevalent, Illinois (and Iowa) grow mostly corn and soy, with Illinois producing the most soy and the second most corn in 2017. Given these crops are often grown in rotation on the same fields, the relative ranking of Illinois and Iowa often swaps.

Meanwhile, Illinois is close to the middle of the distribution in terms of growing season precipitation (April–September). The annual average for Illinois counties, stretching back to 1900, is 579 mm of precipitation. For comparison, counties in California saw just 116 mm during that period. Wetter states, primarily in the southeast, averaged over 650 mm, with Florida,

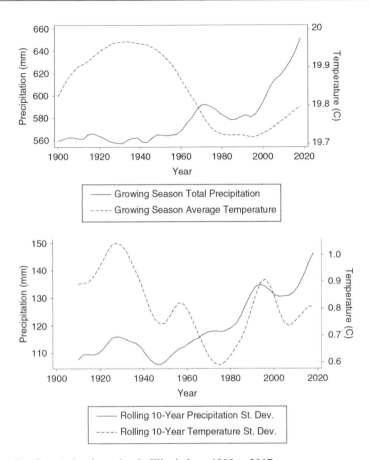

Fig. 4.2   County-level weather in Illinois from 1900 to 2017

Note: Panel A shows the annual growing season total precipitation (solid) and growing season average temperature (dashed) using a local polynomial to plot the state level averages. Panel B shows the standard deviation for the weather variables based on a 10-year rolling calculation. *Source*: Data for both come from PRISM.

the wettest, averaging 888 mm. At 20°C, Illinois is also near the middle of the states in terms of average monthly temperature during the growing season as well. The upshot is that Illinois, relative to the other continental states, is not an extreme case but rather a temperate-humid setting.

Figure 4.2 shows that while temperatures in Illinois have not exhibited a clear trend, precipitation has increased while also becoming more variable. The plots are local polynomial fits for the county-level, annual growing season weather variables across time. Precipitation has exhibited an upward trend since 1960, going from around 565 mm to nearly 650 mm, a 15 percent increase. Temperature is lower now than in 1940, but it is higher compared to more recent baselines like 2000. The shifts are also relatively slight, ranging from 19.7°C to about 19.9°C. We should note, however, the annual averages omit important within season variation that matters for corn production (see Berry, Roberts, and Schlenker 2014, for instance).

Panel b of figure 4.2 shows that the precipitation, although greater on average, has also become more variable. The plots are again local polynomial for county-year measures, but for a rolling, 10-year standard deviation. As the average precipitation began to increase, around 1940, so did the temporal standard deviation, increasing from about 105 mm up to 145 mm. Therefore, the wetter years are interspersed with drier years. Again, no clear trend emerges for temperature given that it is highly sensitive to how far back one looks.

A severe drought occurred in 2012 in Illinois as with much of the Corn Belt. Corn yields in Illinois counties were about 50 percent lower in 2012 than either 2007 or 2017. Berry, Roberts, and Schlenker (2014) predicted corn losses of 20 percent across the Corn Belt. However, the authors do not account for irrigation in their estimates. On average, at the time, this may have been a reasonable assumption, but 15 counties in Illinois that year did irrigate over 5 percent of their harvested crops, with one county topping out at 41 percent. Illinois farmers have drastically increased their irrigated acreage. In 1950, Illinois had just 140 irrigated farms with 1,510 acres irrigated. In 2017, 2,541 farms reported irrigating a total of 607,442 acres, or roughly as much as New Mexico, an arid state associated with irrigation. Illinois irrigation still only amounts to 5 percent of the farms and just 2.3 percent of the acres, meaning many there are yet to adopt irrigation but could potentially do so in the future.

An alternative adaptation to irrigation is crop insurance. Notably, coverage is broader than for only droughts and can include flooding, fire, pests, or even commodity prices. Crop insurance generally comes in two forms: yield protection and revenue protection. Yield protection guarantees a payout—known as "indemnity"—equal to a percentage of a farmer's average crop yield. The average crop yield is measured as the mean of that farmer's previous 10 years of harvest, so if a crop loss event occurs, the insurance will pay a percentage of the difference between the average crop yield and the realized crop yield in the current year's prices. Revenue protection works similarly, but the insurance company pays out a percentage of the farmer's average revenue instead of their average crop yield. The biggest difference is that if farmers opt into revenue protection, they are able to submit insurance claims for drops in crop value (Plastina, Johnson, and Edwards 2021). In addition to the option of yield or revenue protection, farmers are also able to select a coverage level, between 50 percent and 85 percent, which determines the percentage of lost yield or revenue a farmer receives as indemnity.

Uptake of crop insurance in Illinois is extensive and provides substantial protection from droughts. As of 2020, 96 percent of corn acreage is insured and 93 percent of soy acreage, up from previous years (USDA 2021). In 2005, Illinois farmers collected nearly $25 million in indemnity payments due to the drought. More eye opening, in 2012, the most severe drought, they received $2.9 billion.

4.2.2   Analytical Framework

Given that Illinois has increased irrigation capacity despite being humid and getting wetter on average, we provide a formal theoretical model to provide some insight into the decision process for installing a CPIS as a form of drought risk mitigation. We include crop insurance as an alternative mitigation tool given its prevalence and moral-hazard-inducing potential. We also discuss additional factors as extensions to the model that would implicitly affect the functional forms to further generate hypotheses about the cause and effects of CPIS installations in the Corn Belt.

Assume that the farmers' choice is between having insurance without irrigation or having both irrigation and insurance simultaneously. To focus on the decision for drought risk mitigation, also assume that average production is unchanged and set aside the potential losses of too much precipitation. The farmers' profits ($\pi(w_t, p_t)$) can be thought of as a function of irrigation water (w_t) and precipitation (p_t) in a given year, and their expected profits are the sum of annual profits from the present to time T multiplied by the weighted probability of a normal precipitation year ($1 - \alpha$) or below average precipitation year (α).

(1) $$E[\pi] = \sum_{t=1}^{T} E[\pi_t(w_t, p_t)] - g(k),$$

(2) $$\pi_t(w_t, p_t) = (1 - \alpha)\bar{y} + \alpha[y(w_t, p_t) + i(w_t, p_t) - c(w_t)] - \frac{g(k)}{T},$$

$$s.t. \, w_t \leq k \,, 0 \leq \alpha \leq 1.$$

For simplicity, the value of the crop is normalized to 1 and taken as a constant. Precipitation (p_t) is a random variable that follows a stationary process with an average of $\bar{p}$. Irrigation water w_t is constrained by $w_t \geq 0$ and $w_t \leq k_t$ where k_t is installed irrigation capacity. The profit function is composed of crop yield $y(\cdot)$, net insurance payment $i(\cdot)$, the cost of irrigation water $c(w_t)$, and the annualized cost of irrigation capacity [$g(k)$] / T. The yield function is a concave production function for an arbitrary crop where the inputs w_t and p_t are perfectly substitutable. The benchmark crop yield ($\bar{y}$) is the upper limit of the crop yield function achieved when precipitation reaches the average value ($\bar{p}$) where neither insurance nor irrigation water are necessary. Lastly, α is the probability that a given year will have lower than average rainfall and is therefore a value between 0 and 1. The insurance payout in a given period is:

(3)
$$i(y(w_t, p_t)) = b(\bar{y} - y(w_t, p_t)) - f(b),$$

$$s.t.\ 0 \le b \le 1.$$

The insurance pays a guaranteed percentage (b) of the lost crop yield ($\bar{y} - y(w_t, p_t)$) at the cost of the premium for the level of insurance protection that the farmer has opted into ($f(b)$). The greater the quantity of irrigation water w_t, the smaller the gap between the benchmark crop yield and the realized crop yield in the current year. By normalizing the price of crops to one, our model ignores the difference between the specific coverage types. The percentage of guaranteed crop yield or revenue is selected by the farmer, and premiums reflect the difference in this choice by increasing with greater levels of insurance protection.

If CPIS are an effective form of drought mitigation above and beyond that of insurance alone, the expected value of profits in a farm without a CPIS $(0, p_t)$ would be below that of a farm with a CPIS (w_t, p_t), where there is some positive amount of irrigation.

(4)
$$\pi(0, p_t) = \sum_{t=1}^{T}(1 - \alpha)\bar{y} + \alpha[y(0, p_t) + (\bar{y} - y(0, p_t))b - f(b)],$$

(5)
$$\pi(w_t, p_t) = \sum_{t=1}^{T}(1 - \alpha)\bar{y} + \alpha[y(w_t, p_t) + (\bar{y} - y(w_t, p_t))b - f(b) - c(w_t)]$$

$$- \frac{g(k)}{T},$$

(6)
$$\pi(w_t, p_t) - \pi(0, p_t) = \sum_{t=1}^{T}\alpha[(1 - b)\Delta y - c(w_t)] - \frac{g(k)}{T},$$

where $\Delta y = y(w_t, p_t) - y(0, p_t)$.

For CPIS installation to be an effective form of drought mitigation for a farmer, equation 6 must be positive. In other words, the value of the difference in crop yield as a result of irrigation multiplied by the uninsured percentage of the crop must be greater than the cost of water and annualized cost of irrigation capacity for a farmer to consider installing a CPIS. However, this does not give us the full story. Additionally, we see that as the probability of a below average precipitation year increases, the expected value of irrigation also increases which gives us some insight as to why farmers may be installing CPIS more rapidly in recent years as dry spells have gotten more frequent. We can further examine this effect on the margin by creating a Lagrangian from equation 2 for a given year:

(7)
$$\mathcal{L} = (1 - \alpha)\bar{y} + \alpha[y(w_t, p_t) + i(y(w_t, p_t)) - c(w_t)] - \frac{g(k)}{T} + \lambda(k - w_t).$$

Using the definition of $i(y(w_t, p_t))$ from equation 3 and taking the derivative, our first-order conditions are:

(8.a) $\dfrac{\partial \mathcal{L}}{\partial w} = \alpha \left[\dfrac{\partial}{\partial w} y(w_t, p_t) - b \dfrac{\partial}{\partial w} y(w_t, p_t) - \dfrac{\partial}{\partial w} c(w_t) \right] - \lambda \leq 0,$

(8.b) $\dfrac{\partial \mathcal{L}}{\partial k} = -\dfrac{\partial}{\partial k} \dfrac{g(k)}{T} + \lambda \leq 0,$

(8.c) $\dfrac{\partial \mathcal{L}}{\partial \lambda} = k - w_t \geq 0.$

From these first-order conditions, we are mostly interested in equation 8.a, where at the optimal point we may rearrange to find:

(9) $\alpha \left[(1 - b) \dfrac{\partial}{\partial w} y(w_t, p_t) - \dfrac{\partial}{\partial w} c(w_t) \right] \leq \lambda.$

The Lagrange multiplier (λ) is a measure of the shadow price of irrigation capacity. Given our Kuhn-Tucker conditions, farmers either irrigate until the marginal net benefit of water is equal to the Lagrange multiplier or do not irrigate at all. The marginal net benefit of water is positively influenced by the marginal value of the uninsured crop yield and negatively influenced by the marginal cost of water. From this equation, we also see that as the probability of a dry year (α) increases, the marginal net benefit of a unit of water also increases. It follows that as dry spells have been getting more common and intense, the value of irrigation water has increased to farmers in Illinois, incentivizing them to install a CPIS.

To this point, we have largely ignored the factors that may create distinctions in the functional forms across space and time. As mentioned above, the average capital costs, $g(k)$, for a new CPIS are around $150,000 on 160 acres (Sherer 2018). Although this does not explicitly vary across space, variation of the land, both physical and legal, will create differential costs to customize the system appropriately. On the physical side, flatter areas with cheaper access to water will drive down installation costs. Given the relative flatness of Illinois, water access is likely to be much more critical, especially since water rights are based on the riparian doctrine, limiting potential irrigators to those with fields adjacent or above water resources. The field or farm sizes may also matter, exhibiting some economies of scale due to technological aspects or the farm operation and desire or ability to self-insure through CPIS. Finally, public policy in the form of subsidies would influence the farmer's costs. While many programs exist to subsidize water conservation alterations on existing irrigation systems (e.g., USDA's EQIP program), we have not turned up any large-scale programs to help bring rainfed plots under irrigation.

Beyond water access, the pumping costs, $c(w_t)$, will vary by depth to water and saturated thickness as will the price of local energy sources. While we do not have data on the water, we proxy it by the type of water resource (surface, groundwater, or alluvial aquifer). No information on energy prices is available. Finally, although we treated the cost of insurance in the model as constant within a crop, there is variation of $f(b)$, most importantly based on whether irrigation is present or not. Generally, irrigated crops are given a relative discount, meaning the cost of insurance will be reduced if CPIS is installed. Furthermore, should CPIS maintain higher average yields, the payout on insurance claims will also be higher given they rely on field specific average production, offering an additional benefit when considering the interaction with insurance.

In terms of the yield function, $y(w_t, p_t)$, the greater sensitivity to water means larger gaps in profits between irrigated and non-irrigated fields. Although offset by insurance claims, given those are paid out as a percentage, a larger gap will still produce a larger absolute income loss. Relative to soy, corn exhibits both higher and more variable yields, and, although more resistant to extreme heat, greater sensitivity to moisture availability (Zipper, Qiu, and Kucharik 2016). Accordingly, we expect more CPIS installations where more corn is grown and the average yields are higher.

Our model generates few predictions about the effects of CPIS systems, but the model does contain a few assumptions worth testing. At root is that we modeled the CPIS adoption as a mode of self-insurance against drought. If true, we do not expect to see crop switching or an expansion of cropland. However, in more arid regions crop switching to more water-sensitive, but higher-paying, crops has been observed following irrigation investments (e.g., Pfeiffer and Lin 2014). In Illinois, farmers that avoided corn due to its greater sensitivity to precipitation variation may see an opportunity to maintain corn in the rotation more often. Second, we assumed no changes to yields during normal precipitation years. However, yields may trend higher if farmers are able to smooth out dry periods during the year or supplement drier conditions even if the growing season is not a full-blown drought. Furthermore, installing irrigation may reduce the need to use drought-resistant seed variants which often sacrifice yields under non-drought conditions to have higher drought yields (Yu, Miao, and Khanna 2021). Finally, the model assumes farmers will deploy the CPIS in drought years. Accordingly, we expect yield losses to be mitigated in drought years in areas with more CPIS. A corollary to this is that drought-related insurance payouts should also be lower in these areas.

In sum, we aim to test the following hypotheses with empirical data:

i.   Areas with lower cost access to fresh water resources develop more irrigation.

ii.   Flatter and larger farms adopt more irrigation capacity.

iii.   Areas with higher yields and more corn develop more irrigation capacity.

iv.   Areas increase irrigation capacity as the incidence of dry years increases rather than average precipitation trends.

v.   More irrigation capacity does not lead to crop switching or an expansion of cropland.

vi.   In average precipitation years, CPIS does not affect yields, but increases yields in drought years.

vii.   CPIS reduces indemnity (crop insurance) payments.

4.3   Data

Historically, identifying the location of CPIS has been challenging in areas that do not hold publicly accessible records for such things. CPIS identification largely remains a tedious process of visually inspecting aerial or satellite imagery and manually marking their boundaries. This identification method was used to detect CPIS in the Northern Atlantic Coastal Plain (NACP) from satellite imagery in 2013 and indicated that about 271,900 acres were irrigated primarily by CPIS (Finkelstein and Nardi 2016). More relevant for this chapter, the ISWS also manually identified CPIS from aerial imagery of the entire state in 2012 and 2014, which revealed a 14.2 percent increase in CPIS between the two periods (Illinois State Water Survey 2015). One of the reasons our chapter focuses on Illinois is the high quality of the data provided by the ISWS, as it is more complete than other options and shows variation through time rather than being a static snapshot. However, the ISWS data only cover three years of development within which there is only one drought year in 2012, leaving something to be desired.

To overcome this challenge, we utilize deep learning, a type of machine learning that uses neural networks to replicate the learning process of humans. Deep learning is particularly useful for processing unstructured data such as the satellite imagery used for this chapter as it requires very little human input. The goal is to get the predictions of the deep learning model to mirror the state of the real world known as the ground truth by minimizing the difference between the two. The model does this by guessing the correct label of inputs, checking how it performed against the ground truth, then recalibrating itself before repeating the process. For a more expansive and technical treatment of the subject, see Goodfellow, Bengio, and Courville (2016).

While there has been a recent breakthrough in CPIS identification through deep learning methods in arid and semiarid regions where the distinctive crop circles left by CPIS are quite clear, humid regions like Illinois pose a greater difficulty due to the natural precipitation reducing the distinct boundary between the irrigated area and surrounding land cover most years (e.g., Zhang et al. 2018; Deines et al. 2019; Saraiva et al. 2020; Valencia et al. 2020; Tang et al. 2021; Cooley, Maxwell, and Smith 2021). However, we deployed a deep learning method that is able to predict the historical locations of CPIS in Illinois by using the pre-trained model described in Cooley, Maxwell, and Smith (2021) based on the work of Saraiva et al. (2020). This method is particularly useful as it does not rely on the CPIS being a specific size or shape to predict their locations as is the case with previous attempts at detecting CPIS with satellite imagery (Zhang et al. 2018). However, this comes at the cost of a fuzzy border around the CPIS as the model can struggle with determining the exact boundary of the CPIS when utilizing 30 m resolution imagery. Additionally, this method still leaves large time gaps between observations as it is only reliable where the area beyond a CPIS is distinct enough from the irrigated area to be detected.

The model was pre-trained on CPIS over Nebraska, which allowed us to warm-start the process of detecting CPIS in Illinois where the boundaries of CPIS are less distinct through transfer learning. Transfer learning is a training technique that applies the weights and values from a model intended for one purpose to another model in order to expedite the learning process and minimize the loss of the model. The model was then retrained using manually labelled GIS data from Illinois in 2012 and 30 × 30 m resolution top-of-atmosphere (TOA) reflectance satellite imagery collected from Google Earth Engine's Landsat database. The data were randomly divided into three parts for use in training, validation, and testing, with the training set receiving 80 percent of the total data and the remainder being allocated equally between validation and testing sets.

Figure 4.3 provides a comparison of the ground truth (dark gray) and model output (black) in Illinois. To evaluate the performance of the model, four metrics are utilized at the pixel level: accuracy, specificity, precision, and recall. Accuracy is simply the number of correctly identified pixels over the total number of pixels. Precision is the number of correctly identified CPIS pixels over the total number of CPIS pixels predicted by the model regardless of correctness compared to the ground truth. Recall is the number of correctly identified CPIS pixels over the total number of CPIS pixels as given by the ground truth. Lastly, specificity is the number of correctly identified background (non-CPIS) pixels identified by the model over the number of background pixels given by the ground truth.

The model's accuracy and specificity ratings are well over 99 percent, which is due to the very large number of non-CPIS pixels in the state of Illinois that the model correctly identified. The recall rate of the model is 85.4 percent, which is in line with other CPIS detection models (e.g., Zhang et al. 2018; Saraiva et al. 2020; Cooley, Maxwell, and Smith 2021). A good portion of the loss in recall rate at the pixel level can be accounted for by the model's inability to accurately define the boundaries of the CPIS resulting in misidentified pixels near the edges of the crop circle left by the CPIS.

Perhaps the most relevant portion of the score is the precision of the model, which has a rating of 54.5 percent suggesting that the model predicts

Fig. 4.3   Deep learning output for locating CPIS in Illinois, 2012

Note: Black portions are the predicted locations of CPIS from the deep learning model while dark gray portions are the ground truth from the manually labelled CPIS.

a large number of false positives. This may be due to the indistinct nature of the CPIS boundaries in such a humid climate, the massively unbalanced data set in which there are many more negative examples than positive examples of CPIS presence, or some combination of the two. As such, the model is limited in its usefulness for detecting newly installed CPIS.

However, we assume that CPIS are durable through time. While they occasionally change shape or move to a newly dug well, they tend to persist on the same plot of land for many decades. With this in mind, the model was put to use detecting CPIS in previous drought years and the results were compared the CPIS identified via aerial imagery in 2012. If a predicted CPIS fell within an area where there is not one manually labelled in the 2012 data set, it was thrown out of the final data set. This process takes advantage of the model's high recall rate while limiting the impact of its low precision. The key assumption with this method is that if a CPIS was installed prior to 2012, it would still show up in the 2012 imagery even if it is a slightly different shape or size.

In Illinois, CPIS is the dominant form of irrigation. We compared the share of Illinois counties with irrigation during the 2012 growing season from the USDA Census that year to the hand-collected CPIS data from the Illinois State Water Survey from that year and found the relationship between the two incredibly tight. Shown in figure 4A.1 of the appendix, the binary regression yields a coefficient of 1.05, a constant of −0.003, and an R-Square value of 0.94. Furthermore, the 2015 USGS water use data for Illinois shows that 100 percent of the ~600,000 acres of irrigated cropland in Illinois were irrigated by sprinkler, which aligns well with the Illinois State Water Survey's estimate of center pivot irrigation systems irrigating approximately 625,000 acres of cropland in Illinois in 2014 (USGS 2018; ISWS 2015). Therefore, we view CPIS as the primary technology for irrigation and a good measure of the capital constraint for acreage irrigated in any particular year.

CPIS in Illinois can only be accurately predicted by the deep learning model during drought years, so it is not possible to directly identify how CPIS installation is correlated with indemnity payouts and crop yield. In order to work around this complication, gaps between observations were filled in using linear interpolation from one observation to the next. With the dramatic increase in CPIS seen from 2012 to 2014 after the drought in 2012 though, a linear trend may not accurately portray how CPIS have developed in the state through time. So, two other methods were employed to bookend the possibilities: one in which all of the CPIS observed in a drought year were installed immediately following the previous drought year and one in which all of the CPIS observed in a drought year were freshly installed in that year. The latter seems to be the least likely case as it carries with it the implication that CPIS were installed immediately before a drought occurred while the former case suggests that the CPIS were installed immediately after a drought scare, aligning with more general trends in irrigation and recent droughts (Smith and Edwards 2021). Filling the CPIS observation gaps allows for the inclusion of non-drought year data to get a better idea of the baseline for crop yield, precipitation, and temperature through time. The model's remaining inaccuracy is assumed to not be biased in any particular direction as it was trained on drought years and only utilized to predict CPIS locations during other drought years in areas known to have CPIS installed at some point before 2012. A comparison between CPIS presence in 1988 and 2014 is provided in figure 4.4.

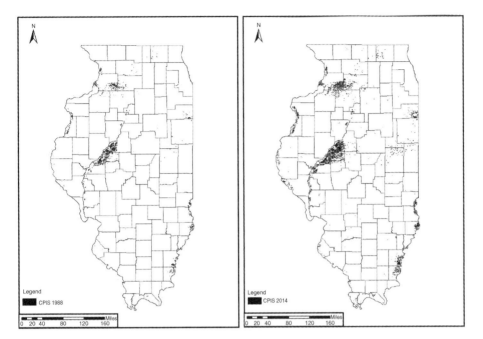

Fig. 4.4   Illinois CPIS in 1988 and 2014

Note: Panel A shows the location of CPIS in 1988 as predicted by the deep learning model. Panel B shows the location of CPIS in 2014 according to a manually labelled data set produced by the Illinois State Water Survey. There was nearly a threefold increase in CPIS between the two periods, but new CPIS were not uniformly distributed across the state, instead being concentrated in a few areas.

The rest of the data collected come from standard sources. USDA censuses provide agriculture and irrigation statistics at the county level roughly every five years, and estimated annual crop data at the county level were gathered from the USDA's National Agricultural Statistics Service (NASS). Illinois crop insurance data regarding the indemnity amount, crop loss, and cause of loss for insurance claims from 1988 to 2020 were collected from the cause of loss and summary of business files from the USDA's Risk Management Agency (RMA). For additional covariates, we gather precipitation and temperature data by county from 1988 to 2012 using the NOAA National Centers for Environmental Information, Climate at a Glance online application. We also draw on annual county-level temperature and precipitation data constructed by Smith and Edwards (2021) from PRISM data. Soil quality and other geographic information was derived from the Gridded National Soil Survey Geographic (gNATSGO) Database for Illinois. Summary statistics for the data may be found in the appendix in table 4A.1.

4.4   Methods

We conduct four empirical analyses. First, we consider drivers of CPIS installation. Second, we consider how CPIS affects cropping patterns. Third, we test for yield effects both in normal and drought years. Fourth, we consider how indemnity payouts in drought years are associated with CPIS.

To test the first four hypotheses concerning irrigation uptake (i through iv), we draw on both our CPIS measures and, given the longer time period available, the agriculture census information on irrigated acres. The latter, however, does not necessarily capture capacity as installed CPIS may not be deployed in a given year. To test the hypotheses, we estimate several versions of the following equation:

$$(10) \qquad Irr_{iy} = \alpha_0 + \bar{W}_i + \bar{X}_i + \bar{C}_{it} + \bar{\delta}_t + \varepsilon_{iy}.$$

Irrigation in county i in year y is measured either by the share of the county irrigated that year according to the census or the share with a CPIS captured in drought years by our machine learning (1988, 2005, 2012). $\bar{W}_i$ is a vector of freshwater availability and their coefficients. These measures include the share of the county over an aquifer, share within a 15-mile buffer of a large stream, and the share over an alluvial aquifer where an alluvial aquifer is one that is closely connected, hydrologically, to a stream. $\bar{X}_i$ is a vector of time-invariant county measures like topography, average weather and variation, 1940 (pre-irrigation) farm characteristics, or, in some specification, just county fixed effects. $\bar{C}_{it}$ is a vector of time varying weather variables. The main ones are locally normalized weather disturbances from Smith and Edwards (2021) measuring how many standard deviations away from the county mean that year's weather is. More attention is given to "severe" years indicated as being more than 1.5 standard deviations drier or hotter than average. Finally, $\bar{\delta}_t$ is a vector of year fixed effects.

For tests of the effects of irrigation in this setting, hypotheses v to vii, we estimate the following equations:

$$(11) \quad y_{it} = \rho_0 + \rho_1 CPIS_{it} + \rho_2 CPIS_{it} * D_t + \rho_3 D_t + \rho_4 P_{it} + \rho_5 T_{it} + \bar{X}_i + \bar{\delta}_t + \varepsilon_{it}.$$

$CPIS_{it}$ is the CPIS presence in county i in year t, where the non-drought years are filled in as previously described. In these specifications, we capture CPIS in a given county by dividing the CPIS area by the area of field crops in the county. Using field crops as the denominator allows for an additional acre irrigated by CPIS to count differently for counties with different crop acreage and composition and excludes crops that couldn't be irrigated via CPIS from being taken into consideration. D_t is a drought indicator equal to one in years determined to qualify for drought insurance payments. In addition, we control for precipitation (P_{it}) and temperature (T_{it}) as linear functions. $\bar{X}_i$ is either a vector of time-invariant controls and their coefficients (elevation, variation in elevation, and soil class) or, in our preferred model, county-level fixed effects. With so few available covariates due to data scarcity, the chance for significant omitted variable bias is high. Grouping at the county level using fixed effects helps to account for those time invariant omitted variables that may be different across counties and correlated with crop yield and CPIS presence. Additionally, while using county fixed effects does reduce the variation that the model has to work with, an inspection of the data reveals that the variables of interest retain at least a third of their variation when comparing within-county standard deviation to between-county standard deviation. Last, $\bar{\delta}_t$ represents year fixed effects.

In order to assess changes in crop patterns, we use the share of the cropland in corn, soy, both, other, or any crop as the outcome of interest. To reduce variation due to changes in the denominator, we use the maximum observed planted cropland in the county to provide a measure of cropland capacity. For this specification, we drop the drought indicator as we are measuring planted acres, not harvested acres, and this ex ante decision by the farmer is not expected to depend on later growing season weather realizations.

Next, we consider crop yield measured as bushels harvested per planted acre. Given the dominance of corn and soy, we focus on the yields of those crops in a county by year. We also log the yields so we can interpret the coefficients as percent changes in yield. We are interested in both the average effect of CPIS (ρ_1) and its interaction in drought years (ρ_2) as a measure of resilience. The sample is from 1988 through 2012.

Finally, to explore the connection between insurance and CPIS we use indemnity payouts as the outcome. Indemnity is logged and defined as the insurance payout in dollars for each drought-related claim in each county (i) during year (t) for a specific crop. We estimate it only for corn and soybean payments. In this case we also remove the drought year indicator as payments for drought losses in non-drought years are uniformly zero. In other words, the sample includes only observations in 1988, 2005, and 2012.

4.5   Results

In terms of *where*, access to an alluvial aquifer is the dominant factor predicting irrigation in Illinois. A series of regressions provided in the appendix (table 4A.2) supports this claim and we will discuss it further, but figure 4.5 first provides main point. It plots the year fixed effects and the year fixed effects interacted with the share of the county overlaying an alluvial aquifer from a simple county-fixed effect model. Irrigation has steadily grown, on average, since 1964, but almost solely where an alluvial aquifer is present.

Additional context is garnered from the additional regressions reported in the appendix. Across all specifications, alluvial aquifer access is a statistically and economically significant predictor of irrigation. In specification 4 (with the most covariates), 100 acres of land over an alluvial aquifer is associated

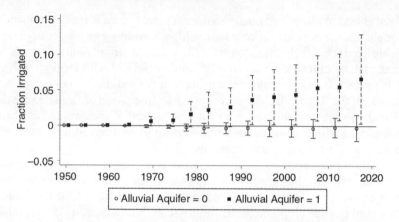

Fig. 4.5   Predicted share of county irrigated by year and aquifer access

Note: Coefficient estimates and their 95th percentile confidence intervals for year-fixed effects are plotted from a two-way fixed effect regression estimating the fraction irrigated, by total county acres. Circles are for counties with no alluvial aquifer and squares are the year-fixed effect interacted with the continuous share of the county overlapping an alluvial aquifer, scaled to 100 percent.

with an additional 4 irrigated acres. Given that just 0.5 acres per 100 are irrigated in Illinois, this is a significant increase. Comparing this to alternative water resources is illuminating: being near a large stream or a non-alluvial aquifer does not increase irrigation in Illinois. In the West, these are significant predictors of irrigation (Edwards and Smith 2018). Although we do not have data on depth-to-water for these alluvial aquifers, they tend to be relatively close to the surface, meaning it is associated with lower costs, both for the drilling of the wells and subsequent pumping. This likely contributes to its role in predicting where CPIS has been installed.

Measured at the county level, soil suitability and slope are not statistically significant predictors, although point estimates are in the direction one would expect. Pre-irrigation farm characteristics (1940) show a slight, but consistent, reduction in irrigation where farm values were higher. This is counter to expectations, but it may be capturing that some early adopters (e.g., Mason County in the 1960s) had less productive soils absent irrigation, incentivizing the initial wave development for average yield effects, not just for resilience benefits. Also counter to expectations, corn production in the 1940s is not predictive of later CPIS installation. Meanwhile, counties with larger average farms are associated with more irrigation. Finally, the weather variables suggest that counties with smaller variations in temperature and that have more precipitation on average are likely to irrigate more. To explore the time-varying component more, we introduce county fixed effects.

Table 4.1 shows that counties are more sensitive to precipitation shocks than temperature shocks in irrigation decisions. Column 1 presents estimates from regressing the fraction of the county irrigated in a census year on the

Table 4.1   CPIS and irrigation uptake timing

	Fraction Irrigated			CPIS Share		
	(1)	(2)	(3)	(4)	(5)	(6)
Average PPT Bin Prior 5 Years	0.00200* (0.00110)			0.00318 (0.00304)		
Average Temp. Bin Prior 5 Years	-0.000898 (0.00194)			-0.00935 (0.00640)		
Severe PPT in Prior 5 years		0.00348* (0.00208)	-0.00810 (0.00562)		0.00552 (0.00397)	-0.0168*** (0.00575)
Severe Temp. in Prior 5 Years		-0.00182 (0.00223)	0.00447 (0.00332)		-0.00453 (0.00342)	0.000715 (0.00259)
Severe PPT × Alluvial Aquifer Share			0.0354 (0.0222)			0.0918*** (0.0245)
Severe Temp. × Alluvial Aquifer Share			-0.0206 (0.0136)			-0.0115 (0.00766)
Constant	0.0695 (0.0524)	0.0684 (0.0494)	0.0558 (0.0412)	0.0232 (0.0228)	0.00414*** (0.00131)	0.00372*** (0.00124)
Observations	1632	1632	1632	303	303	303
Adjusted R-squared	0.123	0.127	0.171	0.824	0.827	0.863
Years	Census, 1950–2017			Recent Droughts (1988, 2005, 2012)		

Note: This table presents the results of estimating equation 10. Measures are at the county level. Columns 1–3 use reported irrigation (as a share of county acres) in USDA Census Years from 1950 to 2017. Columns 4–6 use CPIS capacity (as a share of county acres) from our machine learning output during more recent drought years. Average bins (PPT and Temp) are five discrete bins based on county specific variation from long-run averages, constructed such that higher numbers are drier (lower precipitation) and hotter (higher temperatures). Severe PPT and Severe Temp are indicator variables for experiencing at least one bin-five year in the prior five years. All models include county and year fixed effects. Also, unreported, are controls for current production year precipitation and temperature for columns 1–3 that measure actual irrigation decisions instead of capacity. Robust standard errors in parentheses. ***$p < 0.01$, **$p < 0.05$, *$p < 0.1$.

prior five years' county-specific normalized precipitation and temperature bins. These are constructed as 1–5 with higher numbers indicating "drought" conditions. There is considerable variation both across and within year for these measures (see figure 4A.2 for boxplots). Additional controls are current year precipitation and temperature, and year fixed effects. Experiencing relatively drier years in the past five years increases the share irrigated in a given year. No effect is found for temperature. Column 2 replaces the average bin over the past five years with an indicator variable equal to 1 if any of the past five years fell in the most severe bin (greater than 1.5 standard deviations drier or warmer). Again, a severe dry year in the past five leads to more irrigation. Finally, column 3 interacts the indicators with the share of the county over an alluvial aquifer. Here, the statistical significance is weakened, but the effect appears to be driven by counties with access to the alluvial aquifer.

These irrigation decisions, in any given year, are constrained by the installed capacity. This capacity, meanwhile, need not be deployed in a particular year. Accordingly, we consider similar specifications in columns 4–6 but with CPIS share as the dependent variable and no controls for current year weather given that CPIS installation is not a within-season decision. To accurately capture CPIS from the machine learning model, we use only the statewide drought years, limiting the sample to just three years, straining our ability to pick up statistical significance. Still, the pattern is similar. Particularly in column 6, we find that experiencing at least one severe dry year in the past five years where an alluvial aquifer is present increases CPIS by 0.075 acres per county acre. Overall, it appears county-level CPIS adoption in Illinois is limited to areas with alluvial aquifers and done in response to recently experienced dry years by local standards.

Table 4.2 shows the effect of CPIS investment on crop choices in Illinois. It displays coefficients as the estimated percent change in corn, soybeans, other crops, and cropland land coverage. The strength of the evidence is sensitive to how we fill in the CPIS measures in between drought years and is, generally, imprecisely estimated. Still a few patterns emerge. First, it appears corn acreage has increased with CPIS installations, between 0.13 and 0.17 depending on the CPIS measure. Second, it appears the gain in corn is from a reduction in soybeans. Each version has a negative point estimate on soybean share and positive point estimate on corn share, but statistical significant only when using the CPIS minimum, which we deem the least likely case. Furthermore, other than when CPIS maximum is used, the share of both (combined corn and soy) is relatively close to zero. However, we do note that overall crop shares have positive point estimates, meaning some corn expansion may have occurred.

The expansion of corn may be due to the newly irrigated land being able to support a different crop rotation pattern such as continuous corn or corn-corn-soybeans instead of corn-soybeans or continuous soybeans indicating

Table 4.2   Effects of CPIS presence on crop selection and cropland expansion

	corn share (1)	soybeans share (2)	both share (3)	other share (4)	crop share (5)
CPIS max (%)	0.17	−0.01	0.18	0.03	0.00
	(0.29)	(0.33)	(0.14)	(0.13)	(0.00)
Observations	3162	3162	3162	3162	3162
Adjusted R-squared	0.64	0.43	0.58	0.81	0.75
County FE	Y	Y	Y	Y	Y
Year FE	Y	Y	Y	Y	Y
CPIS trend (%)	0.16	−0.12	0.06	0.04	0.14
	(0.15)	(0.12)	(0.15)	(0.14)	(0.15)
Observations	2529	2529	2529	2529	2529
Adjusted R-squared	0.75	0.36	0.49	0.6	0.22
County FE	Y	Y	Y	Y	Y
Year FE	Y	Y	Y	Y	Y
CPIS min (%)	0.13*	−0.18**	0.04	0.12	0.15
	(0.07)	(0.08)	(0.09)	(0.12)	(0.11)
Observations	2529	2529	2529	2529	2529
Adjusted R-squared	0.55	0.36	0.49	0.61	0.22
County FE	Y	Y	Y	Y	Y
Year FE	Y	Y	Y	Y	Y

Note: This table presents the results of estimating equation 11 when crop share is the dependent variable. Measures are at the county level. Land coverage data was taken from USDA NASS. CPIS share estimates were derived from a deep learning model and represent the share of the maximum cropland observed in a county irrigated via CPIS. The first panel assumes that all CPIS observed in the next drought were built immediately following the previous one. The second panel assumes a linear trend between drought years. The final panel assumes that all of the CPIS observed in a drought year were installed in that year. Columns 1–3 are the share of corn, soybeans, and both corn and soybeans. Column 4 is the share of all other field crops. Column 5 is the share of all cropland relative to the maximum observed cropland in a county. Std. errors in parentheses. *** $p < 0.01$, ** $p < 0.05$, * $p < 0.1$.

that CPIS provide some flexibility in crop rotation patterns. It may also be that farmers installing CPIS are more risk averse than other farmers and take the additional measure of increasing the mix of corn to soybeans as corn is less heat sensitive than soybeans. Lastly, it may be that farmers with newly installed CPIS want to make the most of their irrigation water and increase the mix of corn to soybeans as corn is the more water-efficient crop and more sensitive to water amounts (Dietzel et al. 2015).

Table 4.3 shows the results from estimating equation 11 for corn and soybean yields. The yield values are logged, and CPIS presence is measured as a share of cropland. Therefore, coefficients can be roughly interpreted as the correlation between a percent change in crop yield and a percentage point change in CPIS presence. None of the plausible range for CPIS presence is statistically significantly correlated with average crop yield, although

Table 4.3   Effects of CPIS presence on crop yield (logged)

	corn yield (1)	corn yield (2)	corn yield (3)	soybeans yield (4)	soybeans yield (5)	soybeans yield (6)
CPIS max × drought (%)	0.46** (0.24)			−0.15 (0.13)		
CPIS trend × drought (%)		0.42** (0.19)			−0.16 (0.14)	
CPIS min × drought (%)			0.49** (0.21)			−0.10 (0.15)
CPIS max (%)	0.40 (0.30)			0.14 (0.20)		
CPIS trend (%)		0.20 (0.24)			0.07 (0.17)	
CPIS min (%)			−0.09 (0.21)			−0.11 (0.16)
Observations	2529	2529	2529	2529	2529	2529
Adjusted R-squared	0.73	0.73	0.73	0.65	0.65	0.65
County FE	Y	Y	Y	Y	Y	Y
Year FE	Y	Y	Y	Y	Y	Y

Note: This table presents the results of estimating equation 11 when crop yield is the dependent variable. Measures are at the county level. CPIS estimates are derived from a deep learning model. Crop yields are taken from USDA NASS data. Weather covariates are from NOAA's Climate at a Glance web tool. Yield values are logged, and coefficients may be interpreted as local approximations of the percent change in crop yield when there is a 1 percent change in the share of cropland irrigated via CPIS. The first three rows report the effect of the range of plausible CPIS shares on crop yield during a drought year, and the last three rows report the average effect of CPIS on crop yield. Columns 1 through 3 are logged corn yield, and columns 4 through 6 are logged soybean yield. County and year fixed effects were included in all specifications. Std. errors in parentheses. *** $p < 0.01$, ** $p < 0.05$, * $p < 0.1$.

the point estimates tend to be positive. This leaves open the possibility that average yields do increase, but the effect is hard to detect at the county-level aggregation. However, CPIS presence during a drought year has a significant effect on crop yield for corn, but no significant effect on soybeans. During a drought year, an additional 1 percent of cropland with CPIS is correlated with an approximately 0.46 percent increase in corn yield per acre across the county. Scaling this to a PLSS section level, a single center pivot occupies between 20 percent and 39 percent of the section, so our results imply that the installation of a new CPIS would improve corn yield in a drought year by about 9 percent to 18 percent, depending on the size of the center pivot. Given that average corn yield in a drought year is roughly 99 bushels per acre, this is significant at both a statistical and economic scale. We also find that soybean yield is more sensitive to both heat and precipitation than corn yield.

Finally, in table 4.4 we provide the estimates for equation 11 for indemnity

Table 4.4   **Effects on indemnity payouts in drought years**

	(log) Indemnity Corn (1)	(log) Indemnity Soybeans (2)
Share CPIS	−6.34**	−2.81***
	(2.65)	(0.62)
Observations	2550	2550
Adjusted R-squared	0.60	0.43
County FE	Y	Y
Year FE	Y	Y

Note: This table presents the results from estimating equation 11 when indemnity payouts are the dependent variable. Indemnity quantity and cause are from the USDA's Risk Management Agency. CPIS estimates are derived from a deep learning model. Unreported weather covariates are from NOAA's Climate at a Glance web tool. Indemnity values are logged, and coefficients may be interpreted as local approximations of the percent change in indemnity amount when there is a 1 percentage point increase of CPIS share. Column 1 reports logged indemnity values for corn, and column 2 reports logged indemnity values for soybeans. County and year fixed effects were included in both specifications. Std. errors in parentheses. *** $p < 0.01$, ** $p < 0.05$, * $p < 0.1$.

payments. Indemnity amounts are logged, and the share of CPIS is measured as CPIS acreage in a county divided by the cropland acreage of that county. CPIS presence has statistically significant negative effect on drought indemnity for both corn and soybeans. The coefficients imply that another percentage point of cropland with a CPIS decreases insurance payouts for corn by approximately 6.34 percent and soybeans by about 2.81 percent. This could be because farmers who are disproportionately impacted by drought conditions are more likely to be early adopters of CPIS thus being the ones with the most to benefit from their installation by virtue of relying less on crop insurance payouts. Additionally, the average county only has eight insurance claims filed during a drought year, so a single foregone claim amounts to a large percentage change in indemnity payments.

4.6   Discussion and Conclusion

This chapter has provided insights into the use of irrigation, specifically center pivots, as an agricultural adaptation to climate change in more humid regions. Although just one possible adaptation, the eastern US states have tripled their irrigation since the 1980s, making it an important adaptation to understand. Installing CPIS in Illinois has affected crop rotations, corn yields, and indemnity payments during drought years. The results of this study are significant as they diverge from previous work regarding how improvements in irrigation technology tend to increase average crop yields and expand crop acreage, which suggests that the environmental factors of the setting play a large role in the effect irrigation technology has on both production and a farmer's decision to invest in CPIS. This chapter shows that CPIS provide a measure of drought risk mitigation that goes above and beyond that provided by crop insurance alone, which would provide a reason for Illinois farmers to install them despite the lack of other benefits like those seen in western regions.

These results are most pertinent in the context of investing in irrigation as an adaptation to climate change. Our theoretical model suggests that the probability of a drought event in any given year plays a significant role in a farmer's decision to invest in irrigation even when their crops are insured. We show that, despite the increasing mean precipitation, the variability of precipitation in Illinois is increasing. Our county-level analysis provides evidence that recent local precipitation shocks are correlated with increased shares of irrigation. Moreover, irrigation may not be a viable option for farmers in areas not overlying an aquifer or near a stream. This is important to note, as much of the eastern US remains unirrigated, but further adaptation is likely to continue, especially in similarly endowed regions.

The increase in CPIS presence and resulting pumping could lead to necessary policy decisions being made to prevent excess pumping in the future, especially during times of low water supply when the CPIS will be used the most. At present, groundwater rights in Illinois are dictated through the Reasonable Use Rule established by the Water Use Act of 1983, which provides the right to extract groundwater to the owner of the overlying land if it is put to "reasonable use" (Cain et al. 2017). In a report about the 2012 drought, the ISWS pointed out that Illinois has very limited management authority with no regulation of groundwater sources, no regulation of riparian water use, and few identified alternative municipal water supplies (Fuchs 2021). The latter is important because overextraction of water will produce negative externalities for both other local water users through cones of depression and possibly other riparian water rights holders in the case of alluvial groundwater extraction. For instance, Lake Decatur provides water for nearly 90,000 people and dropped to critical levels in 2012 (Fuchs et al. 2012). This problem is further complicated by the lack of available pumping data for the area, the delay in CPIS identification, and the fact that most CPIS in Illinois are located near rivers where they overlie alluvial aquifers (ISWS 2015). Although these specifics are for Illinois, many of the eastern states also lack oversight of this growing use of water.

This study is limited in its ability to inform policy decisions about pumping limits, spacing rules, or other ways to prevent greater-than-optimal water extraction during drought years as a result of the proliferation of CPIS in Illinois due to the lack of available data. Probably the most relevant of the complications mentioned above is the lack of pumping data, as this makes it extremely difficult to determine an accurate cost function at the farm level. The state has only required that individual irrigators report groundwater pumping since 2015, and the only other pumping data available are in annual aggregate for the state from 1987 onward, which is not fine enough detail to identify anything about where or when the pumping is occurring at the CPIS level (Illinois Department of Natural Resources 2015). However, future studies could attempt to gather or estimate this data through farmer surveys or clever utilization of detailed hydrological and evapotranspiration data which has seen some recent success (e.g., Valencia et al. 2020).

Additional analysis is warranted. Although we identified county-scale factors for CPIS installation, we have said little about what determines CPIS adoption within counties. Some factors may be similar to county-level variation. CPIS require some kind of water source which is most commonly alluvial aquifers created by nearby waterways in Illinois. While there are shallow and deep bedrock aquifers in the state, they are less common, and it remains that not all cropland has access to a sufficient groundwater supply to merit installing a CPIS. For areas that have access to groundwater, differences in CPIS installation across locations could be driven by peer effects (Sampson and Perry 2018), county-level differences in agriculture subsidies (Pfeiffer and Lin 2014; Environmental Working Group 2020), or other factors that may make one area more susceptible to drought than another.

Other adaptations may also warrant consideration; both why farmers choose irrigation rather than another option and how irrigation alters the use of other adaptations. For instance, how does irrigation alter crop insurance choices. Although we found irrigation lowers indemnity payouts and other work has found insurance coverage leads to increasing in water application (Deryugina and Konar 2017), less is known about how transitioning to irrigation affects coverage choices.

We also only consider the drought side of the increasing variability of precipitation in Illinois and the eastern US more broadly, but the variability will also increase flooding events that may influence farmer decisions about investing in irrigation or alternative adaptations. While CPIS installation does seem to provide some assurance during more frequently occurring droughts, it may bring about other concerns about groundwater depletion, water rights, cropland retirement, crop subsidies, and field erosion that are worth additional study. Additionally, further research could be done to determine how generalizable the results of this study are to other humid regions. Illinois was used primarily due to the CPIS ground truth available, but the field of machine learning is rapidly growing and advancing, so it may be possible to quickly identify CPIS in other humid regions in the near future. Still, the preponderance of corn in the state makes it an important case study since corn is the most irrigated crop by acreage (25 percent) in the US (Hrozencik and Aillery 2021). However, 85 percent of corn acreage is not irrigated, leaving plenty of rainfed acres that may consider the adoption of irrigation. These results from Illinois important first steps that show farmers are willing to invest in irrigation as a form of drought mitigation even if average yield enhancements are not present.

Appendix

Raw Data Sources:

- Manually identified CPIS, ISWS (2015)
- Land coverage, USDA NASS (2021)
- County borders, Illinois Geospatial Data Clearinghouse (2003)
- Temperature and precipitation, NOAA (2021); PRISM (2014)
- Indemnity payments, USDA-RMA (Ret. 2021)
- Top-of-atmosphere reflectance satellite imagery, USGS/Google (Ret. 2021)
- County level annual crop yield, USDA NASS (Ret. 2021)
- County level census data, Haines, Fishback, and Rhode (2018), USDA (2019)
- Water resource availability, USGS (2002, 2003, 2014)
- Soil quality and elevation, USDA NRCS (2011)

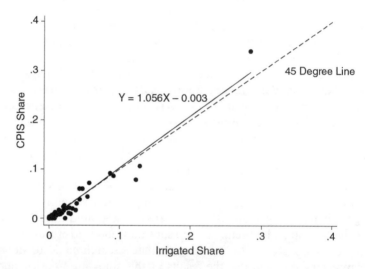

Fig. 4A.1   Comparison of irrigated acreage (USDA Census) to CPIS (ISWS Ground Truth) in 2012 across Illinois counties

Note: Shares are based on total county acres. Linear fit from a binary OLS estimate is compared with the 45-degree line.

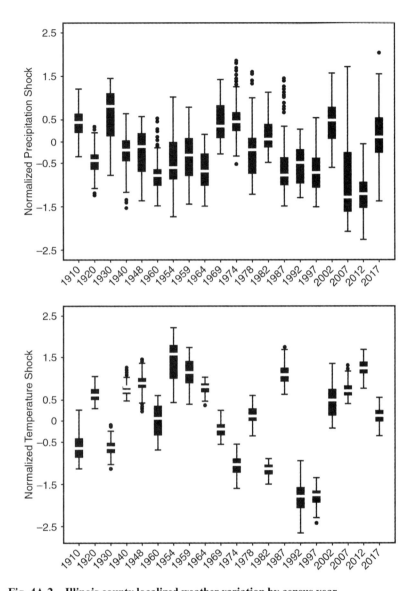

Fig. 4A.2   Illinois county localized weather variation by census year.

Note: Panel A plots the boxplot for the normalized growing season precipitation shock based on historical PRISM data. Each county's average from 1900 to 2017 and standard deviations are utilized to measure the scale of the local shock from "average." Panel B does the same for average growing season temperature.

Table 4A.1   **Summary statistics for the county-level analysis**

Variable	Count	Mean	Std. Dev.	Min	Max	Description and underlying source
Fraction Irrigated	1632	0.00586	0.0225	0	0.380	Share of the county's acres irrigated in a given year (Haines et al. 2018 and USDA 2019)
Fraction with CPIS	303	0.00876	0.0309	0	0.340	Share of the county's acres with a CPIS detected in a drought year (1988, 2005, 2012)
Crop Land (A)	2541	229616.3	130741.7	0	687500	Acreage of major field crops: wheat, winter wheat, soybeans, sorghum, oats, and corn (CDL)
Trend CPIS Share (%)	2550	1.28	4.62	0	58.12	Share of cropland irrigated by CPIS, calculated from predicted acreage, gap years filled using linear interpolation
Min CPIS Share (%)	2550	0.95	3.8	0	55.05	Share of cropland irrigated by CPIS, calculated from predicted acreage, gap years filled by the values in the previous drought
Max CPIS Share (%)	2550	1.61	5.47	0	60.27	Share of cropland irrigated by CPIS, calculated from predicted acreage, gap years filled by the values in the next drought
Corn Share (%)	3187	46.51	15.99	0	100	Share of maximum observed cropland in a county with planted corn (USDA NASS)
Soybeans Share (%)	3187	40.53	10.6	0	100	Share of maximum observed cropland in a county with planted soybeans (USDA NASS)
Both Share (%)	3187	86.04	15.78	0	100	Share of maximum observed cropland in a county with both planted corn and soybeans (USDA NASS)
Other Share (%)	3187	12.73	12.62	0	100	Share of maximum observed cropland in a county with other planted field crops (USDA NASS)
Crop Share (%)	3187	85.47	16.02	0	100	Share of maximum observed cropland in a county with planted field crops (USDA NASS)
Corn Yield (bu/A)	2523	135.11	32.97	19	207	Annual corn yield in bushels per acre (USDA NASS)
Soybeans Yield (bu/A)	2529	41.42	8.41	14.86	63.55	Annual soybeans yield in bushels per acre (USDA NASS)
Corn Indemnity ($)	2550	1069697	6734805	0	135000000	Insurance payment to the insured in dollars for drought claims on corn in a given year (USDA RMA)
Soybeans Indemnity ($)	2550	86159.45	658645.9	0	14000000	Insurance payment to the insured in dollars for drought claims on soybeans in a given year (USDA RMA)
Average Precip (1900–2017)	102	579.1	14.07	544.0	613.9	Total growing season (April - September) precipitation (mm) averaged from 1900 to 2017 (PRISM)
Standard Deviation Precip	102	124.5	8.616	108.5	147.9	Temporal standard deviation of the counties annual precipitation from 1900 to 2017 (PRISM)

Variable	N	Mean	SD	Min	Max	Description
Average PPT Bin Prior 5 Years	1632	2.920	0.431	1.600	4	Average standardized bins in the prior 5 years. Bins based on county specific normalized precipitation distribution with higher bins drier: $1 = [x > 1.5]$, $2 = [0.5 < x < 1.5]$, $3 = [-0.5 < x < 0.5]$, $4 = [-1.5 < x < -0.5]$, $5 = [x < -1.5]$ (Smith & Edwards 2021)
Severe PPT	303	0.122	0.328	0	1	At least one of the prior 5 years had a precipitation bin equal to 5
Average Temperature (1900-2017)	102	19.84	1.211	17.27	22.09	Average growing season (April-September) temperature (C) averaged from 1900 to 2017 (PRISM)
Standard Deviation of Temperature	102	0.834	0.0254	0.755	0.905	Temporal standard deviation of the counties annual temperature from 1900 to 2017 (PRISM)
Average Temp. Bin Prior 5 Years	1632	2.961	0.419	1.600	4.200	Average standardized bins in the prior 5 years. Bins based on county specific normalized temperature distribution with higher bins hotter: $1 = [x < -1.5]$, $2 = [-1.5 < x < -0.5]$, $3 = [-0.5 < x < 0.5]$, $4 = [0.5 < x < 1.5]$, $5 = [x > 1.5]$ (Smith & Edwards 2021)
Severe Temperature	303	0.257	0.438	0	1	At least one of the prior 5 years had a temperature bin equal to 5
Alluvial Aquifer Share	102	0.301	0.239	0	1	Share of the county overlaying an alluvial aquifer
Large Stream Share	102	0.512	0.405	0	1	Share of the county within 15 miles of a strahler order stream 3 or greater (USGS 2014)
Aquifer Share	102	0.270	0.340	0	1	Share of the county overlaying a non-alluvial aquifer (USGS 2003)
Average Soil Suitability	102	2.423	0.796	2	5.947	County's spatial average of gridded soil suitability data, binned 1-8 (USDA 2006)
Elevation	102	184.1	37.16	111.7	266.3	Spatial average of elevation for the county (USGS 2006)
Average Slope	102	12.28	4.525	8	35	Spatial average of the county slope, calculated from the elevation
Slope Range	102	17.05	11.62	0	39	Spatial range of the slope within the county
Longitude	102	−89.18	0.926	−91.19	−87.73	Longitidude of the county centroid
Latitude	102	39.84	1.452	37.19	42.37	Latituted of the county centroid
Farm Value per Acre (1940)	102	180.9	96.20	43.96	595.1	Farm value per farm acre in 1940 (Haines et al. 2018)
Corn Yield per Acre (1940)	102	46.19	11.96	20.65	63.29	Corn yield ber acre in 1940 (Haines et al. 2018)
Ave. Farm Acreage (1940)	102	147.5	32.60	57.45	229.9	Average farm size in 1940 (Haines et al. 2018)
Total Population (1940)	102	77423.9	400170.4	5289	4063342	Total county population in 1940 (Haines et al. 2018)

Table 4A.2 **Irrigation uptake**

			Fraction Irrigated			
	(1)	(2)	(3)	(4)	(5)	(6)
Alluvial Aquifer Share	0.0284**	0.0373**	0.0385***	0.0403***	0.0404***	0.0182***
	(0.0143)	(0.0146)	(0.0125)	(0.0123)	(0.0123)	(0.00670)
Large Stream Share	0.0147***	0.00759*	−0.00280	−0.00235	−0.00242	−0.00104
	(0.00505)	(0.00429)	(0.00448)	(0.00450)	(0.00450)	(0.00298)
Aquifer Share	−0.00401	0.000391	0.00550	0.00545	0.00557	0.00426
	(0.00401)	(0.00346)	(0.00351)	(0.00341)	(0.00341)	(0.00303)
Average Soil Suitability		0.00912	0.00853	0.00661	0.00658	−0.000403
		(0.00719)	(0.00557)	(0.00575)	(0.00572)	(0.00225)
Average Slope		−0.00129	−0.000941	−0.00115	−0.00115	−0.0000440
		(0.00100)	(0.000690)	(0.000718)	(0.000717)	(0.000357)
Slope Range		0.000161	0.000189	0.000217	0.000216	0.000206
		(0.000166)	(0.000158)	(0.000148)	(0.000148)	(0.000132)
Longitude		−0.00496	−0.00188	−0.00404	−0.00409	0.00141
		(0.00321)	(0.00250)	(0.00427)	(0.00428)	(0.00144)
Latitude		0.00419	0.00693**	0.00698**	0.00700**	0.00220
		(0.00294)	(0.00330)	(0.00315)	(0.00316)	(0.00160)
Farm Value per Acre (1940)			−0.0000770***	−0.0000760**	−0.0000752**	−0.0000512***
			(0.0000268)	(0.0000301)	(0.0000298)	(0.0000181)
Corn Yield per Acre (1940)			0.0000909	0.0000204	0.0000160	0.000322
			(0.000276)	(0.000282)	(0.000282)	(0.000221)
Ave. Farm Acreage (1940)			0.000158*	0.000166**	0.000167**	0.0000279
			(0.0000856)	(0.0000815)	(0.0000816)	(0.0000437)
Total Population (1940)			3.63e-09	5.47e-09**	5.46e-09**	2.76e-09*
			(2.21e-09)	(2.63e-09)	(2.62e-09)	(1.56e-09)
Temporal PPT St. Dev.				0.0000212	0.0000173	
				(0.000355)	(0.000355)	
Temporal Temp. St. Dev.				−0.147**	−0.150**	
				(0.0610)	(0.0608)	
Average PPT Bin Prior 5 Years					0.00316**	
					(0.00136)	
Average Temp. Bin Prior 5 Years					−0.00213	
					(0.00246)	
Constant	−0.0150**	−0.596	−0.426	−0.494	−0.499	0.0516
	(0.00624)	(0.375)	(0.315)	(0.435)	(0.433)	(0.147)
Observations	1632	1632	1632	1632	1632	1616
Adjusted R-squared	0.195	0.273	0.313	0.323	0.324	0.248

Note: This table presents the results of estimating equation 10. Measures are at the county level. The outcome is reported irrigation (as a share of county acres) in USDA Census Years from 1950 to 2017. Alluvial aquifer share is the share of the county overlaying an aquifer defined by the USGS (2002). Large stream share is the portion of the county overlaying a 15-mile buffer around a Strahler Order Stream of 3 or greater (USGS 2014). Aquifer share is the share overlaying a non-alluvial aquifer (USGS 2003). All columns include unreported year fixed effects. Columns sequentially add geographical controls (column 2), pre-irrigation farm and demographic attributes (column 3), long-term weather variability (column 4), recent localized weather variation (column 5). Column 6 returns to the column 3 specification but removes Mason County, the most densely irrigated county, as a robustness check. Robust Standard errors in parentheses. *** $p < 0.01$, ** $p < 0.05$, * $p < 0.1$.

References

Anderson, J. 2018. "How Center Pivot Irrigation Brought the Dustbowl Back to Life." *Smithsonian Magazine*, September 10. https://www.smithsonianmag.com/innovation/how-center-pivot-irrigation-brought-dust-bowl-back-to-life-180970243/.

Annan, F., and W. Schlenker. 2015. "Federal Crop Insurance and the Disincentive to Adapt to Extreme Heat." *American Economic Review* 105 (5): 262–66.

Athey, S. 2019. "The Impact of Machine Learning on Economics." In *The Economics of Artificial Intelligence: An Agenda*, edited by Ajay Agrawal, Joshua Gans, and Avi Goldfarb, 507–47. Chicago, IL: University of Chicago Press.

Baerenklau, K. A., and K. C. Knapp. 2007. "Dynamics of Agricultural Technology Adoption: Age Structure, Reversibility, and Uncertainty." *American Journal of Agricultural Economics* 89 (1): 190–201.

Berry, S. T., M. J. Roberts, and W. Schlenker. 2014. "Corn Production Shocks in 2012 and Beyond: Implications for Harvest Volatility." *The Economics of Food Price Volatility*, edited by Jean-Paul Chavas, David Hummels, and Brian D. Wright, 59–81. Chicago, IL: University of Chicago Press.

Bigelow, D. P., and H. Zhang. 2018. "Supplemental Irrigation Water Rights and Climate Change Adaptation." *Ecological Economics* 154: 156–67.

Brent, D. A. 2017. "The Value of Heterogeneous Property Rights on the Costs of Water Volatility." *American Journal of Agricultural Economics* 99 (1): 73–102.

Bureau of Land Management. 2020. BLM National Public Land Survey System Polygons—National Geospatial Data Asset (NGDA).

Burke, M., and K. Emerick. 2016. "Adaptation to Climate Change: Evidence from US Agriculture." *American Economic Journal: Economic Policy* 8 (3): 106–40.

Cain, R. L., M. Goll, T. Hood, C. Lauer, M. McDonough, B. Miller, S. Pearson, S. Rodriguez, and T. Riley. 2017. *Groundwater Laws and Regulations: A Preliminary Survey of Thirteen U.S. States*. Texas A&M University School of Law. Accessed May 8, 2021. https://law.tamu.edu/docs/default-source/faculty-documents/groundwater-laws-reg-13states.pdf?sfvrsn=0.

Christine, H., S. Fuss, J. Szolgayova, F. Strauss, and E. Schmid. 2012. "Investment in Irrigation Systems Under Precipitation Uncertainty." *Water Resource Management* 26: 3113–3137.

Connor, L., and A. L. Katchova. 2020. "Crop Insurance Participation Rates and Asymmetric Effects on US Corn and Soybean Yield Risk." *Journal of Agricultural and Resource Economics* 45 (1): 1–19.

Cooley, D., R. M. Maxwell, and S. M. Smith. 2021. "Center Pivot Irrigation Systems and Where to Find Them: A Deep Learning Approach." *Frontiers in Water* 178. https://doi.org/10.3389/frwa.2021.786016.

Deines, J. M., A. D. Kendall, M. A. Crowley, J. Rapp, J. A. Cardille, and D. W. Hyndman. 2019. "Mapping Three Decades of Annual Irrigation Across the High Plains Aquifer Using Landsat and Google Earth Engine." *Remote Sensing of Environment* 111400.

Deryugina, T., and M. Konar. 2017. "Impacts of Crop Insurance on Water Withdrawals for Irrigation." *Advances in Water Resources* 110: 437–44.

Dietzel, R., M. Liebman, R. Ewing, M. Helmers, R. Horton, M. Jarchow, and S. Archontoulis. 2015. "How Efficiently do Corn- and Soybean-based Cropping Systems Use Water? A Systems Modeling Analysis." *Global Change Biology* 22 (2). doi:10.1111/gcb.13101.

Edwards, E. C., and S. M. Smith. 2018. "The Role of Irrigation in the Development of Agriculture in the United States." *Journal of Economic History* 78 (4). doi: 10.13140/RG.2.2.19247.12965.

Elliott, J., D. Deryng, C. Müller, K. Frieler, M. Konzmann, D. Gerten, M. Glotter, M. Flörke, Y. Wada, N. Best, and others. 2014. "Constraints and Potentials of Future Irrigation Water Availability on Agricultural Production Under Climate Change." *Proceedings of the National Academy of Science* 111 (9): 3239–3244.

Environmental Working Group. 2020. *Illinois Farm Subsidy Information.* Accessed Sep 1, 2021. https://farm.ewg.org/region.php?fips=17000&statename=Illinois.

Evans, R. G. 2001. *Center Pivot Irrigation.* USDA Agricultural Research Service. https://www.ars.usda.gov/ARSUserFiles/21563/center%20pivot%20design%202.pdf.

Finkelstein, J. S., and M. R. Nardi. 2016. "Geospatial Compilation and Digital Map of Center-Pivot Irrigation Areas in the Mid-Atlantic Region, United States." *USGS and University of Delaware Agricultural Extension.*

Ford, T. W., L. Chen, and J. T. Schoof. 2021. "Variability and Transitions in Precipitation Extremes in the Midwest United States." *Journal of Hydrometeorology* 22: 532–45. doi:10.1175/JHM-D-20–0216.1.

Fuchs, B., N. Umphlett, M. S. Timlin, W. Ryan, N. Doesken, J. Angel, O. Kellner, H. J. Hillaker, Knapp, and others. 2012. "From Too Much to Too Little: How the Central U.S. Drought of 2012 Evolved Out of One of the Most Devastating Floods on Record In 2011." National Drought Mitigation Center.

Grady, K. A., L. Chen, and T. W. Ford. 2021. "Projected Changes in Spring and Summer Precipitation in the Midwestern United States." *Frontiers in Water 3.* doi:10.3389/frwa.2021.780333.

Griliches, Z. 1957. "Hybrid Corn: An Exploration in the Economics of Technological Change." *Econometrica 25 (4)*: 501–22.

Goodfellow, I., Y. Bengio, and A. Courville. 2016. *Deep Learning.* Cambridge, MA: MIT Press.

Haines, M., P. Fishback, and P. Rhode. 2018. United States Agriculture Data, 1840—2012. Inter-University Consortium for Political and Social Research. doi: 10.3886/ICPSR35206.v4.

Hrozencik, R. A., and M. Aillery. 2021. "Trends in U.S. Irrigated Agriculture: Increasing Resilience Under Water Supply Scarcity." U.S. Department of Agriculture, Economic Research Service. EIB-22947.

Illinois Department of Natural Resources. 2013. *The Drought of 2012.* https://www2.illinois.gov/dnr/WaterResources/Documents/TheDroughtOf2012.pdf.

Illinois Department of Natural Resources. 2015. *Action Plan for a Statewide Water Supply Planning and Management Program.* Chicago, Illinois: Office of Water Resources.

Illinois Geospatial Data Clearinghouse. 2003. *Illinois County Boundaries, Polygons and Lines.* https://clearinghouse.isgs.illinois.edu/data/reference/illinois-county-boundaries-polygons-and-lines.

Illinois State Water Survey (ISWS). 2015. *Illinois Center Pivot Irrigation.* Illinois Geospatial Data Clearinghouse. Accessed May 7, 2021. http://clearinghouse.isgs.illinois.edu/data/hydrology/illinois-center-pivot-irrigation.

Just, R. E., A. Schmitz, and D. Zilberman. 1979. "Technological Change in Agriculture." *Science* 206 (4424): 1277–1280.

Koundouri, P., C. Nauges, and V. Tzouvelekas. 2006. "Technology Adoption Under Production Uncertainty: Theory and Application to Irrigation Technology." *American Journal of Agricultural Economics* 88 (3): 657–70.

Leonard, B., and G. D. Libecap. 2019. "Collective Action by Contract: Prior Appropriation and the Development of Irrigation in the Western United States." *Journal of Law and Economics* 62 (1): 67–115. doi: 10.1086/700934.

Lau, L. J., and P. A. Yotopoulos. 1989. "The Meta-Production Function Approach to Technological Change in World Agriculture." *Journal of Development Economics 31 (2)*: 241–69.

Mishra, V., and K. A. Cherkauer. 2010. "Retrospective Droughts in the Crop Growing Season: Implications to Corn and Soybean Yield in the Midwestern United States." *Agricultural and Forest Meteorology* 150 (7–8): 1030–1045. doi: doi.org /10.1016/j.agrformet.2010.04.002.

Negri, D. H., N. R. Gollehon, and M. P. Aillery. 2005. "The Effects of Climatic Variability on US Irrigation Adoption." *Climatic Change* 69 (2): 299–323.

NOAA National Centers for Environmental Information. 2021. *Climate at a Glance: County Time Series*. https://www.ncdc.noaa.gov/cag/.

Pfeiffer, L., and C.-Y. C. Lin. 2014. "Does Efficient Irrigation Technology Lead to Reduced Groundwater Extraction? Empirical Evidence." *Journal of Environmental Economics and Management* 67 (2): 189–208. doi: https://doi.org/10.1016 /j.jeem.2013.12.002.

Plastina, A., S. Johnson, and W. Edwards. 2021. *Revenue Protection Crop Insurance*. Accessed Jan 2022. Iowa State University Extension and Outreach Ag Decision Maker. https://www.extension.iastate.edu/agdm/crops/html/a1–55.html.

PRISM Climate Group. 2014. PRISM Climate Data. Oregon State University. http:// prism.oregonstate.edu.

Rosa, L., D. D. Chiarelli, M. Sangiorgio, A. A. Beltran-Peña, M. C. Rulli, P. D'Odorico, and I. Fung. 2020. "Potential for Sustainable Irrigation Expansion in a 3 C Warmer Climate." *Proceedings of the National Academy of Science* 117 (47): 29526–29534.

Ruttan, V. W. 1960. "Research on the Economics of Technological Change in American Agriculture." *Journal of Farm Economics* 42 (4): 735–54.

Sampson, G. S., and E. D. Perry. 2018. "The Role of Peer Effects in Natural Resources Appropriation—The Case of Groundwater." *American Journal of Agricultural Economics* 101 (1): 154–71.

Saraiva, M., E. Protas, M. Salgado, and C. Souza. 2020. "Automatic Mapping of Center Pivot Irrigation Systems from Satellite Images Using Deep Learning." *Remote Sensing* 12 (3): 558.

Schlenker, W., W. M Hanemman, and A. C. Fisher. 2005. "Will US Agriculture Really Benefit from Global Warming? Accounting for Irrigation in the Hedonic Approach." *American Economic Review* 95 (1): 395–406. http://www.jstor.org /stable/10.2307/4132686.

Schlenker, W., and M. J. Roberts. 2009. "Nonlinear Temperature Effects Indicate Severe Damages to US Crop Yields Under Climate Change." *Proceedings of the National Academy of Sciences* 106 (37): 15594–15598.

Schnitkey, G. 2021. *Farmdoc*. University of Illinois. https://farmdoc.illinois.edu /handbook/historic-corn-soybeans-wheat-and-double-crop-soybeans.

Seager, R., N. Lis, J. Feldman, J. Feldman, M. Ting, P. A. Williams, J. Nakamura, L. Haibo, and N. Henderson. 2018. "Whither the 100th Meridian? The Once and Future Physical and Human Geography of America's Arid—Humid Divide. Part I: The Story So Far." *Earth Interactions* 22 (5): 1–22. journals.ametsoc.org/view /journals/eint/22/5/ei-d-17–0011,1,xml.

Sherer, T. 2018. "Selecting a Sprinkler Irrigation System." North Dakota State University Extension. https://www.ag.ndsu.edu/publications/crops/selecting-a -sprinkler-irrigation-system.

Smit, B., and M. W. Skinner. 2002. "Adaptation Options in Agriculture to Climate Change: a Typology." *Mitigation and Adaptation Strategies for Global Change* 7 (1): 85–114.

Smith, S., and E. Edwards. 2021. "Water Storage and Agricultural Resilience to Drought: Historical Evidence from the United States." *Environmental Research Letters* 16 (12): doi:10.1088/1748–9326/ac358a.

Smith, V. H., and B. K. Goodwin. 1996. "Crop Insurance, Moral Hazard, and Agricultural Chemical Use." *American Journal of Agricultural Economics* 78 (2): 428–38.

Soil Survey Staff. 2020. *The Gridded National Soil Survey Geographic (gNATSGO) Database for Illinois*. USDA Natural Resource Conservation Service. Accessed June 9, 2021. https://www.nrcs.usda.gov/wps/portal/nrcs/detail/soils/survey/geo/?cid=nrcseprd1464625.

State Climatologist Office for Illinois. 2015. *Drought Trends in Illinois*. ISWS. Accessed June 14, 2021. www.isws.illinois.edu/statecli/climate-change/ildrought.htm.

State of Nebrasksa Open Data. 2019. *2005 Center Pivots in the Central Platte River Basin*. https://www.nebraskamap.gov/datasets/8e7b99d90da84c82889e00ba8f90ef41_0?geometry=-108.327%2C40.017%2C-90.936%2C42.898.

Storm, H., K. Baylis, and T. Heckelei. 2020. "Machine Learning in Agricultural and Applied Economics." *European Review of Agricultural Economics* 47 (3): 849–92. doi: 10.1093/erae/jbz033.

Stubbs, M. 2016. "Irrigation in U.S. Agriculture: On-Farm Technologies and Best Management Practices." *Congressional Research Service*.

Tack, J., A. Barkley, and N. Hendricks. 2017. "Irrigation Offsets Wheat Yield Reductions from Warming Temperatures." *Environmental Research Letters* 12 (11).

Tang, J., D. Arvor, T. Corpetti, and P. Tang. 2021. "Mapping Center Pivot Irrigation Systems in the Southern Amazon from Sentinel-2 Images." *Water* 13 (3). doi: 10.3390/w13030298.

Torkamani, J., and S. Shajari. 2008. "Adoption of New Irrigation Technology Under Production Risk." *Water Resources Management* 22: 229–37.

Trenberth, K. E., A. Dai, G. Van Der Schrier, P. D. Jones, J. Barichivich, R. Briffa, and J. Sheffield. 2014. "Global Warming and Changes in Drought." *Nature Climate Change* 4 (1): 17–22. doi:10.1038/nclimate2067.

Troy, Tara J., C. Kigpen, and I. Pal. 2015. "The Impact of Climate Extremes and Irrigation on US Crop Yields." *Environmental Research Letters* 10 (5).

Tsur, Y. 1990. "The Stabilization Role of Groundwater When Surface Water Supplies Are Uncertain: The Implications for Groundwater Development." *Water Resources Research* 26 (5): 811–18.

Tsur, Y., and T. Graham-Tomasi. 1991. "The Buffer Value of Groundwater with Stochastic Surface Water Supplies." *Journal of Environmental Economics and Management* 21 (3): 201–24.

United States Department of Agriculture. 2019. Census Data Query Tool. www.nass.usda.gov/Quick_Stats/CDQT/chapter/1/table/1.

United States Department of Agriculture. 2021. *Illinois Crop Insurance*. Risk Management Agency. Accessed May 1, 2021. https://www.rma.usda.gov/en/RMALocal/Illinois/State-Profile.

USDA Natural Resources Conservation Service Illinois. 2011. *Illinois Soils*. Accessed September 1, 2021. https://www.nrcs.usda.gov/wps/portal/nrcs/il/soils/.

USDA NASS. 2021. *Agricultural Land Values*. Economics, Statistics, and Market Information System, U.S. Department of Agriculture. https://usda.library.cornell.edu/concern/publications/pn89d6567.

USDA NASS. 2021. *Data and Statistics*. USDA. Accessed June 9, 2021. https://www
.nass.usda.gov/Data_and_Statistics/index.php.

USDA NASS. 2008–2020. *Cropland Data Layer*. www.nass.usda.gov/Research
_and_Science/Cropland/Release/index.php.

USGS. 1996. USGS EROS Archive—Digital Elevation—Global 30 Arc-Second
Elevation (GTOPO30). U.S. Geological Survey, Washington, DC. doi: /10.5066/
F7DF6PQS.

USGS. 2002. Alluvial and Glacial Aquifers. Groundwater Atlas of the United States.
https://water.usgs.gov/GIS/metadata/usgswrd/XML/alluvial_and_glacial
_aquifers.xml.

USGS. 2003. Principal Aquifers of the 48 Conterminous United States, Hawaii,
Puerto Rico, and the US Virgin Islands: Digital Data. Reston, Virginia.

USGS. 2014. USGS Small-Scale Dataset—1:1,000,000-Scale Hydrographic Geodatabase of the United States—Conterminous United States 201403 FileGDB 10.1.
https://www.sciencebase.gov/catalog/item/581d0551e4b08da350d5273e.

USGS. 2018. Water Use Data for Illinois. National Water Information System: Web
Interface. https://waterdata.usgs.gov/il/nwis/water_use/.

USGS/Google. 2012. *USGS Landsat 5 TM Collection 1 Tier 1 TOA Reflectance*.
Earth Engine Data Catalog. Accessed June 16, 2021. https://developers.google
.com/earth-engine/datasets/catalog/LANDSAT_LT05_C01_T1_TOA.

Valencia, O. M., K. Johansen, B. J. Solorio, T. Li, R. Houborg, Y. Malbeteau,
S. AlMashharawi, M. U. Altaf, E. M. Fallatah, H. P. Dasari, I. Hoteit, and M. F.
McCabe. 2020. "Mapping Groundwater Abstractions from Irrigated Agriculture:
Big Data, Inverse Modeling, and a Satellite-Model Fusion Approach." *Hydrology
and Earth System Sciences* 24: 5251–5277.

Wang, R., R. M. Rejesus, and S. Aglasan. 2021. "Warming Temperatures, Yield Risk
and Crop Insurance Participation." *European Review of Agricultural Economics*
48 (5): 1109–131.

Yu, C., R. Miao, and M. Khanna. 2021. "Maladaption of US Corn and Soybeans
to a Changing Climate." *Scientific Reports* 11 (1): 1–12.

Zaveri, E., and D. B. Lobell. 2019. "The Role of Irrigation in Changing Wheat Yields
and Heat Sensitivity in India." *Nature Communications* 10 (1): 1–7.

Zhang, C., P. Yue, L. Di, and Z. Wu. 2018. "Automatic Identification of Center Pivot
Irrigation Systems from Landsat Images Using Convolutional Neural Networks."
MDPI Agriculture, Special Issue: Remote Sensing in Agricultural System. MDPI.

Zhang, T., X. Lin, and G. F. Sassenrath. 2015. "Current Irrigation Practices in the
Central United States Reduce Drought and Extreme Heat Impacts for Maize
and Soybean, but not for Wheat." *Science of the Total Environment* 508: 331–42.

Zilberman, D. 1984. "Technological Change, Government Policies, and Exhaustible Resources in Agriculture." *American Journal of Agricultural Economics* 66
(5): 634–40.

Zipper, S. C., J. Qiu, and C. J. Kucharik. 2016. "Drought Effects on US Maize and
Soybean Production: Spatiotemporal Patterns and Historical Changes." *Environmental Research Letters* 11 (9): 094021.

5

Perceived Water Scarcity and Irrigation Technology Adoption

Joey Blumberg, Christopher Goemans,
and Dale Manning

5.1 Introduction

Scientists and policy makers are actively exploring efficient and sustainable resource planning under a changing climate (Masson-Delmonte et al. 2021). Climate change describes a shift in the underlying distribution of weather patterns over a long period of time. Shifting temperature and precipitation patterns are expected to contribute to increased water scarcity, which poses a threat to food production (Mancosu et al. 2015). Meeting the needs of expanding populations depends on the ability of industries and governments to adapt. This is particularly relevant in arid regions, where water supplies are expected to be intensely affected by climate change (Lioubimtseva 2004) and agriculture is often dependent on irrigation. For areas that rely on irrigation water derived from snowpack, accelerated snowmelt will change the timing and quantity of water available during the growing season, increasing the risk of costly shortages (Adam, Hamlet, and Lettenmaier 2009). It is estimated that water shortages result in more annual crop loss than all pathogens combined, totaling $30 billion in global production

Joey Blumberg is a PhD student in the Department of Agricultural and Resource Economics at Colorado State University.

Christopher Goemans is chair of the graduate program and an associate professor in the Department of Agricultural and Resource Economics at Colorado State University.

Dale Manning is an associate professor in the Department of Agricultural and Resource Economics at Colorado State University.

This material is based upon work supported by the National Institute for Food and Agriculture under Award No. 2018-69011-28369 and the National Science Foundation under Grant No. 1828902. For acknowledgments, sources of research support, and disclosure of the authors' material financial relationships, if any, please see https://www.nber.org/books-and-chapters/american-agriculture-water-resources-and-climate-change/perceived-water-scarcity-and-irrigation-technology-adoption.

losses over the past decade (Gupta, Rico-Medina, and Caño-Delgado 2020). As overall water availability changes, maintaining agricultural output will depend on how producers adapt. Adopting water conservation strategies is one possible mechanism.

Understanding what motivates producers to conserve water resources is important for future planning. While there exists a rich literature on potential conservation strategies (e.g., Howden et al. 2007), little emphasis has been placed on the role of perceptions that influence the implementation of these strategies. One reason for the lack of research in this area is that identifying events in the natural world that change perceptions about scarcity, and subsequently change behavior, can be difficult. Some literature suggests that personally experiencing an extreme weather event can change climate change perceptions and increase the inclination to adopt conservation strategies (Spence et al. 2011; Wang 2017; Wang and Lin 2018). Maddison (2007) finds that many farmers in Africa perceive climate change to be real, yet some still do not respond in their practices. While these studies provide important insights into attitudes on climate change, they rely on cross-sectional survey data and cannot track behavioral changes over time. Overall, literature investigating how perceptions impact behavior using observational, non-survey data is scant. In this article, we explore how changing perceptions about water scarcity affect conservation investment decisions for agricultural producers. A simple theoretical framework is developed to demonstrate the conditions under which a producer's perception of water availability would incentivize investment in irrigation efficiency. Then, a unique period of extreme drought and institutional reform in Colorado is leveraged as a natural experiment to compare empirical results to simulations from the theoretical model.

Perceptions about water availability play a critical role in decision making for producers who are dependent on irrigation. In the US, western states (the American West) account for 81 percent of total irrigation withdrawals (Dieter et al. 2018), and productivity in many areas is dependent on surface water from snow runoff. In this area, rising temperatures cause more precipitation to fall as rain instead of snow, which reduces snowpack depth and changes the seasonality of runoff (EPA 2016). Several studies have examined recent hydrological changes in the American West, documenting trends in earlier snowmelt-driven streamflows and declines in April snowpack (Mote et al. 2005; Hamlet et al. 2005; Mote 2006; EPA 2016). In addition to increasing temperature and evaporation trends, monthly projections of the Palmer Drought Severity Index (PDSI) suggest that climate change will amplify the length and severity of droughts while also hindering the recovery of macroscale water supplies (Gutzler and Robbins 2011). One mechanism for agricultural producers to adapt to water scarcity is to adopt more water-efficient, pressurized irrigation systems like sprinkler or drip (Howden et al. 2007;

Frisvold and Bai 2016), but gravity systems are still prevalent throughout the American West partially due to high costs of sprinkler investment and relatively low water prices. Carey and Zilberman (2002) use a stochastic dynamic model to demonstrate how uncertainty in water supplies creates an option value and deters irrigation technology adoption unless the expected present value of the investment exceeds the cost by a large margin. However, some empirical evidence has shown that farmers adopt new technologies to hedge against production risk (e.g., Koundouri, Nauges, and Tzouvelekas 2006). The present article provides insights into the disconnect between some theoretical predictions and empirical evidence surrounding investment in irrigation technology.

Presumably, producers have a belief about the probability distribution of input shocks. When considering multiyear investments in water conservation technologies, a farmer likely assesses the probability of a water shortage. Ji and Cobourn (2021) provide an intuitive framework of expectation formation, proposing that perceptions about weather—or supply conditions—develop with an increased bias toward recent events. Therefore, experiencing a disproportionately extreme event triggers a larger revision to expectations, and a subsequent series of events closer to long-run averages would be necessary to decrease the perceived likelihood of another extreme event. They corroborate their theoretical hypotheses empirically, finding that weather shocks significantly impact short-run planting decisions for farmers. Similarly, Cobourn et al. (2021) demonstrate that irrigators anticipating water shortages are more likely to fallow land and plant drought-resilient crops. Complementing these recent studies that focus on short-run responses (i.e., yearly planting decisions), our attention lies on long-run responses (i.e., investment in infrastructure). Our novel data set of over 60 years of water right curtailment recordings alleviates our reliance on weather data in estimating producer expectations of water availability. We are able to pinpoint irrigators that directly experienced shortages, allowing us to identify changes in perceptions and subsequent long-run improvements in water use efficiency via irrigation technology adoption.

The present article contributes to the relevant literature in two aspects. First, we develop a theoretical model to analyze the conditions under which an agricultural producer's perception of a possible water shortage would incentivize investing in a more water-efficient irrigation system. The model framework captures how risk is perceived for farmers operating under a priority-based water allocation institution through two parameters: (1) the probability that water supply will be curtailed in a given year, and (2) if curtailed, the intensity of the water loss. We then consider a range of model parameters to identify when the benefit of investing in more efficient irrigation infrastructure is highest. Since our framework captures the nuances of a priority-based water rights regime, insights on how investment decisions are influenced by perceptions are particularly applicable to the American West, though similar regimes exist throughout the world.

Second, we capture changing perceptions of water shortage risk for farmers in northeast Colorado using a comprehensive panel data set of irrigated cropland, agricultural water rights, and curtailment recordings. For many producers, increased water scarcity will change the perceived reliability of water right portfolios. In Colorado, producers with historically secure water rights are facing increases in curtailment due to institutional changes resulting from litigation and sustained drought in the early 2000s (Waskom 2013). Our empirical context provides us with a unique opportunity to identify a change in perceptions about the reliability of a water supply, which allows us to measure how those changes affect decisions to adapt to increasing scarcity.

Our theoretical model shows that the net benefit of adopting more efficient irrigation technology increases as the probability of curtailment increases, holding all else constant. However, changes in the expected amount of water received (when curtailed) impacts net benefits non-monotonically. Treatment and control groups are determined by water right curtailments during the early 2000s shock relative to historical droughts. Results of a difference-in-difference analysis indicate that the treatment group, those who experienced an unprecedented increase in curtailment, adopted more water-efficient irrigation systems at significantly higher rates than the control group. Additionally, cropland with corn experienced the largest increases in irrigation efficiency improvements in years immediately following the shock, although total corn acreage was reduced. Corn is considered more sensitive to water stress than other popular crops grown in the region, such as alfalfa or wheat, further indicating that the shock incentivized a change in practices to hedge against production risks. Some producers in our study area supplemented their surface water irrigation practices with groundwater, yet we find that changes in groundwater use did not differ substantially between treatment and control groups beyond years immediately following the shock. This is, in part, due to the conjunctive governance of surface water and groundwater in Colorado. These empirical findings provide fresh evidence of the link between updating perceptions and conservation investment behavior.

The remainder of the article is organized as follows. First, we describe a theoretical framework used to analyze the impact of perceptions on irrigation technology adoption in the context of prior appropriation, which is the dominant water allocation system in the American West. We then discuss our study area and the period of extreme drought and institutional reform that we leverage as a natural experiment, followed by a description of the data and modeling approach. In the final sections we present the estimation results, analyze their robustness, and conclude by discussing policy implications.

5.2 Theoretical Framework

When adapting to increasing water scarcity, farmers face a menu of potential strategies to reduce their overall water dependency. Drought-tolerant crop varieties and species can be planted in lieu of water-intensive ones. Deficit irrigation on large quantities of land or increased irrigation on a reduced quantity of land offer opportunities for water savings. Technologies that harvest and store water or reduce conveyance losses can increase average supplies. Improving the application efficiency of an irrigation system can reduce the amount of diverted water necessary to achieve full evapotranspiration. The advent of water markets and water-sharing agreements allow farmers to diversify their income through selling water or to hedge against drought risk by buying water. In general, the actual costs and benefits of different adaptation strategies from this suite of options depend on a farmer's characteristics, such as geographic location. However, farmers' *expected* costs and benefits, which ultimately drive adoption decisions, vary largely according to perceptions about water scarcity. Otherwise similar farmers with different perceptions may exhibit a vastly different willingness to invest in practices that reduce water use. In the following theoretical framework, we focus solely on how perceptions drive the decision to improve irrigation efficiency. Although our theoretical model is presented in the context of a producer evaluating the expected benefits from adopting an irrigation technology, the findings apply more generally with respect to how perceptions of scarcity influence producer decisions to invest in technologies and/or practices that improve water use efficiency.

To examine the impact of perceptions on conservation investment, we develop a theoretical model describing a producer's decision to improve the efficiency of his irrigation infrastructure. We adopt the conceptual framing of a producer's irrigation water supply under a prior appropriation system from Li, Xu, and Zhu (2019), with some simplification. After summarizing prior appropriation, we show a general condition characterizing the net benefit of investment in a conservation technology. We then impose assumptions on the parameters to estimate the impact of perceptions about input availability on the investment decision.

Water allocation in most of the American West is governed by a system of prior appropriation, a legal framework that rules over all water use. To divert water under prior appropriation, one must obtain a water right from a court or purchase an existing right. Water rights are usufructuary, meaning that the rights holder does not own the water itself but the right to divert and use it. Rights are ranked in a hierarchy of priority determined by the date on which a user first appropriated and diverted water for beneficial use, colloquially phrased as "first in time, first in right." Owners of agricultural water rights cannot divert more water for irrigation than what is decreed by their right, and when basin water supplies are insufficient to fulfill all decreed water rights, rights holders with older water rights have priority over users with newer rights. In the state of Colorado, water rights are curtailed through a system of administrative "calls." When inflows are insufficient to satisfy all water rights holders, the state engineer places a "call" on a stream, which curtails the ability for junior water rights holders to divert. The administrative call communicates a priority level required to continue diverting water. In essence, when senior rights are unable to divert their decreed allotment, all junior upstream users must temporarily stop diverting to make more water available (Getches 2009).

Consider a producer operating under a system of prior appropriation who uses water to grow crops. The producer owns a water right with a fixed priority level and a maximum amount of $\bar{w}$ units of water that may be diverted from a specified stream. Irrigation water w available to the producer to grow crops over a growing season is a random variable that takes the form

$$(1) \qquad w = \begin{cases} \bar{w}, & S \geq V \\ \delta\bar{w}, & S < V, \end{cases}$$

where S is a stochastic stream supply term, corresponding to the total quantity of water available for diversion by all water rights holders, and V is the total supply necessary within the stream system for the producer to divert the maximum quantity of water associated with the water right. If $S < V$, the producer's water right is called, and he receives a proportion $\delta \in [0,1)$ of the total allotment. We further assume a relationship between irrigation water and crop yield equal to

$$(2) \qquad y(w, \lambda, \alpha) = \begin{cases} (\lambda w)^\alpha, & 0 \leq \lambda w < w_m \\ y_m, & \lambda w \geq w_m, \end{cases}$$

where $y(w, \lambda, \alpha)$ is the total quantity of output, $\lambda \in (0,1)$ is an irrigation efficiency coefficient, y_m is maximum yield, w_m is the net irrigation requirement for maximum yield, and $\alpha \in (0,1)$ is a shape parameter.[1]

Now consider the case where the producer has an existing low-efficiency flood irrigation system and can invest in a high-efficiency sprinkler system. The producer can pay an annualized cost of the upfront capital investment, c_s, that would increase irrigation efficiency from λ_f to λ_s. Assume this producer's objective is to maximize expected profit by first choosing whether to invest in the new irrigation system, taking prices as given, and then applying water to his fields after w is realized. The profit function after realization is composed of the per unit price of output p, output $y(w, \lambda, \alpha)$, and some fixed cost of production k,

---

1. Our choice of functional form attempts to exhibit the typical relationship between total seasonal irrigation and crop yield as represented on page 4 of Foster and Brozović (2018). We assume no yield when no water is applied, i.e., $y(w = 0, \lambda, \alpha) = 0$, since irrigated crop varieties in Colorado are often not drought tolerant.

(3) $$\pi = py(w,\lambda,\alpha) - k.$$

The producer assumes a probability that his water right will be called in a given year, $P(S < V) = \theta \in [0,1]$, and the magnitude of water loss, $\delta \in [0,1)$, should the call occur. Given his perception of parameters θ and δ, and conditional on efficiency, the producer's expected profit, prior to the realization of w, is

(4) $$\mathbb{E}[\pi] = p[(1 - \theta)y(\overline{w},\lambda,\alpha) + \theta y(\delta\overline{w},\lambda,\alpha)] - k.$$

The decision to invest in the new irrigation system is modeled as binary, so the producer chooses between only two profit functions. For simplicity, we examine the payoff of investing for a single period case. The annualized expected net benefit of investment is

(5) $\mathbb{E}[\pi_s] - [\mathbb{E}\pi_f] =$

$$p\{(1 - \theta)[y(\overline{w},\lambda_s,\alpha) - y(\overline{w}, \lambda_f,\alpha)] + \theta[y(\delta\overline{w},\lambda_s,\alpha) - y(\delta\overline{w},\lambda_f,\alpha)]\} -c_s,$$

and assuming $\overline{w}$ is the amount of water necessary for maximum yield with flood irrigation,[2] i.e., the marginal productivity of water is zero beyond $\overline{w} = y_m^{1/a} / \lambda_f$, equation (5) is reduced to

(6) $$\mathbb{E}[\pi_s] - \mathbb{E}[\pi_f] = p\theta[y(\delta\overline{w},\lambda_s,\alpha) - y(\delta\overline{w},\lambda_f,\alpha)] - c_s.$$

Lastly, we assume that a producer adopts the technology if the net benefit of investment is greater than zero:

(7) $$p\theta[y(\delta\overline{w},\lambda_s,\alpha) - y(\delta\overline{w},\lambda_f,\alpha)] - c_s > 0$$

or after rearranging,

(8) $$\theta[y(\delta\overline{w},\lambda_s,\alpha) - y(\delta\overline{w},\lambda_f,\alpha)] > \frac{c_s}{p}.$$

The left-hand side of equation (8) is the difference in yields when water is called multiplied by the probability that water is called. It represents the expected gross benefit of technology adoption. As the difference increases, producers become more likely to adopt the technology. The right-hand side describes the ratio of cost to output price. As the cost of the investment increases, producers are less likely to invest while as price increases, the net benefit of adoption becomes higher, and producers become more likely to adopt. A key feature of this model is that the benefit of technology adoption comes only from reducing downside risk. The adoption of a more efficient irrigation technology allows the producer to achieve a higher yield per unit of water when he does not receive the entirety of his water right. The highest priority farmer has $\theta = 0$ and an expected gross benefit equal to zero.

---

2. Water allotments under prior appropriation are determined by the historical consumptive use of the activity allowed by the water right, so this assumption is appropriate in this context.

5.3   Parameter Simulations

We now examine how perceptions of θ and δ incentivize adoption of the efficient technology by parameterizing the left-hand side of (8). We are only concerned with identifying where a producer would have the highest likelihood of adopting, so we focus on the range of parameters in which the gross benefits of adoption are highest. When the gross benefits of adoption are highest, we would expect adoption to be more likely.

First, we assume the following parameter values: $\lambda_f = 0.5$, $\lambda_s = 0.9$, $y_m = 6$, and $\alpha = 0.5$. Irrigation application efficiencies are used from Bauder, Waskom, and Andales (2014) and maximum yield can be interpreted as tons of corn per acre. We then calculate the left-hand size of (8) over the range of plausible values of θ and δ to generate a heat map displaying the areas in which the gross benefit of adoption are greatest given the producer's perceptions (figure 5.1). Each point on the heat map corresponds to a possible combination of θ (probability of a call) and δ (magnitude of shortage) for an individual producer. The background shading at each point corresponds to the gross benefit, or increase in expected yield, associated with that combination of θ and δ. Darker (lighter) areas are associated with lower (higher) gross benefits, so a change that results in movement from a dark area to a lighter area would result in an increase in benefit. The impact of θ is straightforward and monotonic. Pick one point along the horizontal axis and hold δ constant, and each point directly above (increasing θ) lies on a lighter area on the map. In other words, as a possible call becomes more likely, the benefit of improving water application efficiency increases monotonically. If the perceived probability of a call is 0, there is no incentive to invest.

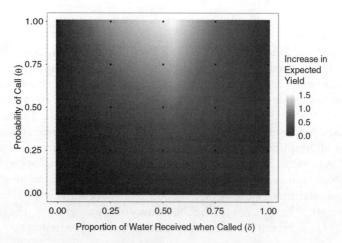

Fig. 5.1   How perceptions impact the gross benefit of investment

The impact of δ is less straightforward. When holding θ constant and moving along points from left to right, the benefit is greatest around $\delta = 0.55$, after which the benefit decreases. When the producer expects to receive nearly all or nearly none of his water during a call, the benefit of improving water application efficiency approaches 0. If a producer were to experience a change in perception that moved him from a dark to light area on figure 5.1, we would expect an increased likelihood of investing in the high-efficiency sprinkler system. The region in which producers are most likely to adopt the new technology occurs when the probability of a call is perceived as high, and the volume of water lost during a shortage is about half of the full right. In the empirical section of this paper, we investigate these theoretical predictions.

5.4   Study Area

In Colorado, the Water Right Determination and Administration Act of 1969, C.R.S. 37-92 et seq. (1969), designated seven water divisions based on drainage characteristics, each staffed with its own division engineer and water judge. Water Division 1 (WD1), the study area for this analysis, is highly dependent on surface water and contains the South Platte River basin (SPRB), Republican River basin, and Laramie River basin. The Colorado Water Plan (CWCB 2015) provides extensive detail on all basins and water divisions, and here we summarize the details relevant to our analysis. The SPRB alone is home to approximately 80 percent of Colorado's population while also having the largest proportion of irrigated agriculture. Irrigated agriculture accounts for approximately 85 percent of total water diversions within the basin, with water supplies originating in mountain snowpack along the Continental Divide. Farmland in WD1 typically receives less than 8 inches of precipitation during the growing season (Schneekloth and Andales 2017). In addition to 1.4 million acre-feet of average annual native flow volume, the basin receives an additional 500,000 acre-feet in transmountain diversions. Overall, the basins in WD1 are over-appropriated, meaning the total allotted volume of water rights exceeds the current average supply, and many irrigation season water rights are continuously out of priority.

WD1 provides a relevant case study of many arid regions that are experiencing water scarcity concerns coupled with irrigation-dependent agriculture and fast growing populations. The 17 states wholly or partially west of the 100th meridian in the conterminous United States all utilize a strict or hybrid prior appropriation water rights regime (Leonard and Libecap 2019) and depend on irrigation water for agricultural production. Of these 17 states, 7 were among the top 10 fastest growing states in percent growth from 2020 to 2021 (US Census Bureau 2021). Increased water scarcity due to climate change combined with increasing demands for urban uses place significant pressures on agricultural production in these areas. While there is considerable heterogeneity in producers across the American West, many face similar problems to those represented in this study.

Agricultural producers in WD1 face uncertainty in water availability from two predominant sources. The first source is the variability in water supplies under a changing climate. The second source is institutional, as water administration is complex and constantly evolving. Colorado is experiencing rapid population growth, with increasing water demands for municipal, industrial, recreational, and environmental uses, and the administration of water law frequently undergoes changes from new legislation and court rulings as new problems emerge (Jones and Cech 2009).

5.5   The Natural Experiment

In addition to designating water divisions, the 1969 act determined that groundwater was to be regulated in conjunction with surface water under prior appropriation. The act introduced "augmentation plans" that allow for out-of-priority diversions so long as sufficient replacement water is supplied to prevent injury to senior users. Such plans are required to be approved through a decree of a district water court,[3] but the state engineer was granted the ability to temporarily approve substitute water supply plans (SWSPs). SWSPs were essentially augmentation plans that could be renewed on an annual basis without official approval from the courts. Consequently, many junior users neglected to formally seek court adjudication and relied on the state engineer for continued water use under SWSPs (Waskom 2013). SWSPs were predominantly utilized by groundwater users who would collectively provide replacement water through recharge ponds or reservoirs. Throughout the 1980s and 1990s, groundwater users in particular were accused of providing inadequate replacement water (Waskom 2013), however exceptional precipitation and snowpack (McKee et al. 2000) veiled potential water shortages. Nearly two decades of abundant water supply meant there was little incentive to impose change within the system.

Then, in 1999–2000, Colorado experienced an unexpected combination of low winter snow accumulation and above average spring and summer temperatures that led to drought conditions across the state (Pielke et al. 2005). This revealed that existing replacement efforts under SWSPs did not adequately cover shortfalls in water availability, and as a result, litigation was launched between two water users over misuse of SWSPs. The result of *Empire Lodge Homeowner's Association v. Moyer*, 39 P.3d 1139 (Colo.

---

3. Colorado water courts are specialized state courts with water judges appointed by the state Supreme Court. Water judges have jurisdiction over all water use and administration within their water division. See https://www.courts.state.co.us/Forms/PDF/JDF%20301W.pdf for the application and detailed requirements for approval of an augmentation plan.

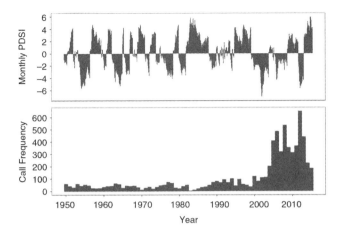

Fig. 5.2   Monthly PDSI and frequency of calls by state engineer, Colorado Water Division 1

2001) declared that the state engineer did not have legal authority to approve SWSPs on an annual basis and shifted more oversight of water replacement plans to the water courts.[4] Although this ultimately led to the permanent curtailment of many groundwater rights, it had a direct impact on surface water. First, producers faced increased dependence on uncertain surface water supplies during the summer months. Additionally, the number of formally decreed augmentation plans that require records of actual diversions increased dramatically in subsequent years (Waskom 2013). Since the basins in WD1 are over-appropriated, net surface water diversions could not increase in practice. As more water rights recorded daily diversions, the state engineer had a better understanding of actual surface water supplies, and the likelihood of calls along mainstream rivers increased.

After the institutional change, drought conditions persisted through 2009 with the most intense period occurring in 2002. In 2002, all of Colorado was in extreme drought conditions, and April snowpack was estimated at 52 percent of the previous 30-year average (Pielke et al. 2005). PDSI levels for WD1 reached −6 (figure 5.2, top panel), a classification of drought categorized by widespread crop losses and severe water shortages that result in water emergencies. The newly increased reliance on surface water, better records for actual diversions, and unprecedented drought conditions resulted in a permanent change to the call regime (figure 5.2, bottom panel). The average number of days under call from 2002 to 2012 was two to four times that of

4. For more information on SWSPs and Empire Lodge Homeowners v. Moyer, see the "Guidance Documents" available at https://dwr.colorado.gov/services/water-administration/water-supply-plans-and-administrative-approvals.

1982–2001 for districts within WD1 (Waskom 2013, 149–152). The change in oversight for out-of-priority diversions, combined with an unprecedented decrease in surface water supply, created an exogenous shock to the distribution of surface water available to relatively junior water rights.

5.6   Data and Modeling Approach

To exploit the exogenous change in surface water availability for some users, we compile an extensive data set for WD1 on irrigated cropland, irrigation technology, agricultural surface water rights, call recordings, and population across seven observation years (1976, 1987, 1997, 2001, 2005, 2010, 2015). County-level population data are available through the Colorado State Demography Office, and the remainder of the data from Colorado's Division of Water Resources HydroBase software.[5] Information on individual water rights includes water source, point of diversion, water use type, maximum flow volume, appropriation date, and priority number. The priority number ranks all water rights in terms of seniority, determined by rights' appropriation and court adjudication dates. Information on irrigated cropland includes acreage, point of diversion, and crop type. Irrigation technology at the field level describes whether a field irrigates using flood or sprinklers. Water rights, irrigation technology, and irrigated acres can be matched to a diversion structure, such as a ditch or canal, however we cannot identify the individual parcels owned by a specific water right holder. Therefore, we aggregate information to the diversion structure as the unit for analysis. Altogether we construct a balanced panel of 411 diversion structures.

Since 1950, all administrative calls by the state engineer have been recorded, which we use for our treatment design. Annual information on the length of curtailment for each water right allows us to define treatment and control groups by losses during the 2000s drought relative to historic droughts in the 1950s and 1970s (McKee et al. 2000). We assume that producers developed a perception about the security of their water rights during drought years from the intensity of their curtailment during the historic droughts. The average number of curtailed days per year in drought period d, C_d, during the growing season (April–October) is calculated over the "historic" drought years (1950–1956 and 1974–1978) and the "recent" drought years (2000–2009) for all water rights sharing a diversion structure. Diversion structures that experienced a considerable increase in average curtailment C_d during the recent drought period are placed into the treatment group at the following cutoff:

$$(9) \qquad \text{Treatment} = \begin{cases} 1, & \Delta C_d \geq 50\% \\ 0, & \Delta C_d < 50\% \end{cases},$$

5. See https://cdss.colorado.gov/software/hydrobase.

Table 5.1   Summary of characteristics of treatment and control diversion structures, 2001

Variables	Control Mean	Control Std Dev	Treatment Mean	Treatment Std Dev
Irrigated Land (Acres)				
Flood Technology	1,110.88	3,814.92	2,755.17	5,721.63
Sprinkler Technology	289.91	1,502.46	894.70	2,830.55
Total	1,400.78	5,142.39	3,649.87	8,028.84
Groundwater Supplemented[a]	372.12	1,776.11	2,064.06	4,713.05
Crop Varieties (Acres)				
Corn	433.15	1,868.73	1,628.41	3,820.95
Alfalfa	483.02	2,055.93	1,149.95	2,572.18
Grass Pasture	250.54	514.26	351.02	701.01
Wheat	107.37	450.89	192.18	572.25
Other[b]	42.23	228.76	109.44	405.71
Water Rights Data[c]				
Appropriation Year	1880	13.14	1892	24.11
Number of Rights	6.13	11.46	2.90	3.52
County Population	211,001.8	136,786.1	147,749.5	122,041.8
Number of Structures	339		72	

Note:

[a] HydroBase provides estimates of surface water irrigated acreage that is supplemented with groundwater.

[b] Other crops include sugar beets, dry beans, and assorted vegetables.

[c] Refers only to water rights with decreed agricultural uses.

where $\Delta C_d = (C_{recent} - C_{historic}) / C_{historic} * 100$. Robustness to the 50 percent cutoff is examined in the first section of the appendix.[6]

In table 5.1, we summarize the sample characteristics of treatment and control groups. Statistics for 2001 are reported to provide a snapshot of the sample just before the natural experiment, and it is used as the reference year for our regression analysis. From the data presented it is apparent that larger diversion structures with slightly more junior water rights were disproportionately impacted by the shock. To investigate if treatment structures are correlated spatially, we present a map of treatment and control structures in figure 5.3. The location of treatment structures provides evidence that the shock was not localized to a specific area. We find treatment structures in both urban and rural areas and along a variety of different streams.

To examine the impact of the shock on the number of irrigated acres at diversion structure i in year t with technology j, y_{it}^j, we estimate the following difference-in-difference models:

6. See http://www.nber.org/data-appendix/c14698/appendix.pdf

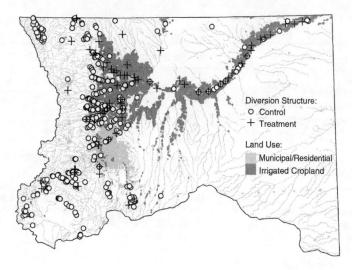

Fig. 5.3   **Treatment and control diversion structure map, Colorado Water Division 1**

$$(10) \qquad \left\{ y_{it}^j = \sum_{t}^{\tau} \beta_t^j D_i T_t + \omega^j x_{it} + \alpha_t^j + \gamma_i^j + \varepsilon_{it}^j \right\},$$

where j denotes the technology-specific model (i.e., sprinkler or flood), $D_i = 1$ if structure i is in the treatment group and 0 otherwise, and T_t is an indicator equal to 1 if t = year T and 0 otherwise. The term x_{it} is county population, and α_t and γ_i are year and diversion structure fixed effects to control for time trends and omitted variables. Lastly, ε_{it} is the error term clustered at the diversion structure. As a placebo test, $D_i T_t$ includes all panel years, excluding the reference year of 2001, to investigate differences prior to and after the natural experiment. Hereinafter we will refer to years 1976, 1987, and 1997 as "pre-treatment" and years 2005, 2010, and 2015 as "post-treatment."

5.7   Empirical Results

Coefficient estimates from (10) with corresponding cluster-robust standard errors are reported in table 5.2. We estimate four iterations of the model with different dependent variables: the number of irrigated acres with flood technology, the number of irrigated acres with sprinkler technology, sprinkler acres as a percentage of total irrigated acres, and total irrigated acres. We include the percentage of sprinkler acres to ensure that estimates in the first two columns are not biased by the behavior of larger diversion structures in our sample. Insignificant estimates for the pre-treatment variables in the first three columns indicate that differences in the outcome variables between treatment and control groups are not statistically distinguishable from zero prior to the shock. This provides suggestive evidence that the

Table 5.2   **Difference-in-difference estimations of the impact of drought and institutional change on irrigation practices**

Variables:	Flood Acres (1)	Sprinkler Acres (2)	% Sprinkler (3)	Total Acres (4)
Treatment*1976	467.0	−251.5	−0.016	215.5*
	(276.2)	(214.3)	(0.022)	(94.6)
Treatment*1987	218.3	−175.7	−0.006	42.6
	(164.8)	(131.9)	(0.013)	(70.5)
Treatment*1997	139.1	−114.1	0.012	25.0
	(78.5)	(78.8)	(0.008)	(36.8)
Treatment*2005	−401.3**	254.5**	0.048***	−146.8
	(151.8)	(87.2)	(0.012)	(82.0)
Treatment*2010	−557.6**	516.7**	0.077***	−40.9
	(196.5)	(163.1)	(0.020)	(76.6)
Treatment*2015	−843.4**	723.3**	0.112***	−120.1
	(305.1)	(237.5)	(0.025)	(102.5)
	−0.001	−0.0007	-2.11×10^{-7}*	−0.002***
County Population	(0.001)	(0.001)	(1.07×10^{-7})	(0.0004)
Fixed effects:				
Diversion Structure	✓	✓	✓	✓
Year	✓	✓	✓	✓
Observations	2,877	2,877	2,877	2,877
Adjusted R^2	0.920	0.792	0.701	0.993

Note: Diversion Structures: 411, Time Periods: 7, Reference Year: 2001. Standard errors (in parentheses) are clustered at the diversion structure level. Signif. Codes: ***: 0.001, **: 0.01, *: 0.05.

treatment and control groups have parallel trends. We present coefficients for the treatment variables graphically in figure 5.4, with 95 percent confidence intervals, to check for the existence of differential pre-trends visually. Dashed confidence intervals indicate overlap with zero. In years after the shock, estimates become significant and increase in magnitude, suggesting that a change in behavior persisted for over a decade. By 2015, the average treatment structure adopted sprinkler technology on 723 more acres than the average control structure. This amounts to 11.2 percent more land converted from flood to sprinkler irrigation on average. Applying this estimate to the entire treatment group, the shock incentivized an increase of over 52,000 sprinkler-irrigated acres in our study area as of 2015.

Surprisingly, there is no statistically significant impact on total irrigated acreage. Although WD1 is experiencing an overall decline in irrigated acreage (CWCB 2015), the rate at which land is leaving production is comparable between the treatment and control groups. This suggests that the treatment group responded to the shock to water availability through more efficient use of the input on the intensive margin. The overall decline in irrigated acres across the basin is perhaps partially explained by the negative and significant coefficient for population, suggesting that a population increase reduces

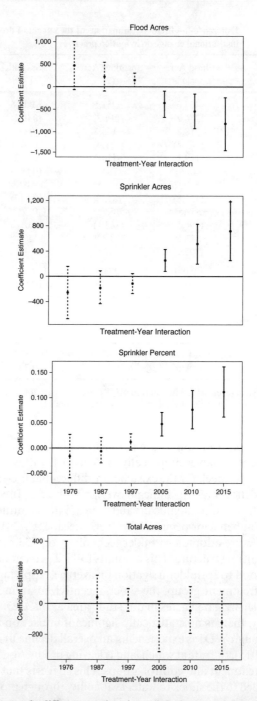

Fig. 5.4 Difference-in-difference estimations of the impact of drought and institutional change on irrigation practices

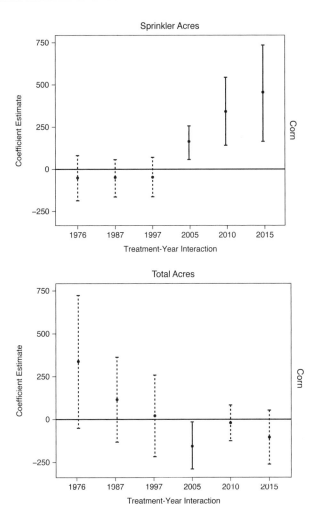

Fig. 5.5 **Difference-in-difference estimations of the impact of drought and institutional change on crop choice**

irrigated acres within a county. This finding is consistent with large cities in Colorado buying agricultural water rights to meet increasing municipal demands (Pritchett, Thorvaldson, and Frasier 2008).

In addition to irrigation technology, agricultural producers can respond to water scarcity by planting less water-intensive crops. We estimate crop-specific models using the same specification as (10) while limiting the dependent variable to total and sprinkler irrigated acres with corn, alfalfa, and wheat. We exclude results for grass pasture as there is very little sprinkler irrigated pasture in our sample. Results from the crop-specific models are presented graphically in figure 5.5, again with 95 percent confidence inter-

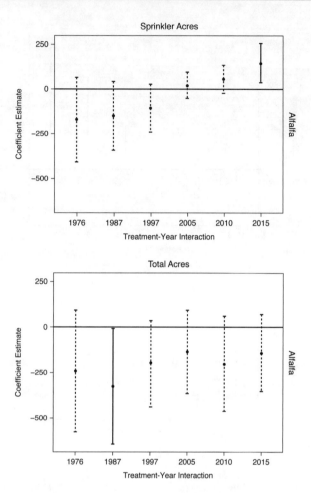

Fig. 5.5   (cont.)

vals, and cluster-robust standard errors are available in table 5.3. Regression results indicate that corn was the crop that experienced the biggest increase in sprinkler-irrigated land as a result of the shock. On average, corn acreage accounted for 60–65 percent of the increase in sprinkler acreage for all post-treatment years.[7] This result holds in 2005 despite the significant average decrease of 149 total corn acres, which was a potential short-run response to the shock. Between the three crops, corn is generally more sensitive to drought than alfalfa or wheat (Lobell et al. 2014), making this

7. This was estimated by dividing the coefficient estimates from the corn-specific Sprinkler Acres model (table 5.3, column 2) by the coefficient estimates from the total Sprinkler Acres model (table 5.2, column 2).

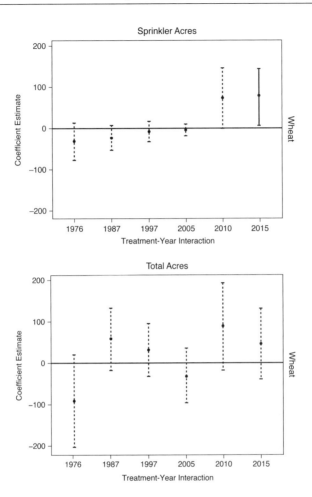

Fig. 5.5   (cont.)

result consistent with risk-mitigating behavior. By 2015, we find significant and positive differences for alfalfa and wheat in addition to corn for the Sprinkler Acres specification. With the exception of corn in 2005, we find no significant differences in total acres post-treatment for each crop. This suggests that adjustments to the change in relative scarcity were made on the intensive margin (adjusting water application per acre) rather than the extensive margin (retiring cropland).

We also investigate the potential impacts on irrigated acreage that is supplemented with groundwater. The average number of estimated acres supplemented with groundwater in table 5.1 indicates that the treatment group utilizes more groundwater to augment their irrigation practices. Since the institutional change in the early 2000s resulted in the curtailment of many

Table 5.3   Difference-in-difference estimations of the impact of drought and institutional change on crop-specific irrigation practices

Variables:	Corn		Alfalfa		Wheat	
	Total Acres (1)	Sprinkler Acres (2)	Total Acres (3)	Sprinkler Acres (4)	Total Acres (5)	Sprinkler Acres (6)
Treatment*1976	338.2	−54.9	−244.0	−169.3	−90.4	−32.2
	(195.8)	(68.4)	(171.1)	(119.9)	(56.4)	(23.2)
Treatment*1987	115.1	−56.8	−328.4*	−149.4	59.0	−23.4
	(125.7)	(55.5)	(161.8)	(97.5)	(38.2)	(15.5)
Treatment*1997	18.8	−49.3	−203.5	−107.5	32.0	−8.34
	(120.3)	(59.7)	(122.4)	(67.8)	(31.9)	(12.7)
Treatment*2005	−149.4*	155.8**	−134.2	21.1	−28.8	−4.65
	(68.9)	(51.2)	(119.5)	(37.1)	(33.7)	(7.29)
Treatment*2010	−23.5	340.7**	−204.2	54.4	88.6	72.0
	(53.8)	(103.3)	(135.4)	(40.3)	(53.2)	(37.4)
Treatment*2015	−103.5	453.2**	−144.1	144.3*	46.3	74.6*
	(79.9)	(145.7)	(109.3)	(56.5)	(42.7)	(35.1)
County Population	−0.001**	−0.0005	−0.0009**	−0.0002	−0.0001	1.5×10^{-5}
	(0.0004)	(0.0005)	(0.0003)	(0.0005)	(0.0001)	(9.58×10^{-5})
Fixed effects:						
Diversion Structure	✓	✓	✓	✓	✓	✓
Year	✓	✓	✓	✓	✓	✓
Observations	2,877	2,877	2,877	2,877	2,877	2,877
Adjusted R^2	0.954	0.781	0.915	0.717	0.798	0.522

Diversion Structures: 411, Time Periods: 7, Reference Year: 2001. Standard errors (in parentheses) are clustered at the diversion structure level. Signif. Codes: ***: 0.001, **: 0.01, *: 0.05.

groundwater rights, it is important to scrutinize what changes in sprinkler adoption can be attributed to changes in surface water versus groundwater availability. We first control for groundwater supplemented acreage in the Sprinkler Acres and Sprinkler % models to check for loss of significance and magnitude of the treatment effects, and then estimate one additional model with groundwater supplemented acres as the dependent variable. Estimates of groundwater supplemented acreage were omitted from the primary regressions due to endogeneity concerns and potential measurement error, particularly because attenuation bias due to measurement error is amplified in fixed effects estimations (Johnston and DiNardo 2009, 404). Regression results for the groundwater models are presented in table 5.4, where columns (2) and (4) correspond to the models with the added groundwater control variable and column (5) to the model with groundwater supplemented acres (GW Acres) as the dependent variable. The only qualitative change to the results from the primary regressions is the loss of significance of the Treatment*2005 variable for the Sprinkler Acres model. Otherwise, the longer-term trends and Sprinkler % results remain largely unaffected. For

Table 5.4   Difference-in-difference estimations, controlling for groundwater use

Variables:	Sprinkler Acres (1)	Sprinkler Acres (2)	Sprinkler % (3)	Sprinkler % (4)	GW Acres (5)
Treatment*1976	−251.5	−190.9	−0.016	−0.014	47.09
	(214.3)	(193.8)	(0.022)	(0.022)	(37.78)
Treatment*1987	−175.7	−207.9	−0.006	−0.007	−24.35
	(131.9)	(146.5)	(0.013)	(0.013)	(36.94)
Treatment*1997	−114.1	−134.0	0.012	0.011	−15.09
	(78.8)	(96.2)	(0.008)	(0.009)	(19.67)
Treatment*2005	254.5**	−94.8	0.048***	0.033**	−269.00**
	(87.2)	(124.7)	(0.012)	(0.013)	(100.50)
Treatment*2010	516.7**	354.5*	0.077***	0.070***	−125.36
	(163.1)	(162.9)	(0.020)	(0.020)	(85.00)
Treatment*2015	723.3**	624.5**	0.112***	0.108***	−76.31
	(237.5)	(219.5)	(0.025)	(0.025)	(58.41)
County Population	−0.0007	−0.002	-2.11×10^{-7}*	-2.46×10^{-7}*	−.0007**
	(0.001)	(0.001)	(1.07×10^{-7})	(1.07×10^{-7})	(.0002)
GW Acres	—	−1.32***	—	-5.38×10^{-5}***	—
	—	(0.232)	—	(7.96×10^{-6})	—
Fixed effects:					
Diversion Structure	✓	✓	✓	✓	✓
Year	✓	✓	✓	✓	✓
Observations	2,877	2,877	2,877	2,877	2,877
Adjusted R2	0.792	0.820	0.701	0.707	0.989

Diversion Structures: 411, Time Periods: 7, Reference Year: 2001. Standard errors (in parentheses) are clustered at the diversion structure level. Signif. Codes: ***: 0.001, **: 0.01, *: 0.05.

the GW Acres model, we find a significant decrease in groundwater supplemented acreage for the treatment group in 2005, which reflects the immediate curtailment of groundwater rights after the shock. However, estimates for Treatment*2010 and Treatment*2015 are not statistically distinguishable from 0, indicating that long-term changes in groundwater use did not differ significantly between the treatment and control groups. It is therefore likely that the significant increases in sprinkler adoption in the treatment group was a mechanism to adapt to long-run changes in surface water availability due to the shift in the call regime.

In summary, we observe a short-run response to the shock in the reduction of total corn acreage and a long-run response in the increased and consistent adoption of sprinkler technology. To examine what this implies for potential water use, we use seasonal crop-water demands for corn, alfalfa, and wheat to make a back-of-the-envelope calculation of the reduction in water required for full crop yields for treatment structures. First, we multiply the 2015 coefficient estimates in table 5.2 for corn, alfalfa, and wheat by the number of treatment structures. Next, we calculate the difference in the seasonal net-irrigation requirement, accounting for precipitation and soil moisture typical to northeastern Colorado, for an acre of each crop with flood irrigation versus sprinkler irrigation.[8] The difference for each crop is then multiplied by the values from the first step. In total, we estimate a potential reduced seasonal irrigation demand for water diversions of 85,000 acre-feet or 28 billion gallons of water across WD1 by 2015 attributable to the change in expectations of water availability. The average Colorado household needs about 0.5 acre-feet of water per year (Waskom and Neibauer 2014), so the demand reduction is roughly equivalent to the yearly water demands of 170,000 households.[9]

In figure 5.4, there is some evidence of pre-trends given the direction of coefficient estimates across time, particularly for the Sprinkler % model. One might attribute these trends to the difference in the average appropriation year (table 5.1) between the treatment and control water rights. Given our theoretical results, it is reasonable to assume that junior water right holders would invest more in water-efficient technologies than senior water rights holders, regardless of the shock to surface water availability in the 2000s. If that is the case, then our coefficient estimates could be biased. We test this supposition by limiting our sample to similar treatment and control structures and re-estimating the Sprinkler % model. We use a propensity score matching algorithm using the minimum, median, and maximum appropriation year for the water rights associated with a structure to make the distribution of all water rights between treatment and control groups as similar as possible.[10] Results from the matching exercise are presented in figure 5.6. The treatment group is smaller than the control group, so we first match every treatment structure with two similar control structures (second column) and then one-to-one (third column). The first row of figure 5.6 displays a smoothed density curve for the total sample and the two matched samples, and the second row displays coefficient estimates corresponding to the sample directly above. Although the densities do not completely overlap in the two-to-one matched sample, any evidence of pre-trends in the resulting coefficient estimates is virtually eliminated, and post-shock estimates remain positive and significant. The one-to-one matching results in a nearly perfect overlap between densities, but the regression suffers from a small sample and estimates are not statistically significant until 2015. This exercise provides evidence that our main results are not driven by differences in seniority among the water rights at treatment and control diversion structures. Additional robustness checks and analysis relating to treatment

8. Net crop water requirements are calculated from data presented in Schneekloth and Andales (2017).

9. This comparison is made only to provide perspective on the volume of water. According to Colorado water law, water "saved" via irrigation efficiency gains cannot be reused or sold.

10. We use the MatchIt package in R to perform a greedy nearest neighbor matching algorithm. Details can be viewed at https://cran.r-project.org/web/packages/MatchIt/MatchIt.pdf.

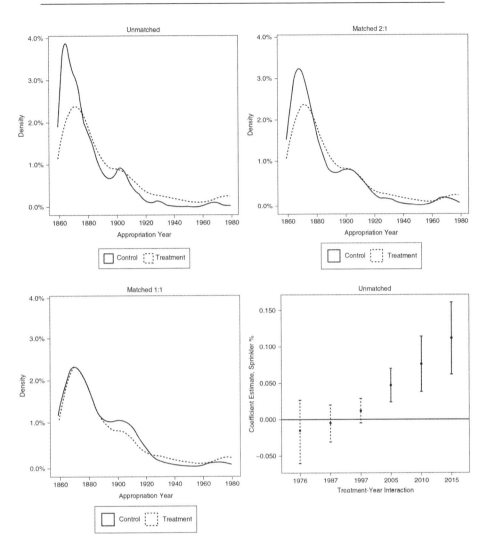

Fig. 5.6   Difference-in-difference estimations before and after propensity score matching, Sprinkler %

design, model specification, and nonlinear impacts from the shock can be found in the first three sections of the appendix, respectively.[11]

5.8   Conclusion and Policy Implications

In this article, we explore the impact of perceived input scarcity on conservation investment decisions. We develop a theoretical model to examine

11. See http://www.nber.org/data-appendix/c14698/appendix.pdf.

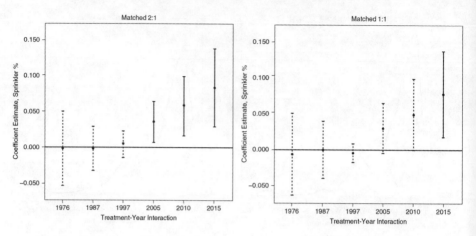

Fig. 5.6   (cont.)

the conditions under which an agricultural producer's perception of water shortages would incentivize investment in a more efficient irrigation technology. A numerical exercise is used to demonstrate a range of perceptions that maximize the gross benefit of investing in irrigation efficiency, and we test our theoretical predictions empirically. A period of severe drought and institutional change in Colorado that led to a change in expectations about the availability of irrigation water is leveraged as a natural experiment. Results suggest that agricultural producers who experienced an unprecedented shock to their irrigation water supply transitioned more land from low- to high-efficiency irrigation systems in the following decade. Our analysis provides evidence that input shocks can trigger investment in efficiency due to changes in perceptions.

This research has limitations that must be acknowledged. Subsidy programs such as the Environmental Quality Incentives Program (EQIP) that can significantly reduce the costs of investment may affect conservation decisions. Although we can observe general rates of adoption through land use changes, we do not know producer-specific costs of a sprinkler system. We also cannot observe conservation practices beyond irrigation technology and crop choice in our data. For example, when evaluating EQIP enrollment, Wallander et al. (2013) found that many drought-facing producers adopted tillage practices that conserve soil moisture. Lining or replacing irrigation ditches to reduce seepage is another practice identified as water saving by EQIP that we are unable to detect.

In Colorado, water rights can be bought and sold, and a distinct feature of our study area is the presence of active water markets. Most market activity consists of municipal and industrial buyers and agricultural sellers.

Some rights were undoubtedly traded during our study period, which we are unable to track. We can only observe the decreed uses of a particular water right as they exist today, and although we limited our analysis to water rights that have a decreed agricultural use, some have gained additional uses through previous transactions. It is possible that not all water right owners with an agricultural water right are using their water for agricultural production in a given year. Although we cannot identify which water rights were sold, we find that changes in total irrigated acreage did not differ substantially between treatment and control groups. This provides some evidence that agricultural water rights are being sold at similar rates across all diversion structures, regardless of the heterogeneous impacts of the shock.

Concerning water right transactions, improving irrigation efficiency does not generally reduce the value of a water right. One aspect of prior appropriation is that water rights may be forfeited if the owner consistently fails to apply the water to a beneficial use, otherwise known as "use it or lose it." This component however only applies to the consumptive use determined by the water right. In the case of a farmer, the consumptive use of his water right is determined by the annual documented evapotranspiration of his crops, not the total amount of water diverted. Since improving irrigation efficiency only reduces the amount necessary for diversion and not the beneficial, consumptive use, the water right's value should not be affected. In the case of a water right transfer, the transferee buys only the right to the consumptive use regardless of the transferor's former diversion amounts. In general, the "use it or lose it" rule is not a true barrier to improving irrigation efficiency, although it is potentially perceived that way by some (Waskom et al. 2016).

Another important characteristic of our study area is that all surface water and most groundwater resources are administered similarly under prior appropriation, which is not uniformly the case across the American West. Their conjunctive governance effectively limits their substitutability, so agricultural producers cannot rely on increased groundwater pumping when surface water supplies are low during drought. This lack of substitutability certainly affected producers' willingness to invest in technology to use surface water more efficiently. Groundwater aquifers are often exhaustible in practice, since they can take long periods of time to replenish naturally. Inhibiting the ability to excessively pump groundwater during drought may prompt an earlier adoption of water conserving technologies. Improving the use efficiency of renewable surface water supplies before exhausting limited groundwater resources may increase the longevity of agricultural production under climate change.

Lastly, it is worth noting that hydrological systems are exceptionally complex, and any changes to how and when water is diverted has common property resource implications. Water is considered a public good under prior appropriation, and water rights are usufructuary. If downstream users in a basin are reliant on return flows, i.e., the water that returns to the system after human use, reducing upstream flows by improving irrigation efficiency can impact their water availability. In some cases, it may not be clear if the adoption of efficient application technologies improves system-level performance. An area of future research that warrants attention is evaluating how uncertainty in return flows impacts the overall efficiency of a basin. Return flows are difficult to track and can vary in their amounts depending on the crop being grown, soil type, weather conditions, and when the water is applied. This added uncertainty can make a system more difficult to manage, all else equal. High efficiency irrigation technologies however increase the control that a producer has to target water to a crop, which reduces the uncertainty in the value added from a unit of water that could have otherwise been applied with a low efficiency technology. Scrutinizing these uncertainties and understanding how incentives for efficient water use are aligned across producers within a basin are crucial for agricultural sustainability.

Despite some limitations, our results are generally informative and have important policy and water management implications. First, drought in arid regions is expected to worsen under a changing climate, and perceptions will play a critical role in the future adoption of conservation practices in agriculture. Neglecting how costs and benefits are perceived when assessing the effectiveness of programs designed to encourage conservation efforts could provide policy makers with misleading information. For example, if a policy maker is considering the implementation of a subsidy program to promote the adoption of water-conserving technologies, it is important to understand whether non-adoption is driven by conventional cost hurdles or perceptions about necessity. If the latter is the driving factor, efforts to accelerate revisions to perceptions to align with actual shortage distributions before the realization of costly weather disruptions could bolster a more efficient path to adoption. This may be an opportunity for agricultural extension to address and build perceptions about water scarcity in arid climates. Surveys and qualitative interviews can be administered to local farmers to gauge perceptions about climate change, drought risk, and the efficacy and necessity of adaptation strategies. If climate change risk is perceived as negligible, awareness campaigns tailored to communicating water scarcity concerns in localized areas may be effective at accelerating changes. If climate change is perceived as a real risk, communicating the benefits of increasing water use efficiency and providing better information on the possibility of future water shortages can enable producers to minimize their downside risk. Highlighting the conservation practices of local farming operations may also facilitate changes in perceptions, as the behavior of neighbors has been found to be influential in adoption behavior (Case 1992). Once climate change perceptions align with a need to improve water use efficiency, disseminating opportunities that reduce costs of implementation, such as EQIP participation, can hasten the path to adoption.

References

Adam, J. C., A. F. Hamlet, and D. P. Lettenmaier. 2009. "Implications of Global Climate Change for Snowmelt Hydrology in the Twenty-First Century." *Hydrological Processes: An International Journal* 23 (7): 962–972.

Bauder, T. A., R. M. Waskom, and A. Andales. 2014. "Nitrogen and Irrigation Management." Colorado State University Extension Fact Sheet (0.514).

Callaway, B., and P. H. Sant'Anna. 2021. "Difference-in-Differences with Multiple Time Periods." *Journal of Econometrics* 225 (2): 200–230.

Carey, J. M., and D. Zilberman. 2002. "A Model of Investment under Uncertainty: Modern Irrigation Technology and Emerging Markets in Water." *American Journal of Agricultural Economics* 84 (1): 171–183.

Case, A. 1992. "Neighborhood Influence and Technological Change." *Regional Science and Urban Economics* 22 (3): 491–508.

Cobourn, K. M., X. Ji, S. Mooney, and N. F. Crescenti. 2021. "The Effect of Prior Appropriation Water Rights on Land-Allocation Decisions in Irrigated Agriculture." *American Journal of Agricultural Economics* 104 (3): 947–975.

CWCB. 2015. "Colorado's water plan." Colorado Division of Natural Resources.

Dieter, C., M. Maupin, R. Caldwell, M. Harris, T. Ivahnenko, J. Lovelace, N. Barber, and K. Linsey. 2018. "Estimated Use of Water in the United States in 2015." US Geological Survey.

EPA. 2016. "Climate Change Indicators in the United States." United States Environmental Protection Agency.

Foster, T., and N. Brozović. 2018. "Simulating Crop-Water Production Functions Using Crop Growth Models to Support Water Policy Assessments." *Ecological Economics* 152: 9–21.

Frisvold, G., and T. Bai. 2016. "Irrigation Technology Choice as Adaptation to Climate Change in the Western United States." *Journal of Contemporary Water Research & Education* 158 (1): 62–77.

Getches, D. H. 2009. *Water Law in a Nutshell*, 4th edition. St. Paul, MN: West Academic Publishing.

Gupta, A., A. Rico-Medina, and A. I. Caño-Delgado. 2020. "The Physiology of Plant Responses to Drought." *Science* 368 (6488): 266–269.

Gutzler, D. S., and T. O. Robbins. 2011. "Climate Variability and Projected Change in the Western United States: Regional Downscaling and Drought Statistics." *Climate Dynamics* 37 (5): 835–849.

Hamlet, A. F., P. W. Mote, M. P. Clark, and D. P. Lettenmaier. 2005. "Effects of Temperature and Precipitation Variability on Snowpack Trends in the Western United States." *Journal of Climate* 18 (21): 4545–4561.

Howden, S. M., J.-F. Soussana, F. N. Tubiello, N. Chhetri, M. Dunlop, and H. Meinke. 2007. "Adapting Agriculture to Climate Change." *Proceedings of the National Academy of Sciences* 104 (50): 19691–19696.

Ji, X., and K. M. Cobourn. 2021. "Weather Fluctuations, Expectation Formation, and Short-Run Behavioral Responses to Climate Change." *Environmental and Resource Economics* 78 (1): 77–119.

Johnston, J., and J. DiNardo. 2009. *Econometric Methods*, 4th edition. New York: McGraw-Hill.

Jones, P. A., and T. Cech. 2009. *Colorado Water Law for Non-Lawyers*. Denver, CO: University Press of Colorado.

Koundouri, P., C. Nauges, and V. Tzouvelekas. 2006. "Technology Adoption under Production Uncertainty: Theory and Application to Irrigation Technology." *American Journal of Agricultural Economics* 88 (3): 657–670.

Leonard, B., and G. D. Libecap. 2019. "Collective Action by Contract: Prior Appropriation and the Development of Irrigation in the Western United States." *The Journal of Law and Economics* 62 (1): 67–115.

Li, M., W. Xu, and T. Zhu. 2019. "Agricultural Water Allocation under Uncertainty: Redistribution of Water Shortage Risk." *American Journal of Agricultural Economics* 101 (1): 134–153.

Lioubimtseva, E. 2004. "Climate Change in Arid Environments: Revisiting the Past to Understand the Future." *Progress in Physical Geography* 28 (4): 502–530.

Lobell, D. B., M. J. Roberts, W. Schlenker, N. Braun, B. B. Little, R. M. Rejesus, and G. L. Hammer. 2014. "Greater Sensitivity to Drought Accompanies Maize Yield Increase in the US Midwest." *Science* 344 (6183): 516–519.

Maddison, D. 2007. *The Perception of and Adaptation to Climate Change in Africa*, vol. 4308. World Bank Publications.

Mancosu, N., R. L. Snyder, G. Kyriakakis, and D. Spano. 2015. "Water Scarcity and Future Challenges for Food Production." *Water* 7 (3): 975–992.

Masson-Delmonte, V., P. Zhai, A. Pirani, S. Connors, C. Péan, S. Berger, N. Caud, Y. Chen, L. Goldfarb, M. Gomis, et al. 2021. "IPCC 2021: Climate Change 2021: The Physical Science Basis: Contribution of Working Group I to the Sixth Assessment Report of the Intergovernmental Panel On Climate Change."

McKee, T. B., N. J. Doesken, J. Kleist, C. J. Shrier, and W. P. Stanton. 2000. "A History of Drought in Colorado: Lessons Learned and What Lies Ahead," no. 9, 2nd edition. Colorado Climate Center, Colorado State University.

Mote, P. W. 2006. "Climate-Driven Variability and Trends in Mountain Snowpack in Western North America." *Journal of Climate* 19 (23): 6209–6220.

Mote, P. W., A. F. Hamlet, M. P. Clark, and D. P. Lettenmaier. 2005. "Declining Mountain Snowpack in Western North America." *Bulletin of the American Meteorological Society* 86 (1): 39–50.

Pielke, R. A., N. Doesken, O. Bliss, T. Green, C. Chaffin, J. D. Salas, C. A. Woodhouse, J. J. Lukas, and K. Wolter. 2005. "Drought 2002 in Colorado: An Unprecedented Drought Or a Routine Drought?" *Pure and Applied Geophysics* 162 (8): 1455–1479.

Pritchett, J., J. Thorvaldson, and M. Frasier. 2008. "Water As a Crop: Limited Irrigation and Water Leasing in Colorado." *Review of Agricultural Economics* 30 (3): 435–444.

Schneekloth, J., and A. Andales. 2017. "Seasonal Water Needs and Opportunities for Limited Irrigation for Colorado Crops." Colorado State University Extension Fact Sheet (4.718).

Spence, A., W. Poortinga, C. Butler, and N. F. Pidgeon. 2011. "Perceptions of Climate Change and Willingness to Save Energy Related to Flood Experience." *Nature Climate Change* 1 (1): 46–49.

US Census Bureau. 2021. "New Vintage 2021 Population Estimates Available for the Nation, States And Puerto Rico." RELEASE NUMBER CB21-208.

Wallander, S., M. Aillery, D. Hellerstein, and M. Hand. 2013. "The Role of Conservation Programs in Drought Risk Adaptation." Economic Research Service ERR 148.

Wang, X. 2017. "Understanding Climate Change Risk Perceptions in China: Media Use, Personal Experience, and Cultural Worldviews." *Science Communication* 39 (3): 291–312.

Wang, X., and L. Lin. 2018. "The Relationships among Actual Weather Events, Perceived Unusual Weather, Media Use, and Global Warming Belief Certainty in China." *Weather, Climate, and Society* 10 (1): 137–144.

Waskom, R. M. 2013. "HB12-1278 Study of the South Platte River Alluvial Aquifer." Colorado Water Institute.

Waskom, R., and M. Neibauer. 2014. "Water Conservation in and around the Home." Colorado State University Extension 9.952.

Waskom, R., K. Rein, D. Wolfe, and M. Smith. 2016. "How Diversion and Beneficial Use of Water Affect the Value and Measure of a Water Right." Colorado Water Institute Special Report 25. Fort Collins, CO.

6

Climate, Drought Exposure, and Technology Adoption
An Application to Drought-Tolerant Corn in the United States

Jonathan McFadden, David Smith,
and Steven Wallander

6.1 Introduction

Ongoing climate change is causing complex and potentially irreversible changes to crop growing conditions across the world. In the US, increases in average temperatures, accompanied by increases in the frequency of warm days and nights, and higher precipitation variability over the central US are amplifying drought risk (Intergovernmental Panel on Climate Change 2014). Widespread drought prevalence and risk are projected to increase through 2050, with the Rocky Mountain states and the southwestern US experiencing greater drought frequency (Strzepek et al. 2010). These regions, along with the Central Plains, are at high risk of a multi-decadal drought this

Jonathan McFadden was an assistant professor of economics at the University of Oklahoma when this paper was written, and is currently a research economist in the Resource and Rural Economics Division of the USDA-Economic Research Service.

David Smith is an economist at the National Center for Environmental Economics of the US Environmental Protection Agency.

Steven Wallander is an Economist in the Conservation and Environment Branch of the Resource and Rural Economics Division of the USDA-Economic Research Service.

We thank Ryan Williams and Kevin Hunt for assistance with the geocoded data, as well as Andrew Bryant for valuable research assistance. Ariel Dinar, Gary Libecap, James MacDonald, Roger Claassen, and Seth Wechsler are thanked for their feedback on earlier versions of this research. Seminar participants at USDA-ERS, USDA-FAS, Oklahoma State University, and the University of Oklahoma, as well as participants at the 2018 AAEA conference, 2019 AERE annual meetings, and 2022 NBER conference "Economic Perspectives on Water Resources, Climate Change, and Agricultural Sustainability" are thanked for comments and useful discussions. The views expressed in this working paper are those of the authors and do not necessarily represent the views or policies of the US Department of Agriculture or the US Environmental Protection Agency. For acknowledgments, sources of research support, and disclosure of the authors' material financial relationships, if any, please see https://www.nber.org/books -and-chapters/american-agriculture-water-resources-and-climate-change/climate-drought -exposure-and-technology-adoption-application-drought-tolerant-corn-united-states.

century (Cook, Ault, and Smerdon 2015), even as groundwater sources for irrigation in certain areas continue to decline (Steward et al. 2013).

Using historical crop yield data, several applied econometric studies have found little evidence of long-run adaptation to climate change in the production of major US row crops. For example, through a comparison of long difference and panel regression coefficients, Burke and Emerick (2016) conclude that adaptation has offset no more than half of the negative short run impacts of extreme heat on US corn yields. Their median estimate of future climate change impacts suggests that such limited adaptation will contribute to decreases in annual yields by 15 percent in 2050. Much of this is driven by the finding that corn yields decline sharply at temperatures above 28–29°C (82.4–84.2°F) (Schlenker and Roberts 2009; Burke and Emerick 2016). However, past agricultural adaptations are imperfectly captured, and the prospects of future adaptation-inducing technical progress (Heisey and Day Rubenstein 2015) or large-scale price feedbacks (Auffhammer and Schlenker 2014) cannot be precisely forecasted over long time horizons.

In the short run, crop farmers have few potential tools for reducing downside production risk from changes in drought frequency or intensity. Adoption of irrigation equipment, increased irrigation, or adoption of more efficient equipment are only viable for farmers with access to sufficient irrigation water. Regardless of irrigation availability, no-till crop management and conservation tillage practices can be effective adaptation tools (Powlson et al. 2014) because they reduce soil moisture evaporation and help improve soil water-holding capacity and infiltration. US federal working lands conservation programs, such as the Environmental Quality Incentives Program, pay farmers to use such practices. Federal conservation easement programs like the Conservation Reserve Program can also help certain farmers adapt to drought risk by retiring land from agricultural production (Wallander et al. 2013). Apart from these technologies and management practices, one recent and increasingly available option is adoption of drought-tolerant (DT) crop varieties.

Non-genetically engineered (non-GE) DT corn hybrid varieties were commercialized in 2011, while GE DT varieties were released on a limited basis in 2012. Adoption has been rapid in the first five years since introduction: by 2016, just over 22 percent of US corn acreage was planted to DT varieties (McFadden et al. 2019). Currently, most DT corn acreage is concentrated in drought-prone areas of the western Corn Belt—particularly in Nebraska and Kansas. However, DT corn seed adoption is also significant in the eastern Corn Belt and in regions of relatively lower corn productivity. Climate and drought risk, local growing conditions, and seed company marketing strategies likely factor into farmers' decisions about which corn varieties to grow.

Corn presents a particularly attractive opportunity for studying the economic dimensions of DT crop adoption. First, corn has been a major target of substantial US breeding research into drought tolerance for several decades, culminating in the commercial release of multiple hybrid varieties. Despite the extensive challenges of identifying and manipulating plant genetic material that mediates the plant's complex response to drought, DT varieties have been developed for other crops like soybeans (e.g., Nuccio et al. 2018), though US regulatory approval and commercialization have only occurred very recently, and the full extent of uptake is unknown. Second, corn is a water-intensive crop that is sensitive to high temperatures (e.g., Schlenker and Roberts 2009), thus making its on-farm management intrinsically interesting—especially within the context of worsening climate conditions. Third, corn is among the most economically important crops in the US. Since 2006, annual gross cash receipts for corn have comprised at least 10 percent of receipts for all US agricultural commodities (US Department of Agriculture, Economic Research Service 2022), and its share of total annual US planted crop acres has been well over 25 percent during this same period (US Department of Agriculture, National Agricultural Statistics Service 2022). Given the sizeable physical and financial value at risk—risk with an increasingly large downside for the US agricultural economy as climate change deepens—an examination of the determinants surrounding this particular drought adaptation tool would seem compelling.

The objective of this study is to determine how drought risk and recent drought exposure impact the adoption of DT corn in the US. We begin by detailing research and development (R&D) of DT corn seeds and the subsequent use of these varieties across the US. Next, we motivate our empirical analysis of adoption trends through a state-contingent framework that accommodates farmers' beliefs about future drought based on objective drought risk and exposure. Although nationally representative microdata of farmers' fields are available, they do not track the same field (or farm operation) over time, which generates concerns of potential bias arising from unobserved heterogeneity. To alleviate this source of potential bias, we implement a new and intuitive econometric method, spatial first differences, which is designed to eliminate time-invariant confounds in cross-sectional regression models.

Our discrete choice analysis suggests that adoption is responsive to long-run drought risk and climate conditions and not the severity or duration of recent droughts. We further find that irrigation reduces the likelihood of adoption, while high erodability increases this likelihood, both of which are consistent with the location-based marketing of these varieties that initially targeted the western Corn Belt.

6.2   Research, Development, and Uptake of Drought-Tolerant Corn Varieties

Private-sector research on drought tolerance in corn dates back to at least the 1950s (Cooper et al. 2014). Initial research focused on varietal selection for yield performance under drought conditions. Public-sector research began in the mid-1970s (Edmeades 2012). In contrast to the private sector, non-profit institutions like the International Maize and Wheat Improvement Center (CIMMYT) began selecting for drought tolerance in tropical varieties using an index of traits. The methods (e.g., conventional breeding, molecular breeding, and genetic engineering) and germplasm (i.e., seeds and plants used in crop breeding research) used in both sectors have improved over the past several decades. To date, the international research community has introduced several hundred varieties of DT corn that are adapted to growing conditions in the US and abroad.[1]

By 2012, both non-GE and GE DT corn hybrid varieties had been commercialized in the US. Non-GE varieties were developed through the use of molecular breeding, which entailed analysis of field trial data and subsequent selection based on statistical predictions of yield performance and other traits under drought conditions. Non-GE DT corn acreage has increased following its 2011 commercial release (Minford 2015). By contrast, GE drought tolerance involves insertion of a certain soil bacterium gene into the corn plant's genetic material, which causes expression of a specific protein that helps the plant mitigate drought damages. GE DT corn acreage has also increased since commercialization in 2012 (Castiglioni et al. 2008; Waltz 2014), but at a slower rate than acreage to corn varieties with conventional drought tolerance (McFadden et al. 2019).[2]

The percent of US corn acreage planted to DT corn varieties increased swiftly in the years following commercialization (figure 6.1). In 2012, DT corn (all varieties) accounted for a little over 2 percent of national corn acreage. By 2016, roughly 22 percent of US corn acres were drought tolerant, a 20 percentage point increase over five years. This growth is similar to that of insect-resistant (Bt) and herbicide-tolerant (HT) corn varieties in their first five years (1996–2000). Comparisons of adoption trends for these three technologies must be interpreted cautiously because each is used to manage different factors that cause yield loss. Bt varieties are mainly used to manage two insect pests common throughout the US Corn Belt. HT traits are used to manage weeds. In contrast, there are several areas of the eastern Corn Belt (e.g., eastern Indiana, Ohio, and western Pennsylvania) where

1. The Drought Tolerant Maize for Africa (DTMA) project (2007–15) was an international partnership that developed and released DT corn varieties adapted for sub-Saharan growing conditions. Wossen et al. (2017) found average yields were 12.6 percent higher among DTMA DT corn adopters in Nigerian villages under mild droughts than non-adopters. The Water Efficient Maize for Africa (WEMA) project (2013–18) is a public-private partnership that has also developed and commercialized DT corn varieties for certain sub-Saharan nations (Edmeades 2012). A successor, the Stress Tolerant Maize for Africa project (2016–19), had the goal of developing 70 varieties with more resistance to non-drought stresses, e.g., low soil fertility, diseases, and pests (CIMMYT 2018).

2. As our study relies on data at the initial stage of the DT corn adoption path, it is infeasible to perform an ex post analysis of the returns to drought tolerance R&D for the commercialized varieties.

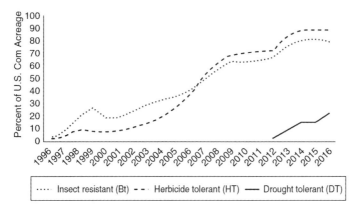

Fig. 6.1   Adoption of insect-resistant, herbicide-tolerant, and drought-tolerant corn in the US since 1996

Note: The HT and Bt genetically-engineered (GE) traits were commercially introduced in corn in 1996. The year in which GE varieties of drought-tolerant (DT) corn were introduced was 2012. For year 2016, the DT adoption rate is calculated from Phase II ARMS data. Since Phase II of ARMS does not survey corn each year, estimates of adoption rates are unavailable prior to 2016. For years 2012–15, DT corn acreage is taken from various industry sources and then divided by total harvested acreage of corn for grain. Adoption rates for Bt and HT corn include minimal acreages of stacked varieties (1996–2000), whereas adoption rates for DT corn include varieties that are overwhelmingly stacked with Bt and/or HT traits (2012–2016).

Source: Fernandez-Cornejo and McBride (2002) for Bt and HT corn adoption rates; USDA, National Agricultural Statistics Service (2018b) and various industry estimates for DT (2011–15); USDA, Economic Research Service and National Agricultural Statistics Service, 2016 Agricultural Resource Management Survey.

drought is less common. In addition, there was no universal expectation that HT and Bt traits would subsequently become "breakthrough" and widely adopted crop production technologies. Nonetheless, the pace of early DT corn adoption has been comparable to that of other major recent innovations in corn varieties.[3]

Aggregation of national acreage trends masks significant regional variation in 2016 DT corn adoption (figure 6.2). Roughly 16–21 percent of corn in 2016 for the traditional Corn Belt (e.g., Iowa, Illinois, and Indiana) was drought tolerant, with 14–20 percent shares of corn acreage in the northern Great Lakes states of Minnesota, Wisconsin, and Michigan. The Great Plains exhibit significant within-state variability, particularly in Nebraska,

---

3. It should be noted that these traits are often bundled in the same variety. For example, 54 percent of corn fields in our sample were planted to varieties with HT and Bt traits, while 20 percent were planted to varieties with DT, HT, and Bt traits. Despite such bundling, it does not appear that climate, drought risk, and other regressors of interest have larger effects on adoption of HT or Bt corn than DT corn. This conclusion is supported by complementary multivariate econometric analysis, which we exclude here for space considerations (estimates available upon request).

Fig. 6.2   DT shares of US state corn acreage, 2016

Note: In 2016, total US corn acreage planted to DT varieties was 18.6 million acres.

Source: USDA, Economic Research Service and National Agricultural Statistics Service, 2016 Agricultural Resource Management Survey.

Kansas, and Texas, though adoption is highest (39–42 percent) for cropland generally over or near the Ogallala Aquifer. Certain high-latitude regions, like North Dakota, had minimal adoption but also tend to have relatively less corn production.

It is expected that adoption is correlated with incidences of recent and severe droughts (figure 6.3). The US Drought Monitor indicates that substantial swaths of the Corn Belt experienced a "severe" (category D2) or worse drought in July 2012. One year later, droughts were concentrated in western states, with many counties in Colorado, Nebraska, Kansas, and Texas experiencing "extreme" (category D3) or "exceptional" (category D4) droughts. These drought patterns are confirmed by spatial variation in July drought risk, as measured by the standard deviation of the Palmer Modified Drought Index (PMDI). Counties that had one or more severe-or-worse droughts in 2011–15 tended to have higher risk. However, most counties in South Dakota, North Dakota, and Minnesota did not experience significant drought over 2011–15, despite being higher risk. Conversely, many Texas counties experienced more frequent and/or more severe droughts during 2011–14 than might be implied by their long-run risk. These discrepancies in short-run exposure and long-run risk partially motivate our economic framework, which we discuss next.

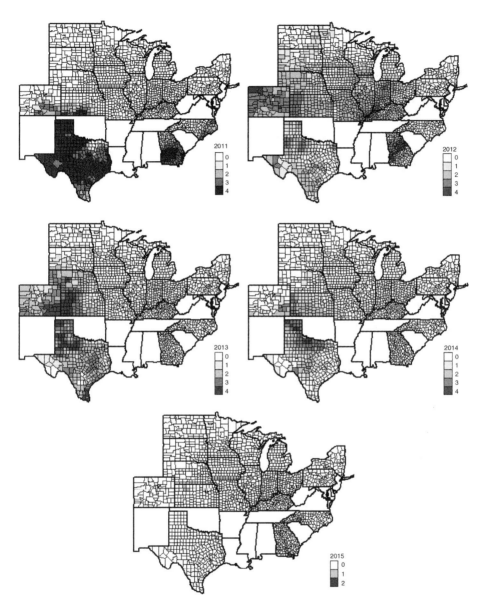

Fig. 6.3   US Drought Monitor index of droughts, 2011–15

Note: The US Drought Monitor sorts drought into five categories: abnormally dry (a drought precursor, D0), moderate (D1), severe (D2), extreme (D3), and exceptional (D4). These can be thought of in terms of potential impacts; short-term dryness (D0), some damage to crops (D1), crop losses likely (D2), major crop losses (D3), and exceptional and widespread crop losses (D4), although actual agricultural impacts will vary by crop and irrigation use.

Source: US Drought Monitor 2019.

6.3   Economic Framework

We define the farmer's state-contingent and partial (drought and drought-abating) per-acre profit (π) function as

$$(1) \qquad \pi_s = P Y_s(\mathbf{X^D}, \mathbf{X_s^D}) - \mathbf{P^X}\mathbf{X^D} - \mathbf{P^{X_s}}\mathbf{X_s^D},$$

where P is the output price,[4] Y_s is yield, $\mathbf{X^D}$ is a vector of state-independent drought-abating inputs, $\mathbf{X_s^D}$ is a vector of state-contingent drought-abating inputs, $\mathbf{P^X}$ is a vector of prices for these inputs, and s is an index for a state of the world.

We assume that yields are a multiplicative function of drought-free yields (Y) and a drought abatement function ($d_s(.) \in \{0,1\}$), defined over the unit interval. The drought abatement function is the percentage of the drought-free yields that are not damaged by the drought,

$$(2) \qquad Y_s = Y d_s(\mathbf{X^D}, \mathbf{X_s^D}; D_s, \mathbf{I}).$$

The drought abatement function is dependent on drought damage (D_s), drought abatement ($\mathbf{X^D}, \mathbf{X_s^D}$), and the ability to irrigate ($\mathbf{I}$). In this framework, $\mathbf{I}$ is a set of exogenous parameters indicating: (1) whether irrigation equipment has been previously installed on a field, and (2) whether irrigation water is available. This reflects the idea that within a single growing season, farmers cannot choose whether their fields are irrigable or if water is available. In other words, farmers must make their drought abatement choices contingent on the capacity to irrigate.[5] When there is no drought or if the inputs fully abate damages from drought, then $d_s = 1$ and yields are equivalent to drought-free yields. If the drought is devestating, then $d_s = 0$ and the crop is lost ($Y_s = 0$).

The drought-abating inputs can be separated into a vector of inputs contingent on drought $\mathbf{X_s^D}$ and inputs that are not contingent on drought $\mathbf{X^D}$. Importantly, seed decisions occur well before planting, and thus drought-tolerant (DT) seed use is not state contingent (i.e., farmers must make this decision before observing the state of drought during the growing season). Other examples of state-independent drought-abating inputs are adjustments to planting dates, row spacing, seeding rates, and the installation or upgrade of irrigation equipment. These also include conservation practices that have long-term effects on soil organic matter and water-holding capacity, such as conservation tillage and cover crops. The set of state-contingent drought-abating inputs is somewhat limited but include increases in irrigation application rates, changes in late season herbicide use to reduce competition from weeds for water, and reductions in fertilizer applications to reduce water demand.

---

4. We assume that farmers consider output and input prices not to be state contingent.
5. To economize on notation in the remaining exposition, we do not explicitly index farmers' optimal choices by these irrigation parameters.

The farmer's objective is to maximize state-contingent profits, choosing whether or not to use drought-abating inputs

$$(3) \qquad\qquad \max_{x^D, x_s^D} W(\boldsymbol{\pi}),$$

where $W : \mathbb{R}^s \mapsto \mathbb{R}$ is a non-decreasing continuous function of a profit vector $(\boldsymbol{\pi} = \pi_1, \pi_2, \ldots, \pi_s)$ indexed by state $s \in S$. Farmers will choose to use a drought-abating input if

$$(4) \qquad\qquad W(\boldsymbol{\pi}^1) > W(\boldsymbol{\pi}^0),$$

where $\boldsymbol{\pi}^1 = \boldsymbol{\pi}(x^D = 1 \mid \mathbf{X^D}, \mathbf{X_s^D}; \mathbf{I})$ and $\boldsymbol{\pi}^0 = \boldsymbol{\pi}(x^D = 0 \mid \mathbf{X^D}, \mathbf{X_s^D}; \mathbf{I})$.

A number of preference functions, $W : \Pi^s \mapsto \mathbb{R}$, defined on the set of profit outcomes, Π with elements, π_s, have been used in the economics literature on choice under risk and uncertainty. We present two commonly used preference functions to motivate our empirical models. Such preferences give rise to empirical specifications suggesting that short-run drought shocks and long-run drought risk can influence farmers' economic decision making. Given that farmers' underlying preference structures are unknown and difficult to determine empirically, we estimate separate regression models.[6]

6.3.1   Mean-Variance Utility

The objective of a risk-facing farmer with risk-neutral preferences is simply to maximize expected profits. However, some empirical agricultural production models have shown utility-maximization models provide better fit than expected profit maximization (e.g., Lin, Dean, and Moore 1974). We choose the mean variance model with preference function

$$(5) \qquad\qquad W(\boldsymbol{\pi}) = v(\mu(\boldsymbol{\pi}), \sigma^2(\boldsymbol{\pi})).$$

Note that $v : (\mu, \sigma^2) \mapsto \mathbb{R}$ is a utility function, $\mu[\boldsymbol{\pi}] = \sum_{s=1}^{S} \rho_s \pi_s$, where π_s occurs with probability ρ_s and $\sigma^2[\boldsymbol{\pi}] = \sum_{s=1}^{S}(\pi_s - \sum_{s=1}^{S} \rho_s \pi_s)^2$. Substituting in the production function, $Y_s = Yd_s$, we can then choose a preference function such that

$$(6) \quad v^1 - v^0 = PY \exp\{\mu[\pi_s^1(d_s)] - \mu[\pi_s^0(d_s)] - (\sigma[\pi_s^1(d_s)] - \sigma[\pi_s^0(d_s)])\} - p^x.$$

The preference function defines the marginal utility of drought abatement.[7] With this preference function, a farmer will choose to use drought-abating

---

6. In earlier versions of this research that used drought severity indicators, we estimated the structural regression models implied by the below derivations. We found many of the structural parameters were individually significant and implied marginal effects similar to those from reduced-form weighted probit regressions of adoption on the drought severity indicators.

7. The preference function is increasing in output prices, drought free-yields, and mean drought abatement. The marginal utility of drought abatement is decreasing in the variance of drought abatement. As Chambers and Quiggin (2000) note, the variance can be replaced by any index for riskiness. Without loss of generality, we replace the variance with the standard deviation.

inputs if $v^1 - v^0 > 0$, i.e., if her utility from adopting these inputs exceeds her utility from not adopting them. Rearranging this inequality and applying the logarithmic transformation implies that

(7) $\ln P + \ln Y + \mu[\pi_s^1(d_s)] - \mu[\pi_s^0(d_s)] - (\sigma[\pi_s^1(d_s)] - \sigma[\pi_s^0(d_s)]) > \ln p^x.$

If the mean of utility (of profits) from adopting drought-abating inputs relative to the utility from not adopting is approximately zero, i.e. $\mu[\pi^1(d_s)] \approx \mu[\pi^0(d_s)]$, which would be expected to occur in long-run equilibrium, then (7) leads directly to an estimating equation once an operationalized measure of long-run drought risk is assumed. Using the standard deviation of the Palmer Modified Drought Index (PMDI) as our measure of drought risk, we can estimate

(8) $\beta_0 + \beta_Y \ln Y + \beta \sigma_{PMDI} > \varepsilon,$

where β_0 subsumes prices, β captures the effects of variability in profits from adopting drought-abating inputs relative to the variability in profits from not adopting drought-abating inputs, as underlying functions of drought (conceptualized here as states), and ε is an econometric error term.

6.3.2   Prospect Theory

Mean-variance models and other expected utility models have theoretical shortcomings and do not always reconcile with empirical evidence (e.g., Schoemaker 1982). For example, individuals often over-weight low probability states and under-weight high probability states (e.g., Kahneman and Tversky 1979). This can have important implications for farmers' decisions to use drought-abating inputs. In some regions, severe or extreme drought is rare, but if some farmers attach substantial importance to these rare but damaging occurrences, their adoption rates could be systematically higher than expected, assuming that farmers did not over-weight such events.[8] For these reasons, we also use the following preference function

(9) $W(\pi) = \sum_{s=1}^{S} h(\rho_s) u(\pi_s),$

where $h(.)$ is a probability weight function, and $u(.)$ is a utility function. Note this model assumes farmers know the probability of each possible state and systematically weight these probabilities. This model can be simplified by assuming that the marginal rate of substitution between two state-dependent profits is $MRS_{ss'} = [u(\pi_s^1) - u(\pi_s^0)] / [u(\pi_{s'}^1) - u(\pi_{s'}^0)] = w_s(\pi_s^1 - \pi_s^0) / w_{s'}(\pi_{s'}^1 - \pi_{s'}^0)$, where w_s is the preference weight a farmer assigns to profits in each state. The probability-weighted marginal utility of drought-abating inputs is $\omega_s(\pi_s^1 - \pi_s^0)$

---

8. Similarly, if the frequency, duration, and severity of droughts are expected to increase under unmitigated climate change, then adoption rates for those who systematically underweight such events may be lower, even as the objective probability of such occurrences increases over time.

$= h(\rho_s)w_s(\pi_s^1 - \pi_s^0)$, where ω_s is a weight that combines the probability and preference weights. Farmers will adopt drought-abating inputs if

$$(10) \qquad \sum_{s=1}^{S} \omega_s(\pi_s^1 - \pi_s^0) > 0.$$

Substituting in the production function, $Y_s = Yd_s$, farmers will adopt if

$$(11) \qquad PY\sum_{s=1}^{S} \omega_s(d_s^1 - d_s^0) > p^x,$$

where $d_s^1 \in (0,1)$ is the percent of yields not damaged by drought when using a drought-abating input, and $d_s^0 \in (0,1)$ is the percent of yields not damaged by drought when not using a drought-abating input. We do not observe farmers' preference weights on state-dependent outcomes and probabilities when making their abatement decisions. A common approach is thus to use the historically observed states to proxy for the states that farmers face when making such decisions. Here, we assume that farmers use their experience of realized states from previous years to form their set of possible states. States can be defined according to the drought severity categories from the US Drought Monitor, for example, such that

$$(12) \qquad PY\sum_{s=1}^{S}\sum_{t=1}^{T} \omega_s(d_s^1 - d_s^0)D_{st} > p^x,$$

where D_{st} is an indicator for drought in state s and year t. Assume that drought-related yield losses are reduced according to the function, $\exp(\sum_{s=1}^{S}\sum_{t=1}^{T}\omega_s(d_s^1 - d_s^0)D_{st})$.[9] Substituting this into (12) and applying a logarithmic transformation, we get the following inequality

$$(13) \qquad \ln P + \ln Y + \sum_{s=1}^{S}\sum_{t=1}^{T}\omega_s(d_s^1 - d_s^0)D_{st} - \ln p^x > 0.$$

This leads to an empirical equation that resembles

$$(14) \qquad \alpha_0 + \alpha_Y \ln Y + \sum_{s=1}^{S}\sum_{t=1}^{T}\alpha_{st}D_{st} > \varepsilon.$$

6.4   Empirical Strategy: Identification and Data

The motivating economic framework suggested that producers could consider short-run shocks and/or long-run risk when making drought abatement adoption decisions.[10] However, owing to the standard *ceteris paribus*

---

9. If there is no drought in any state ($D_{st} = 0$), then yields in all states are drought free (i.e., $Y_s = Y$).

10. Farmers may also consider short-run or seasonal climate forecasts when making agricultural decisions. Appendix B in the Supplementary Information (http://www.nber.org/data-appendix/c14693/SupplementaryInformation.pdf) contains a discussion of this point.

assumption in regression models, it would not be suitable to include both measures in the same specification. This is because long-run drought risk will not be plausibly constant if drought frequency, duration, and severity change annually during a half-decade period.[11] We therefore estimate separate specifications.

Using weighted linear probability (LPM) regressions, we first estimate the "mean-variance empirical model" of DT corn variety adoption on field i in year 2016 as

$$(15) \qquad DT_i = \beta_0 + \beta_{DR}\sigma_{PMDI,i} + \beta_{\bar{W}}\bar{W}_i + \beta_{Irr}Irr_i + \beta_{SL}SL_i + \varepsilon_i,$$

where DT_i is an indicator of DT seed variety adoption; $\sigma_{PMDI,i}$ is the standard deviation of the PMDI measure in July for field i in 2016; $\bar{W}_i$ is a vector of average temperature and precipitation conditions, and their standard deviations, over the 30 previous years (as a proxy for climate); Irr_i is a vector of variables denoting field irrigation capacity and their interaction terms; and SL_i is a vector of field-level soil characteristics, land attributes, and corn basis from nearby grain elevators for February 2016. Note that $\sigma_{PMDI,i}$ is a normalized measure of variability in natural soil moisture over the past century (Wallander et al. 2013). For purposes of model assessment, we compare estimated coefficients to average marginal effects from corresponding weighted probit regressions.[12]

Similarly, we estimate the following "prospect-theoretic empirical model" of DT corn variety adoption using weighted LPM regressions

$$(16) \qquad DT_i = \alpha_0 + f(\alpha_{st}, D_{st,i}) + \alpha_{\bar{W}}\bar{W}_i + \alpha_{Irr}Irr_i + \alpha_{SL}SL_i + \varepsilon_i,$$

where $f(\alpha_{st}, D_{st,i})$ is a simple linear function of drought incidence with severity s in year t.[13] We do not use drought severity indicators as suggested by our motivating framework because of multicollinearity concerns. Rather, we make use of two variables designed to capture the duration and severity of farmers' recently-experienced droughts: (1) the total number of months of severe, extreme, or exceptional droughts during years 2011–2015, and (2) the most intense drought experienced during the growing season (May–September) throughout 2011–2015.

11. In contrast, it is plausible to hold long-run drought risk constant while varying the first two moments of long-run average temperature and precipitation (and vice versa), which is done in our empirical application.

12. Under the assumptions of the linear probability model for binary response variables, the ordinary least squares (OLS) estimates of the coefficients are unbiased and consistent. Moreover, the LPM may be preferred to standard probit models when there are several discrete covariates, each with a limited number of values (Wooldridge 2010).

13. Farmer demographics, like education, as well as other dimensions that are more difficult to measure (e.g., risk preferences, broader human capital) are also expected to influence adoption (e.g., Wozniak 1987). We can link our surveyed fields to their operators using data from the 2016 ARMS Phase III, but the benchmark sample size declines by roughly 40 percent. However, our estimates of the impacts of long-run drought risk change only minimally after controlling for years of farming experience, indicators for education level (high school diploma, some college, and four-year college degree), and gender.

6.4.1 Identification and Causal Impacts of Climate in Cross-Sectional Regressions

Over the past two decades, a surge in the number of empirical studies relating weather and climate to economic outcomes has brought methodological improvements in econometric applications. For agricultural outcomes, some studies have analyzed cross-sectional data (e.g., Mendelsohn, Nordhaus, and Shaw 1994), while others make use of panel data (e.g., Deschênes and Greenstone 2007; Schlenker and Roberts 2009), or long [temporal] differences (Burke and Emerick 2016). Under certain conditions, Hsiang (2016) shows that, locally, the marginal effect of climate is equivalent to the marginal effect of weather in linear regression models estimated on data with repeat observations over time.

Credible identification of climate effects in cross-sectional analyses is particularly challenging because of omitted variables bias. Unmodeled factors that are correlated with outcomes (e.g., adoption of seed technologies, crop yields, farmland values, revenues) and one or more covariates will produce biased estimates. For example, cross-sectional regressions of farmland values on temperatures that omit irrigation access will have biased estimates because irrigation access is capitalized into valuation of farmland and is positively correlated with temperature (Schlenker, Hanemann, and Fisher 2005). Generally, climate is not random over large areas. In our application, there are several unmodeled, time-invariant factors correlated with climate and DT variety adoption that could confound identification of causal impacts: (i) input dealer recommendations, (ii) agricultural cooperative guidelines, (iii) ethanol plant contract terms, (iv) university extension guidance, (v) local USDA office practices, and (vi) other location institutions that could induce correlations in farmers' behavior over short distances.

One potential solution is to saturate the regression model with as many economically relevant covariates as possible. However, it is impossible to know if all important variables have been included; further, saturation can lead to overfitting and high standard errors in models with severe multicollinearity. For some economic outcomes (including many in agriculture), accumulated evidence from a variety of studies can provide useful guidance on whether or not important covariates have been excluded (Hsiang 2016).

Spatial first differences (SFD) have been proposed as a method for reducing concerns of omitted variables bias (Druckenmiller and Hsiang 2018). The framework rests on a fundamental premise of spatial statistics: observations closer together are more similar than those farther apart from one another. That is, the presence of unobserved location effects, c_i, may drive most or all of the similarities among observations at short distances. As such, the SFD estimator results from an OLS regression of differenced outcomes (y_i) on differenced covariates x_i between spatially-adjacent observations i and $i - 1$:

$$(17) \qquad y_i - y_{i-1} = (x_i - x_{i-1})\beta + (c_i - c_{i-1}) + (\varepsilon_i - \varepsilon_{i-1}),$$

where ε is a mean-zero i.i.d error term, β is a vector of parameters, and α is a vector that measures the influence of fixed effects.

Identification in the SFD framework rests on a Local Conditional Independence Assumption: $E[y_i \mid x_{i-1}] = E[y_{i-1} \mid x_{i-1}]$ for all $\{i,i\text{-}1\}$, i.e., the expected outcome is the same for two neighbors if they receive the same treatment. This identification condition is nearly equivalent to that assumed to hold in first-differenced time series models and is similar to those for certain differences-in-differences panel estimators and regression discontinuity designs. Importantly, the Local Conditional Independence Assumption is weaker than the Conditional Independence Assumption underlying a cross-sectional regression in which spatial fixed effects have been omitted: $E[y_i \mid x_j] = E[y_j \mid x_j]$ for all $i \neq j$, i.e., distant units are comparable in that the same expected outcome would result if both observations were treated with x_j—even though only j received x_j. In practice, the SFD model eliminates spatially correlated unobserved heterogeneity by filtering out the influence of factors that vary at short distances (not immediately adjacent observations) and differencing out common idiosyncratic influences at adjacent observations (Druckenmiller and Hsiang 2018).

To control for potentially important location fixed effects that cannot be estimated in the cross-sectional analysis, we re-estimate LPM versions of (15) and (16) using differenced data on DT corn seed variety adoption and differenced data for all covariates. Distinct from the application in Druckenmiller and Hsiang (2018), which differences county crop yields and covariates in the north-south direction (and then analyzes to what extent results are robust to differencing in the east-west direction), we define spatially adjacent fields to be fields that are closest to each other, regardless of direction. This assumes that the angle between neighboring fields is unimportant, which is plausible for our measures of adoption, weather, and climate.[14] In particular, adoption of drought-tolerant seed varieties appears to be spatially correlated, though there are not pronounced spatial effects in any cardinal direction (figure 6.4).

6.4.2   Data: Field-level Agricultural Production and Gridded Weather-Climate Observations

The Agricultural Resource Management Survey (ARMS) is our primary source of data.[15] ARMS is a survey conducted each year on farms' production practices (e.g., crop choice and rotations, tillage operations, conservation program enrollment and assistance), input use (e.g., GE trait choices, chemical applications, and irrigation use), soil and land characteristics, farm

---

14. To our knowledge, neither drought incidence (fig. 6.3) nor seed adoption (fig. 6.4) fall along strict geometric patterns. Although major precipitation events captured in our sample generally moved west-east, this directional component would be less important for our 30-year weather average calculations.

15. Appendix A in the Supplementary Information (http://www.nber.org/data-appendix /c14693/SupplementaryInformation.pdf) contains an overview of the data used in this analysis.

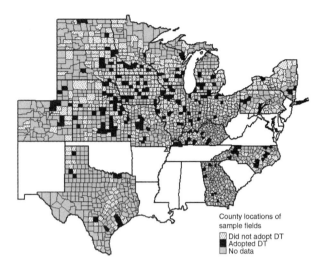

Fig. 6.4   US county adoption of DT corn, 2016

Note: All surveyed states in 2016 are depicted (Colorado, Georgia, Illinois, Indiana, Iowa, Kansas, Kentucky, Michigan, Minnesota, Missouri, Nebraska, New York, North Carolina, North Dakota, Ohio, Pennsylvania, South Dakota, Texas, and Wisconsin). Multiple fields per county appear in the data set frequently, and within these counties, there can be several fields planted with DT varieties. The maximum number of DT fields per county was six.

Source: USDA, Economic Research Service and National Agricultural Statistics Service, 2016 Agricultural Resource Management Survey.

business and household financial characteristics, and operator household demographics (US Department of Agriculture, Economic Research Service 2016). We use data from Phase II of the 2016 ARMS, which randomly sampled one corn field per farm in the survey.

The year 2016 is the first year in which ARMS Phase II surveyed farmers about their use of DT corn seeds, for both the 2015 and 2016 seasons. Using a "base" weight provided by USDA's National Agricultural Statistics Service (NASS), our sample expands to 73.3 million corn acres, representing just over 78 percent of 2016 US corn acreage (US Department of Agriculture, National Agricultural Statistics Service 2018b). Roughly 20 percent of 2016 sample corn acres were planted with DT varieties (table 6.1), similar to the national average of 22 percent (McFadden et al. 2019).

We merge drought-related weather and climate data with the ARMS data for the regression analysis. The US Drought Monitor produces expert-adjusted observational data through a collaboration of USDA, the National Drought Mitigation Center at the University of Nebraska-Lincoln, and the National Oceanic and Atmospheric Administration (NOAA).[16] The main

16. In order to produce drought estimates, the US Drought Monitor first relies on data from other drought and precipitation indices; fire risk data; satellite imagery of vegetation health; soil moisture data and model estimates; and hydrologic data. These are synthesized to develop

Table 6.1   Summary statistics (n = 1,768 fields)

	Units	Sample Mean	Weighted Mean	Standard Error	Min	Max
2016 DT adoption rate	{0,1}	0.21	0.20	0.02	0	1
Severe-or-greater drought duration	Months	4.78	4.45	0.44	0	45
Maximum drought intensity	Index	3.67	3.57	0.11	−0.56	7.57
Drought risk	[0,∞)	2.27	2.26	0.03	1.60	3.08
30-year temp. mean	°C	20.82	20.50	0.19	16.57	28.03
30-year temp. std. dev.	°C	1.42	1.43	0.01	0.84	1.80
30-year precip. mean	In.	4.02	4.11	0.04	1.91	5.69
30-year precip. std. dev.	In.	1.96	2.03	0.03	0.95	2.92
Irrigation	{0,1}	0.10	0.06	0.02	0	1
Irrigation × non-irr. corn share	[0,1]	0.03	0.02	0.005	0	0.99
Clay	{0,1}	0.13	0.15	0.02	0	1
Irrigation × clay	{0,1}	0.01	0.004	0.002	0	1
Irrigation × non-irr. corn share × clay	[0,1]	0.002	0.001	0.001	0	0.94
Highly erodible	{0,1}	0.14	0.17	0.02	0	1
Corn-soy soil index–mean	[0,10]	5.72	6.01	0.16	0.35	9.59
Corn-soy soil index–std. dev.	[0,,∞)	1.09	1.18	0.04	0	3.38
February 2016 basis	$(USD)	−0.12	−0.15	0.03	−0.80	1.13

Note: Estimates are expanded to the population of US corn fields in 2016 using a base expansion factor from NASS. Standard errors are clustered at the crop reporting district (CRD) level. There are 142 CRDs in the full sample with 12.5 fields per CRD, on average.

benefits of the US Drought Monitor are that it provides: (i) drought categories that are easy to interpret, and (ii) policy relevance.[17] However, the US Drought Monitor is subject to expert review and revision, which could introduce noise and reduce comparability across locations. We therefore base our classifications of severe, extreme, and exceptional droughts (the Drought Monitor's severity classes) on values of the PMDI, also produced by NOAA.[18] To construct field-level values of the drought measures, we interpolated PMDI data at the 12 nearest stations within 100 miles of sample fields (within the continental US) using inverse distance weighting. This

---

an initial weekly assessment of the percent area of each county that is in one or more of five drought categories, including an "abnormally dry" category. This initial assessment is sent to over 350 observers across the US. These observers include meteorologists, climatologists, hydrologists, and extension agents whose weather expertise and knowledge of local conditions inform their reports of drought impact. The initial assessment is subsequently revised to incorporate these experts' collective judgment (US Drought Monitor 2019).

17. USDA uses the US Drought Monitor as a guide to making disaster declarations and to help determine eligibility for certain types of loans. The index is also used by USDA's Farm Service Agency to help determine eligibility for the Livestock Forage Program. This program compensates eligible livestock producers for certain grazing losses, including losses resulting from qualifying drought conditions.

18. PMDI values are mapped into Drought Monitor severity categories as follows: −1.0 to −1.99 (D0), −2.0 to −2.99 (D1), −3.0 to −3.99 (D2), −4.0 to −4.99 (D3), and −5.0 to −5.99 (D4) (US Drought Monitor 2019).

assigns more weight to stations closer to our sample fields than more distant stations.

The PMDI itself is an "operational" version of the Palmer Drought Severity Index (PDSI), which quantifies the duration and intensity of long-term drought patterns. The PMDI uses hydrologic models to estimate net soil moisture based on precipitation recharge and losses from evapotranspiration, infiltration, and runoff. The PMDI introduced the concepts of severe, moderate, and extreme drought based on PDSI values (Heddinghaus and Sabol 1991; Wallander et al. 2013). The fields in our sample experienced abnormally dry conditions (D0) much more frequently than droughts across all years. However, nearly two-thirds of fields experienced a severe or extreme drought at some point in 2012. More broadly, the 2012 drought affected two-thirds of all land in the US, causing agricultural losses estimated at $30 billion (Rippey 2015). By contrast, the 2013 drought was mainly severe for fields located in western counties. On average, fields experienced roughly four to five months of severe-or-worse droughts, in total, across the corn growing seasons of 2011–15. Similarly, many of the fields were at risk of experiencing a drought of any severity in any given year; the weighted mean of the standard deviation of long-run PMDI was 2.26 (table 6.1).

Yearly temperature and precipitation are taken from Oregon State University's PRISM Climate Group, from which we construct 30-year averages. The PRISM data use point observations, a digital elevation model, and other spatial data sets to generate estimates of climatic parameters (e.g., temperature and precipitation) at 4×4 km grids (Daly, Neilson, and Phillips 1994). For each of the 4 km grids, we aggregate to the county based on distance to county centroids, clipped and weighted by cropland density. To avoid the possibility that 2015 weather averages might correlate with our measures of drought, drought risk, and other idiosyncratic effects in 2015, we calculate yearly averages from monthly observations between 1985 and 2014. Only months in the corn growing season were used. Nationally, the average 30-year growing season temperature was 20.5°C (69°F), with mean precipitation of 4.1 inches (table 6.1).

Our ARMS-based field-level measure of one aspect of irrigation capacity, whether the field is irrigable, is a variable indicating if the corn crop was irrigated in that year. Only 6 percent of corn acres in our sample are irrigated, consistent with the fact that the vast majority of DT corn was grown on non-irrigated cropland in 2016, even among states that are major corn irrigators (McFadden et al. 2019). Our second measure of irrigation capacity is irrigation water availability, though we must proxy for this in the absence of spatially detailed, representative data on surface water and groundwater availability. We represent this as the share of each county's harvested corn acreage that was not irrigated, using data from the 2012 Census of Agriculture (U.S. Department of Agriculture, National Agricultural Statistics Service 2013). We expect large shares of non-irrigated acreage to signal a general lack of water availability, but because availability only matters for irrigable fields, we prefer to include the [irrigable × water availability] interaction term and omit the proxy in levels. Nationally, approximately 86 percent of harvested corn acreage in 2012 was not irrigated.

We also take certain field-level indicators of soil type and quality from the ARMS data. Roughly 15 percent of the sample acreage had a primarily clay soil type. Given that clay soils are better at retaining water than other soil types, we might expect lower DT corn adoption under adequate soil moisture. Further, the effect of soil type on DT corn adoption may depend on both measures of irrigation capacity, and so we include both interaction terms, [irrigable × clay] and [irrigable × water availability × clay]. In contrast, adoption on highly erodible soils might be higher due to a combination of steeper slopes and greater potential for soil detachment, which can contribute to soil moisture loss. Roughly 17 percent of sample acres are highly erodible.

We control for other unobserved countywide land productivity attributes through use of USDA's National Commodity Crop Productive Index (NCCPI) for corn and soybeans (Dobos, Sinclair Jr., and Robotham 2012). After re-scaling, these index values lie in [0,10], with higher values indicating higher inherent soil suitability for growing corn and soybeans.[19] The NCCPI values are available at the geographic level of a "map unit" polygon, of which there are several thousand per US state. These polygon values are then aggregated to 30 m cells as part of USDA's Gridded Soil Survey Geographic Database (US Department of Agriculture, Natural Resources Conservation Service 2018). Our corn-soy soil productivity index is created by averaging over values of NCCPI for all 30 m cells within 3 km of each field's location. As expected, most fields are suitable for growing corn, with an average county value of 6.01, though there is significant variation in soil quality within our sample.

National seed premiums for DT corn variety traits were roughly $10 per bag of 80,000 seeds in 2016 (McFadden et al. 2019). Although some evidence suggests that DT seed premiums are lower in eastern states, where drought is less common, farm- or state-level data for DT corn premiums are not widely available (Farmers Business Network 2018). To proxy for the price of corn, we use a field-matched measure of basis, which can be thought of as a location-adjusted net corn price (Barr et al. 2011). In particular, we subtract the February 2016 cash price from the March futures price for the December contract. February is chosen as this is the time operators of most US corn

---

19. USDA calculates this index as the product of ratings from five input category subrules: chemical (e.g., soil pH, cation exchange capacity, organic matter); water (e.g., available water-holding capacity, precipitation during the growing season); physical (e.g., saturated hydraulic conductivity, rock fragments); climate (e.g., frost-free days, total precipitation); and landscape (e.g., slope gradient). Interested readers should consult Dobos, Sinclair Jr., and Robotham (2012) for more information.

acreage make choices of seed varieties to plant. We assign cash spot prices to the nearest ARMS fields using USDA-collected cash grain bids data.[20] The weighted average basis in our sample is −$0.15.

6.5   The Role of Drought Exposure and Climate on Adoption of Drought-Tolerant Corn Varieties

We first discuss the LPM regression results for the models motivated by both behavioral assumptions and compare them with benchmark probit results. Next, we examine the extent of spatial variation in the data and the effects of controlling for potentially confounding locality factors through SFD estimation. Last, we provide an illustrative discussion of the role of climate change attitudes, beliefs, and potential barriers to adoption.

6.5.1   Benchmark Adoption Estimates

We find that long-run drought risk is positively associated with farmers' adoption of DT varieties, consistent with expectations and evidence about the geographic regions in which these technologies were first introduced (table 6.2, column 1a). Controlling for climate effects, a one-standard-deviation increase in 100-year drought risk leads to increased adoption by 10.6 percentage points, a large but insignificant effect.[21] Thus, long-run drought risk does not appear to have significant predictive content after controlling for 30-year temperature and precipitation means and standard deviations. However, an additional 1°C increase in average climate conditions increases adoption by 2.4 percentage points. Operators of fields that are of inherently lower productivity—those that are highly erodible—are more likely to have planted DT varieties.

Irrigation has an insignificant main-level effect with the expected negative sign, though its impact varies by soil type. Specifically, the effect of irrigation on adoption is roughly 66 percentage points higher on fields with primarily clay soils located in areas with very high shares of irrigated acreage. But as irrigation water becomes incrementally less available (as proxied by a marginal increase in the county's share of irrigated acreage), DT corn adoption on irrigable fields is higher for non-clay fields but lower for clay fields. However, both interaction terms involving the water availability proxy are insignificant.

The LPM results are robust to the exclusion of the climate controls

---

20. We first removed outliers from the basis data and then fit an inverse distance weighted surface to the remaining points. We looked for the nearest five purchasers (e.g., grain elevators) within a maximum of 70 miles from the field. This is generally the greatest distance growers will truck grain. Roughly 97 percent of fields were within 70 miles of one or more grain purchasers. Among these fields, only 146 had fewer than five purchasers within the 70-mile radius. The average distance to a purchaser was just over 16 miles.

21. For our sample, this would be roughly similar to an area with moderate drought risk becoming an area with severe drought risk.

Table 6.2 Mean-variance-based model: weighted linear probability model estimates and probit average marginal effects

	LPM (marginal effects)			Probit (average marginal effects)		
	Drought risk & climate (1a)	Drought risk only (2a)	Climate only (3a)	Drought risk & climate (1b)	Drought risk only (2b)	Climate only (3b)
Drought risk	0.106 (0.076)	0.197** (0.078)		0.103 (0.077)	0.198** (0.079)	
30-year temp. mean	0.024** (0.011)		0.026** (0.011)	0.024** (0.010)		0.026** (0.010)
30-year temp. std. dev.	0.097 (0.214)		0.165 (0.205)	0.134 (0.208)		0.193 (0.201)
30-year precip. mean	-0.075 (0.059)		-0.097 (0.062)	-0.072 (0.061)		-0.095 (0.060)
30-year precip. std. dev.	0.103 (0.082)		0.138* (0.081)	0.096 (0.083)		0.134* (0.081)
Irrigation	-0.117 (0.075)	-0.076 (0.069)	-0.091 (0.071)	-0.092* (0.055)	-0.068 (0.058)	-0.076 (0.057)
Irrigation × non-irrigated corn share	0.105 (0.126)	0.073 (0.122)	0.073 (0.122)	0.092 (0.132)	0.067 (0.127)	0.067 (0.130)
Clay	0.011 (0.037)	0.011 (0.036)	0.008 (0.037)	0.011 (0.037)	0.010 (0.036)	0.008 (0.037)

Irrigation × clay	0.659**	0.629**	0.675**	0.639***	0.613***	0.653***
	(0.269)	(0.270)	(0.268)	(0.205)	(0.225)	(0.191)
Irrigation × non-irrigated corn share × clay	-0.747	-0.674	-0.749	-0.569	-0.505	-0.579
	(0.591)	(0.588)	(0.592)	(0.478)	(0.481)	(0.480)
Highly erodible	0.073*	0.078**	0.075*	0.072*	0.077**	0.073*
	(0.038)	(0.038)	(0.038)	(0.037)	(0.038)	(0.038)
Corn-soy soil index mean	-0.006	-0.001	-0.006	-0.005	-0.001	-0.006
	(0.008)	(0.007)	(0.008)	(0.008)	(0.007)	(0.008)
Corn-soy soil index std. dev.	0.016	0.019	0.006	0.017	0.019	0.008
	(0.027)	(0.027)	(0.025)	(0.027)	(0.027)	(0.025)
February 2016 basis	0.073	0.109**	0.049	0.086	0.112**	0.061
	(0.082)	(0.049)	(0.082)	(0.082)	(0.050)	(0.081)
Constant	-0.541	-0.257	-0.423			
	(0.384)	(0.187)	(0.369)			
Observations	1,768	1,768	1,768	1,768	1,768	1,768
Correctly classified (%)	79	79	79	79	79	79
F-statistic	2.37***	2.75***	2.33***	2.37***	2.61***	2.36***
R-squared	0.03	0.02	0.03			

Note: Estimates are expanded to the population of US corn fields in 2016 using a base expansion factor from NASS. Standard errors in parentheses are clustered at the crop reporting district (CRD) level. For the LPM estimates, all predicted probabilities were within the unit interval. Goodness-of-fit statistics for the probit regressions are with respect to the fitted model. Significance is denoted as *** $p < 0.01$, ** $p < 0.05$, and * $p < 0.10$.

(table 6.2, column 2a) with few exceptions. The marginal effect of long-run drought risk nearly doubles to 0.197 and becomes significant at the 5 percent significance level, absorbing much of the influence of the excluded moments of the 30-year weather distribution. Somewhat surprisingly, February basis (our output price control) becomes significant at the 5 percent level. This suggests a degree of collinearity between basis and our gridded climate data.

Our set of results are also robust to dropping the long-run drought risk index entirely (column 3a), though the quality of fit somewhat changes. The 30-year temperature average and precipitation variability increase the probability of adoption, with the latter having a large 13.8 percentage point effect. That average precipitation remains insignificant (though with the expected sign) echoes a major finding in the climate econometrics literature on crop yields: average precipitation effects are relatively small and variable (e.g., Schlenker and Roberts 2009; Burke and Emerick 2016). One distinction, however, is that our study relies on drought exposure, climate, and adoption data at a finer spatial scale than the county.

As expected, the average marginal effects from the probit model (columns 1b–3b) are very similar to the partial effects identified in the LPM regressions. Differences between the two sets of estimates imply differences in impacts on adoption rates that are less than 1 percentage point, on average. Given that our focus is on identification of drought exposure and climate effects, rather than out-of-sample prediction, it is of less importance if predicted probabilities from the LPM regressions lie outside [0,1], although this did not occur for any of the regression specifications.

The duration and severity of recent droughts do not appear to affect seed technology adoption (table 6.3). Although the total duration (in months) of severe, extreme, or exceptional droughts has the expected sign, its magnitude is near zero and insignificant. Similarly, the severity variable suggests that adoption is roughly 2 percentage points higher as the five-year maximal PMDI increases by one index value (columns 2a and 2b), though these estimates are also insignificant. The switch in signs on this variable between the "drought shocks & climate" regressions relative to the "shocks only" regressions is suggestive of substantial collinearity with the temperature and precipitation climate variables. Though to some extent surprising, this null result is in agreement with other work suggesting climatic effects of temperature and rainfall explain most of the variation in corn yields, though with some remaining effects due to drought (e.g., Kuwayama et al. 2019).

Consistent with expectations, increased average temperatures and variability of rainfall lead to higher adoption, though the latter is statistically insignificant. Similarly, adoption rates are higher on highly erodible land. As with the estimates in table 6.2, irrigation leads to lower adoption, with point estimates that are comparable but roughly 1–2 percentage points lower and again insignificant. Irrigation effects are also moderated by the dominance

Table 6.3 Prospect-theory-based model: weighted linear probability model estimates and probit average marginal effects

	LPM (marginal effects)		Probit (average marginal effects)	
	Drought shocks & climate (1a)	Drought shocks only (2a)	Drought shocks & climate (1b)	Drought shocks only (2b)
Severe-or-greater drought duration	0.002	0.002	0.002	0.002
	(0.004)	(0.004)	(0.004)	(0.004)
Maximum drought intensity	−0.006	0.022	−0.006	0.022
	(0.018)	(0.017)	(0.017)	(0.017)
30-year temp. mean	0.027**		0.027**	
	(0.011)		(0.011)	
30-year temp. std. dev.	0.183		0.208	
	(0.220)		(0.213)	
30-year precip. mean	−0.096		−0.094	
	(0.060)		(0.060)	
30-year precip. std. dev.	0.136		0.131	
	(0.085)		(0.083)	
Irrigation	−0.093	−0.079	−0.080	−0.073
	(0.083)	(0.074)	(0.064)	(0.061)
Irrigation × non-irrigated corn share	0.077	0.059	0.074	0.063
	(0.125)	(0.122)	(0.134)	(0.132)
Clay	0.007	0.010	0.007	0.009
	(0.038)	(0.037)	(0.038)	(0.036)
Irrigation × clay	0.679**	0.649**	0.656***	0.633***
	(0.270)	(0.272)	(0.189)	(0.207)
Irrigation × non-irrigated corn share × clay	−0.757	−0.664	−0.586	−0.511
	(0.597)	(0.585)	(0.479)	(0.484)
				(continued)

Table 6.3 (cont.)

	LPM (marginal effects)		Probit (average marginal effects)	
	Drought shocks & climate (1a)	Drought shocks only (2a)	Drought shocks & climate (1b)	Drought shocks only (2b)
Highly erodible	0.073*	0.080**	0.072*	0.080**
	(0.038)	(0.039)	(0.037)	(0.039)
Corn-soy soil index mean	−0.007	−0.001	−0.006	−0.001
	(0.008)	(0.008)	(0.008)	(0.008)
Corn-soy soil index std. dev.	0.006	0.007	0.009	0.008
	(0.026)	(0.027)	(0.026)	(0.027)
February 2016 basis	0.057	0.036	0.070	0.037
	(0.084)	(0.045)	(0.083)	(0.045)
Constant	−0.456	0.107		
	(0.401)	(0.067)		
Observations	1,768	1,768	1,768	1,768
Correctly classified (%)	79	79	79	79
F-statistic	2.04**	1.60	2.06**	1.50
R-squared	0.03	0.02		

Note: Estimates are expanded to the population of US corn fields in 2016 using a base expansion factor from NASS. Standard errors (not shown) are clustered at the crop reporting district (CRD) level. Goodness-of-fit statistics for the probit regressions are with respect to the fitted model. Significance is denoted as *** $p < 0.01$, ** $p < 0.05$, and * $p < 0.10$.

of clay in the topsoil profile, though interactions with the proxy for water availability remain insignificant.

6.5.2   Spatial Differences Estimates: Controlling for Unobserved Location-Specific Heterogeneity

As a precursor to SFD estimation, we first examine variation in the differenced data (table 6.4). We choose thresholds corresponding to the 75th percentile (14 miles), 50th percentile (9 miles), and 25th percentile (6 miles) of the distribution of distance between fields. Since we are differencing data for each field from those at the next closest field, many of the means are zero or close to zero.[22] There is no broadly meaningful interpretation of small positive or negative means because of this spatial, rather than temporal, differencing.

Relative to their means, many of the differenced drought and climate data have large standard deviations. Across the three distance quartiles, the range of the absolute value of the coefficient of variation (CV) for the drought duration measure is [49, 142], with a similar CV range for the intensity measure: [31, 182]. Variability in the 30-year climate regressors is somewhat lower. Clay content and the indicator of high erodibility have comparatively large variation, as expected, while the interaction terms, corn-soy soil productivity index, and basis have among the least variation in the differenced data.

We also computed correlation coefficients for each covariate for observations matched with its closest other observation. If we observe relatively high correlations in all of the covariates of interest at close distances, this could suggest (but does not prove) that location fixed effects—if uncontrolled for—could bias estimation in levels. However, even if such effects are absent, spatially differencing the data may help to reduce codependence among fields in our data. Consistent with the summary statistics in table 6.4, many of the weather and climate measures are highly correlated, even when the nearest-neighbor field is roughly 15 miles away. At this distance threshold, correlations in drought risk and 30-year temperature averages are in excess of 0.97, and this correlation increases to over 0.99 for fields located to their closest other fields less than five miles away. A similar pattern holds for basis. For these reasons, we do not estimate the SFD model with these highly spatially dependent data.

There is significant variation in distances between our sample points (figure 6.5), and there are few fields whose nearest field exceeds 30 miles, generally the average north-south distance of major US corn growing coun-

---

22. With first-differenced time series data, a single observation in levels generally appears in differencing calculation exactly *twice* (e.g., subtract year 2014 from year 2015, subtract year 2015 from year 2016). There are 30 fields in our levels data set that are used more than twice to construct the differenced data (e.g., for fields A, B, and C, the closest field is Field D). Our results are robust to their exclusion.

Table 6.4   Summary statistics, spatial first differenced data

	Mi. < 14		Mi. < 9		Mi. < 6	
	Mean	Std. Dev.	Mean	Std. Dev.	Mean	Std. Dev.
2016 drought-tolerant corn adoption	0.014	0.516	0.027	0.532	0.045	0.534
Severe-or-greater drought duration	−0.042	2.037	−0.013	1.846	0.011	1.465
Maximum drought intensity	0.013	0.403	0.002	0.363	−0.005	0.278
Drought risk	0.001	0.066	0.002	0.053	−0.002	0.039
30-year temp. mean	−0.003	0.360	−0.003	0.284	0.008	0.208
30-year temp. std. dev.	0.001	0.022	0.001	0.020	0.001	0.020
30-year precip. mean	−0.004	0.114	−0.004	0.102	0.003	0.097
30-year precip. std. dev.	−0.004	0.091	−0.003	0.082	0.001	0.075
Irrigation	0.006	0.246	0.007	0.249	0.008	0.231
Irrigation × non-irrigated corn share	0.008	0.160	0.008	0.157	0.011	0.155
Clay	−0.005	0.449	−0.031	0.435	−0.056	0.439
Irrigation × clay	−0.002	0.080	−0.007	0.080	−0.011	0.105
Irrigation × non-irrigated corn share × clay	0.0001	0.040	−0.002	0.029	−0.003	0.038
Highly erodible	0.002	0.445	0.010	0.453	0.011	0.443
Corn-soy soil index mean	−0.007	0.080	−0.010	0.073	−0.009	0.065
Corn-soy soil index std. dev.	0.001	0.042	−0.0003	0.041	0.0003	0.038
February 2016 basis	0.002	0.039	−0.001	0.029	0.0002	0.023

Note: For all regressors except the two soil index variables and February 2016 basis, the sample sizes are the following: 929 fields (mi. < 14), 615 fields (mi. < 9), and 358 fields (mi. < 6).

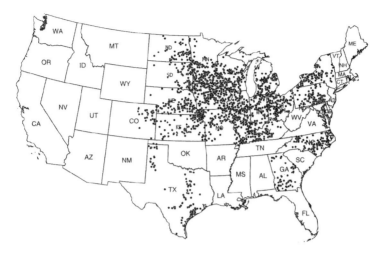

Fig. 6.5   Surveyed fields' locations, 2016

Note: To preserve privacy and meet USDA data disclosure requirements, dots representing field locations are disproportionately large relative to map scale.

Source: USDA, Economic Research Service and National Agricultural Statistics Service, 2016 Agricultural Resource Management Survey.

ties (Druckenmiller and Hsiang 2018). The differencing process results in samples that get increasingly smaller (925, 613, and 356 fields), with an accompanying decline in the models' degrees of freedom. For purposes of comparison, we estimate the same model in level terms, adding back in the climate and price variables that were removed in the SFD estimation, for the samples that would have resulted had we not spatially differenced (with comparable sample sizes). As in the LPM and probit results, all standard errors have been clustered at the level of USDA's crop reporting districts.[23]

For the SFD model estimated at the 14-mile distance, the drought-specific variables are again insignificant (table 6.5, columns 1a and 1b). At this distance, it is unclear whether or not the location effects have been completely differenced out, while the model is estimated with fewer observations—and that exhibit less variation. Regarding other attributes, irrigation and its interaction with clay are significant, while the presence of highly erodible land is insignificant. The corn-soy soil productivity index is now significant at the 5 percent level and suggests that relative to fields which are completely unsuitable for growing corn, DT varieties are adopted at rates 77 percentage

23. Crop reporting districts (CRDs) are groupings of counties that have similar geography, climate, and cropping practices (US Department of Agriculture, National Agricultural Statistics Service 2018a). Most states in our sample are each divided into nine adjacent CRDs. Clustering at this level is designed to mitigate the effects of spatial autocorrelation in the econometric errors.

Table 6.5   Spatial first difference estimates at the 75th, 50th, and 25th percentiles of distance between fields

	SFD			Levels		
	Mi. < 14 (1a)	Mi. < 9 (2a)	Mi. < 6 (3a)	Mi. < 14 (1b)	Mi. < 9 (2b)	Mi. < 6 (3b)
Severe-or-greater drought duration	−0.008 (0.018)	−0.022 (0.028)	−0.042 (0.049)	−0.0003 (0.005)	−0.002 (0.005)	−0.003 (0.005)
Maximum drought intensity	−0.025 (0.057)	−0.095* (0.057)	−0.072 (0.121)	0.009 (0.028)	−0.002 (0.027)	0.026 (0.032)
-year temp. mean				0.039** (0.017)	0.044** (0.018)	0.050** (0.021)
-year temp. std. dev.				0.338 (0.321)	0.460 (0.380)	0.680* (0.372)
-year precip. mean				0.045 (0.078)	0.031 (0.092)	0.052 (0.100)
-year precip. std. dev.				−0.028 (0.107)	−0.007 (0.122)	0.038 (0.148)
Irrigation	−0.293** (0.148)	−0.318 (0.232)	−0.150 (0.286)	−0.067 (0.139)	−0.192** (0.083)	−0.116 (0.112)
Irrigation × non-irrigated corn share	0.315 (0.244)	0.279 (0.305)	0.091 (0.373)	−0.009 (0.186)	0.052 (0.145)	−0.103 (0.130)
Clay	−0.095 (0.057)	−0.111 (0.078)	−0.263** (0.126)	−0.040 (0.029)	−0.020 (0.041)	−0.023 (0.048)

Irrigation × clay	0.714*	0.964***	1.083**	0.375		0.912***
	(0.366)	(0.318)	(0.483)	(0.481)		(0.130)
Irrigation × non-irrigated corn share × clay	−0.883	−1.661**	−1.582	0.358		
	(0.610)	(0.669)	(0.988)	(0.560)		
Highly erodible	0.058	0.111	0.071	0.059	0.119**	0.046
	(0.063)	(0.074)	(0.071)	(0.045)	(0.054)	(0.064)
Corn-soy soil index mean	0.768**	0.406	0.219	−0.036	−0.057	−0.035
	(0.307)	(0.326)	(0.604)	(0.117)	(0.147)	(0.165)
Corn-soy soil index std. dev.	0.509	0.016	0.485	−0.267	−0.421	−0.899***
	(0.457)	(0.427)	(0.675)	(0.283)	(0.305)	(0.335)
February 2016 basis				0.061	0.083	
				(0.123)	(0.139)	
Constant	−0.027	−0.031	−0.032	−1.120*	−1.400*	−1.845**
	(0.023)	(0.027)	(0.050)	(0.651)	(0.720)	(0.702)
Observations	925	613	356	926	614	357
Correctly classified (%)	79	81	83	80	80	83
R-squared	0.03	0.04	0.08	0.03	0.06	0.07
Clusters	113	99	91	113	99	91

Note: Estimates are expanded to the population of US corn fields in 2016 using a base expansion factor from NASS. Standard errors (not shown) are clustered at the crop reporting district (CRD) level. Linear probability models in (1b), (2b), and (3b) are estimated for the regressions in level terms. The R-squared values for the SFD estimates are analogous to the within R-squared from a fixed effects panel estimator, while the R-squared for the levels models (LPMs) are overall R-squared values. Irrigation-related interaction terms are dropped in the specifications of the last two columns due to multicollinearity. Significance is denoted as *** $p < 0.01$, ** $p < 0.05$, and * $p < 0.10$.

points higher among fields that are deemed to have the highest suitability for corn.

Several of the estimated coefficients in the SFD model for the data set at the 9-mile threshold (2a) are qualitative similar to those obtained in the 14-mile sample (1a), as well as level estimates at the 9-mile threshold (2b). The drought duration variable remains small and insignificant, although the drought intensity variable is now marginally significant. There is broad consistency in the signs and sizes of the soil coefficients between (1a) and (2a), though the larger effect of the irrigation-clay interaction term is more precisely estimated in (2a). Less variation in the corn-soy soil productivity index at smaller distances contributes to the smaller coefficient estimate and insignificance.

Many of the SFD point estimates for fields with neighboring observations that are at a distance of 6 miles or less (3a) are similar to counterparts at greater distances. By differencing out potentially bias-inducing spatial fixed effects, the impacts of the weather and climate effects are magnified, especially as seen in (3b). The restricted sample size is also contributing to some of the increase in magnitude of these estimates. Given the novelty of the spatial first differences approach, the consequences of restricting samples to observations that could be "too close" (e.g., using other measures of spatial dependence, like Moran's I statistic) have not been fully explored.

6.5.3   Climate Change Beliefs, Attitudes, and Barriers to Adoption

Recent evidence suggests that the 2012 drought did not significantly affect agricultural advisors' climate change beliefs or adaptation attitudes. However, advisers indicated greater concern about risks from pests and drought arising from 2012 yield damages (Carlton et al. 2016). Against the current backdrop of learning requirements for progressing technologies and evolving drought risk (Lybbert and Bell 2010), further research is needed to assess the potential of drought-tolerant crop varieties to serve as adaptation mechanisms. Although substantial field trial evidence confirms that DT corn has higher average yields relative to non-DT controls under water-stressed growing conditions (e.g., Gaffney et al. 2015; Nemali et al. 2015; Adee et al. 2016), this is not necessarily proof of its adaptation efficacy. Adoption of DT corn is an adaptation tool only if the economic gains from these varieties relative to conventional varieties under a changing climate are larger than the gains under a constant climate (Lobell 2014).

Since the adoption data are only available for 2016, we cannot completely isolate short-run drought effects from circumstances surrounding the 2012 or adjacent-year droughts. This is of less concern if shocks similar in magnitude, timing, and duration to the 2012 drought are not exceptionally rare events, which seems unlikely under climate change. The ideal data set, though, would contain repeated field-level observations over time, which would allow us to distinguish short-run effects from recent-year effects.

However, to the extent that long-run drought risk will not lessen in coming years, our analysis suggests that widespread disadoption of DT corn under current economic, technology, and policy environments is unlikely.

Moreover, we do not formally identify barriers to early DT corn adoption. In the context of sub-Saharan Africa, Fisher et al. (2015) find that poor market access, seed and labor availability, and lack of information about the new varieties hampered adoption of DT corn in 2012–13. Such factors are unlikely to significantly impact adoption in the US, where there are far fewer impediments to flows of seed, labor, capital, and information. Participation in federal crop insurance is also unlikely to affect adoption in the US (Weber, Key, and O'Donoghue 2016; McFadden et al. 2019). However, seed premiums and farmers' perceptions of possible yield penalties could have decreased early adoption rates. We do not formally test this claim due to lack of price data. Rather, our results suggest US DT corn adoption is mainly influenced by—and is likely to continue to be influenced by—drought exposure and risk.[24]

6.6   Conclusions and Policy Implications

Controlling for average weather conditions and the severity and duration of recent droughts, in addition to potentially confounding location fixed effects, we find that adoption of DT corn is generally increasing in long-run average temperatures and drought risk. By contrast, adoption does not appear to be significantly affected by the severity of the worst drought experienced in the field in the previous five years, nor the duration of severe-or-worse droughts during this same period. Farmers are more likely to plant DT corn on highly erodible fields, and they are also less likely to irrigate on these fields, though the impact of irrigation on adoption is affected by soil type (i.e., clay-dominant particles).

There are three main caveats of this research. First, use of cross-sectional data has limited our ability to fully tease out the effects of field- and farm-specific factors on adoption. We have included a large set of covariates suggested by theory and past empirical evidence, though some relevant predictors may be absent from our regressions—as is the case with nearly all cross-sectional analysis. Second, there are no widely available data at fine spatially varying scales for DT corn price premiums relative to prices of other corn varieties. Several empirical studies of climate change and agriculture assume an underlying market equilibrium such that prices are constant

---

24. The pace of DT corn adoption could increase if private seed companies increasingly combine drought tolerance with HT and/or Bt traits. Increased stacking could result from companies selling additional varieties with all three traits, or through a gradual market withdrawal of HT and/or Bt varieties that are not drought tolerant. The latter type of marketing strategy has been used for some types of farm machinery and technologies that—initially introduced as options—gradually become standard equipment.

across space (Blanc and Reilly 2017). Such an assumption is not unrealistic for integrated markets in the US, but the existence of thousands of agricultural cooperatives, seed retailers, and other agricultural input dealers suggests the likelihood of at least modest variability in seed prices across the country (e.g., Shi, Chavas, and Stiegert 2010).[25] Third, the form of our analysis does not provide rigorous evidence about how uptake of drought-tolerant crop varieties could evolve under climate change. This type of analysis would make use of drought severity projections based on output from appropriately downscaled general circulation models (Hsiang and Kopp 2018). Much progress has been made recently to forecast drought severity into the 21st century using climatological models (e.g., Cook, Ault, and Smerdon 2015), but more research is necessary to generate a set of robust drought projections for use in climate economics studies.

A number of general policy implications emerge from our analysis. We find a negative relationship—though not generally statistically significant—between irrigation use and DT corn adoption, which is consistent with the notion that these technologies may not be complementary. Although irrigable fields are currently less likely to be planted to DT corn varieties, this relationship could fundamentally change as irrigation water sources become scarcer under a worsening climate. Faced with increasingly drier conditions, there could be some threshold beyond which irrigators would choose to adopt DT varieties if they perceive them to be a cost-effective way of limiting yield losses from mild-to-moderate droughts. However, if such a case were to occur, it is unknown if DT varieties would be chosen as a way to help preserve irrigation water availability (via the crop's improved water use efficiency) or if they would be chosen as a "last resort" option due to perceptions of yield penalties. In either case, decision makers could consider availability and use of crops with greater water use efficiency when designing agricultural water policy.

In a similar vein, it is possible that US farmers are using DT corn varieties as a kind of shallow loss insurance. That these varieties can serve as shallow loss insurance that complements conventional crop insurance policies is supported by the fact that: (1) drought tolerance does provide some yield loss protection against mild-to-moderate droughts, and (2) greater shares of DT corn fields than non-DT corn fields are insured under the US federal crop insurance program (McFadden et al. 2019). Regardless of mechanism, our results suggest adoption increases with average temperature and drought

25. Although evidence suggests that some spatial variability in drought-tolerant seed prices exist (e.g., Farmers Business Network 2018), the extent to which this input price variability is tightly linked to underlying agricultural productivity is unclear. By leaving out input prices, we minimize the risk of bias from bad control (Angrist and Pischke 2008). However, we present results of a robustness check of our levels estimates to inclusion of state-level DT corn premiums in appendix C of the Supplementary Information (http://www.nber.org/data-appendix /c14693/SupplementaryInformation.pdf).

risk, which are both projected to increase under multiple climate scenarios. In the short run, optimal crop insurance premiums may need to be adjusted to account for the crop's greater resistance to drought, though these adjustments would likely need to be revisited as droughts intensify and become more prolonged.

Despite the positive linkages between temperature, drought risk, and adoption, farmers' learning about the net benefits of DT corn varieties could be hampered by random weather shocks during the growing season (Lybbert and Bell 2010). For example, a farmer who chooses to plant DT corn, faces a substantially water-stressed growing season, and then suffers crop failure may believe that the DT variety or all similar varieties are ineffective even though they were not bred to thrive in such conditions. Conversely, a farmer who plants a DT corn variety could face routine weather conditions and a subsequently poor yield realization due to an unrelated factor (e.g., nutrient deficiency, pest infestation) but misattribute the outcome to a DT yield penalty. These scenarios suggest farmers' learning about DT seed efficacy is more challenging than that for other variable inputs (e.g., labor, fertilizer, energy, pesticides), reiterating the importance of extension programs. While these characteristics do not demonstrate a new role for extension, they highlight the need for clear communication of the idea that several years of data on DT seed use and experimentation may be required before farmers are able to make a fully informed "final" determination of appropriateness of the technology for their operation.

A prime area for further research is a causal analysis of how weather shocks impact DT corn yields and net returns relative to their non-DT counterparts. The main private-sector companies selling DT corn varieties have published field trials demonstrating efficacy of the seeds under water-limited growing environments. Many but not all of these results have been replicated by university researchers on test plots, but there have been no studies that analyze the impacts of weather shocks on DT varieties' yields and economic returns using data on farmers' marketplace behavior across the US. If implemented carefully, such a study would be able to inform policy makers and other market participants of the potential of drought tolerance, in isolation, to serve as a climate change adaptation mechanism.

References

Adee, E., K. Roozeboom, G. R. Balboa, A. Schlegel, and I. A. Ciampitti. 2016. "Drought-Tolerant Corn Hybrids Yield More in Drought-Stressed Environments with No Penalty in Non-Stressed Environments." *Frontiers in Plant Science* 7: 1–9.

Angrist, J. D., and J. S. Pischke. 2008. *Mostly Harmless Econometrics: An Empiricist's Companion*. Princeton, NJ: Princeton University Press.

Auffhammer, M., and W. Schlenker. 2014. "Empirical Studies on Agricultural Impacts and Adaptation." *Energy Economics* 46: 551–61.

Barr, K. J., B. A. Babcock, M. A. Carriquiry, A. M. Nasser, and L. Harfuch. 2011. "Agricultural Land Elasticities in the United States and Brazil." *Applied Economic Perspectives and Policy* 33 (3): 449–62.

Blanc, E., and J. Reilly. 2017. "Approaches to Assessing Climate Change Impacts on Agriculture: An Overview of the Debate." *Review of Environmental Economics and Policy* 11 (2): 247–57.

Burke, M., and K. Emerick. 2016. "Adaptation to Climate Change: Evidence from U.S. Agriculture." *American Economic Journal: Economic Policy* 8 (3): 106–40.

Carlton, J. S., A. S. Mase, C. L. Knutson, M. C. Lemos, T. Haigh, D.P. Todey, and L. S. Prokopy. 2016. "Adaptation to Climate Change: Evidence from U.S. Agriculture." *Climatic Change* 135 (2): 211–26.

Castiglioni, P., D. Warner, R. J. Bensen, D. C. Anstrom, J. Harrison, M. Stoecker, M. Abad, G. Kumar, S. Salvador, R. D'Ordines, S. Navarro, S. Back, M. Fernandes, J. Targolli, S. Dasgupta, C. Bonin, M. H. Luethy, and J. E. Heard. 2008. "Bacterial RNA Chaperones Confer Abiotic Stress Tolerance in Plants and Improved Grain Yield in Maize under Water-Limited Conditions." *Plant Physiology* 147 (2): 446–55.

Chambers, R. G., and J. Quiggin. 2000. *Uncertainty, Production, Choice, and Agency: The State-Contingent Approach.* Cambridge, UK: Cambridge University Press.

Cook, B. I., T. R. Ault, and J. E. Smerdon. 2015. "Unprecedented 21st Century Drought Risk in the American Southwest and Central Plains." *Science Advances* 1 (1): 1–7.

Cooper, M., C. Gho, R. Leafren, T. Tang, and C. Messina. 2014. "Breeding Drought-Tolerant Maize Hybrids for the U.S. Corn Belt: Discovery to Product." *Journal of Experimental Botany* 65 (21): 6191–6204.

Daly, C., R. P. Neilson, and D. L. Phillips. 1994. "A Statistical-Topographic Model for Mapping Climatological Precipitation over Mountainous Terrain." *Journal of Applied Meteorology* 33 (2): 140–58.

Deschênes, O., and M. Greenstone. 2007. "The Economic Impacts of Climate Change: Evidence from Agricultural Output and Random Fluctuations in Weather." *American Economic Review* 97 (1): 354–85.

Dobos, R., H. Sinclair Jr., and M. Robotham. 2012. User Guide for the National Commodity Crop Productivity Index. Website. U.S. Department of Agriculture. Accessed 04/29/2018. https://www.nrcs.usda.gov/wps/portal/nrcs/site/national/home.

Druckenmiller, H., and S. Hsiang. 2018. "Accounting for Unobservable Heterogeneity in Cross Section Using Spatial First Diffferences." NBER Working Paper No. 25177. Cambridge, MA: National Bureau of Economic Research.

Edmeades, G. O. 2012. "Progress in Achieving and Delivering Drought Tolerance in Maize—An Update." In *Global Status of Commercialized Biotech/GM Crops: 2012, Brief 44*, edited by C. James, 239–72. Ithaca, NY: The International Service for the Acquisition of Agri-biotech Applications (ISAAA).

Farmers Business Network. 2018. Zone Pricing in Corn Seed Report. Website. Accessed 04/09/18. https://emergence.fbn.com.

Fernandez-Cornejo, J., and W. D. McBride. 2002. "Adoption of Bioengineered Crops." AER-810. U.S. Department of Agriculture, Economic Research Service.

Fisher, M., T. Abate, R. W. Lunduka, M. Asnake, Y. Alemayehu, and R. B. Madulu. 2015. "Drought Tolerant Maize for Farmer Adaptation to Drought in Sub-Saharan Africa: Determinants of Adoption in Eastern and Southern Africa." *Climatic Change* 133 (2): 283–99.

Gaffney, J., J. Schussler, C. Löffler, W. Cai, S. Paszkiewicz, C. Messina, J. Groeteke, J. Keaschall, and M. Cooper. 2015. "Industry-Scale Evaluation of Maize Hybrids Selected for Increased Yields in Drought-Stress Conditions of the U.S. Corn Belt." *Crop Science* 55 (4): 1608–1618.

Heddinghaus, T. R., and P. Sabol. 1991. "Review of the Palmer Drought Severity Index and Where Do We Go from Here?" Proceedings, 7th Conference on Applied Climatology, 242–46. Boston, MA: American Meterological Society.

Heisey, P. W., and K. Day Rubenstein. 2015. "Using Crop Genetic Resources to Help Agriculture Adapt to Climate Change: Economics and Policy." EIB-139. U.S. Department of Agriculture, Economic Research Service.

Hsiang, S. 2016. "Climate Econometrics." *Annual Review of Resource Economics* 8: 43–75.

Hsiang, S., and R. E. Kopp. 2018. "An Economist's Guide to Climate Change Science." NBER Working Paper No. 25189. Cambridge, MA: National Bureau of Economic Research.

Intergovernmental Panel on Climate Change. 2014. Climate Change 2014: Synthesis Report. Contribution of Working Groups I, II, and III to the Fifth Assessment Report of the Intergovernmental Panel on Climate Change. Geneva, Switzerland: ICPP.

International Maize and Wheat Improvement Center (CIMMYT). 2018. Project Profiles. Website. Accessed 05/02/18. https://www.cimmyt.org/project-profile.

Kahneman, D., and A. Tversky. 1979. "Prospect Theory: An Analysis of Decision under Risk." *Econometrica* 47 (2): 263–92.

Kuwayama, Y., A. Thomspon, R. Bernknopf, and B. Zaitchik. 2019. "Estimating the Impact of Drought on Agriculture Using the U.S. Drought Monitor." *American Journal of Agricultural Economics* 101 (1): 193–210.

Lin, W., G. W. Dean, and C. V. Moore. 1974. "An Empirical Test of Utility vs. Profit Maximization in Agricultural Production." *American Journal of Agricultural Economics* 56 (3): 497–508.

Lobell, D. B. 2014. "Climate Change Adaptation in Crop Production: Beware of Illusions." *Global Food Security* 3 (2): 72–76.

Lybbert, T. J., and A. Bell. 2010. "Stochastic Benefit Streams, Learning, and Technology Diffusion: Why Drought Tolerance Is Not the New Bt." *AgBioForum* 13 (1): 13–24.

McFadden, J. R., D. J. Smith, S. J. Wechsler, and S. Wallander. 2019. "Development, Adoption, and Management of Drought-Tolerant Corn in the United States." EIB-204. U.S. Department of Agriculture, Economic Research Service.

Mendelsohn, R., W. D. Nordhaus, and D. Shaw. 1994. "The Impact of Global Warming on Agriculture: A Ricardian Analysis." *American Economic Review* 84 (4): 753–71.

Minford, M. 2015. Farmers Test Drought-Tolerant Corn Hybrids. Website. Accessed 04/12/18. Corn and Soybeans Digest. http://www.cornandsoybeandigest.com /corn.

Nemali, K. S., C. Bonin, F. G. Dohleman, M. Stephens, W. R. Reeves, D. E. Nelson, P. Castiglioni, J. E. Whitsel, B. Sammons, R. A. Silady, D. Anstrom, R. E. Sharp, O. R. Patharkar, D. Clay, M. Coffin, M. A. Nemeth, M. E. Leibman, M. Luethy, and M. Lawson. 2015. "Physiological Responses Related to Increased Grain Yield under Drought in the First Biotechnology Derived Drought-Tolerant Maize." *Plant, Cell, and Environment* 38 (9): 1866–1880.

Nuccio, M. L., M. Paul, N. J. Bate, J. Cohn, and S. R. Cutler. 2018. "Where Are the Drought Tolerant Crops? An Assessment of More than Two Decades of Plant Biotechnology Effort in Crop Improvement." *Plant Science* 273: 110–19.

Powlson, D. S., C. M. Stirling, M. L. Jat, B. G. Gerard, C. A. Palm, P. A. Sanchez, and K. G. Cassman. 2014. "Limited Potential of No-Till Agriculture for Climate Change Mitigation." *Nature Climate Change* 4: 678–83.

Rippey, B. R. 2015. "The U.S. Drought of 2012." *Weather and Climate Extremes* 10 (A): 57–64.

Schlenker, W., W. M. Hanemann, and A. C. Fisher. 2005. "Will U.S. Agriculture Really Benefit from Global Warming? Accounting for Irrigation in the Hedonic Approach." *American Economic Review* 95 (1): 395–406.

Schlenker, W., and M. J. Roberts. 2009. "Nonlinear Temperature Effects Indicate Severe Damages to U.S. Crop Yields under Climate Change." *Proceedings of the National Academy of Sciences* 106 (37): 15594–15598.

Schoemaker, P. J. H. 1982. "The Expected Utility Model: Its Variants, Purposes, Evidence, and Limitations." *Journal of Economic Literature* 20 (2): 529–63.

Shi, G., J. P. Chavas, and K. Stiegert. 2010. "An Analysis of the Pricing of Traits in the U.S. Corn Seed Market." *American Journal of Agricultural Economics* 92 (5): 1324–1338.

Steward, D. R., P. J. Bruss, X. Yang, S. A. Staggenborg, S. M. Welch, and M. D. Apley. 2013. "Tapping Unsustainable Groundwater Stores for Agricultural Production in the High Plains Aquifer of Kansas, Projections to 2110." *Proceedings of the National Academy of Sciences* 110 (37): E3477–E3486.

Strzepek, K., G. Yohe, J. Neumann, and B. Boehlert. 2010. "Characterizing Changes in Drought Risk for the United States from Climate Change." *Environmental Research Letters* 5 (4): 1–9.

U.S. Department of Agriculture, Economic Research Service. 2016. Agricultural Resource Management Survey: Overview. Website. Accessed 06/05/19. https://www.ers.usda.gov/data-products.

———. 2022. U.S. Farm Income and Wealth Statistics. Website. Accessed 03/30/22. https://www.ers.usda.gov/data-products/farm-income-and-wealth-statistics/.

U.S. Department of Agriculture, National Agricultural Statistics Service. 2013. Census of Agriculture, 2012. Website. Accessed 06/23/22. https://www.nass.usda.gov/AgCensus/index.php.

———. 2018a. County Data FAQs. Website. Accessed 07/01/19. https://www.nass.usda.gov/Data and Statistics/index.php.

———. 2018b. Quick Stats. Website. Accessed 02/11/19. https://quickstats.nass.usda.gov.

———. 2022. Acreage. Website. Accessed 03/30/22. https://www.nass.usda.gov/Publications/Todays Reports/reports/acrg0621.pdf.

U.S. Department of Agriculture, Natural Resources Conservation Service. 2018. Gridded Soil Survey Geographic Database for the Coterminous United States. Website. Accessed 10/16/18. https://gdg.sc.egov.usda.gov/.

U.S. Drought Monitor. 2019. Drought Classification. Website. Accessed 03/28/19. http://droughtmonitor.unl.edu/Data.aspx.

Wallander, S., M. Aillery, D. Hellerstein, and M. Hand. 2013. "The Role of Conservation Programs in Drought Risk Adaptation." ERR-144. U.S. Department of Agriculture, Economic Research Service.

Waltz, E. 2014. "Beating the Heat." *Nature Biotechnology* 32 (7): 610–13.

Weber, J. G., N. Key, and E. O'Donoghue. 2016. "Does Federal Crop Insurance Make Environmental Externalities from Agriculture Worse?" *Journal of the Association of Environmental and Resource Economists* 3 (3): 707–42.

Wooldridge, J. M. 2010. *Econometric Analysis of Cross Section and Panel Data*. Cambridge, MA: MIT Press.

Wossen, T., T. Abdoulaye, A. Alene, S. Feleke, A. Menkir, and V. Manyong. 2017. "Measuring the Impacts of Adaptation Strategies to Drought Stress: The Case of Drought Tolerant Maize Varieties." *Journal of Environmental Management* 203 (1): 106–13.

Wozniak, G. D. 1987. "Human Capital, Information, and the Early Adoption of New Technology." *Journal of Human Resources* 22 (1): 101–12.

Cover Crops, Drought, Yield, and Risk
An Analysis of US Soybean Production

Fengxia Dong

7.1 Introduction

Climate change has caused increasing frequency and severity of drought stress in the US. Water scarcity has become one of the most severe constraints to agricultural production, adversely affecting crop yields and presenting a major challenge to sustainable food production. Along with prescribed grazing, mulching, micro-irrigation, and conservation tillage, cover cropping is among the five short- and long-term strategies for dealing with drought conditions that farmers can receive financial assistance for from the Natural Resources Conservation Service (NRCS) (USDA Climate Hubs 2021). Cover crops are defined by USDA NRCS in Cover Crop Termination Guidelines Version 4 (2019) as *"crops including grasses, legumes and forbs for seasonal cover and other conservation purposes. Cover crops are primarily used for erosion control, soil health improvement, weed and other pest control, habitat for beneficial organisms, improved water efficiency, nutrient cycling, and water quality improvement. A cover crop managed and terminated according to these Guidelines is not considered a 'crop' for crop insurance purposes."*

Legumes and grasses are currently the two most popular cover crop types

Fengxia Dong is a Research Agricultural Economist in the Agricultural Policy and Models Branch of the Market and Trade Economics Division at USDA's Economic Research Service (ERS).

The findings and conclusions in this paper are those of the author and should not be construed to represent any official USDA or US Government determination or policy. For acknowledgments, sources of research support, and disclosure of the author's material financial relationships, if any, please see https://www.nber.org/books-and-chapters/american-agriculture-water-resources-and-climate-change/cover-crops-drought-yield-and-risk-analysis-us-soybean-production.

(SARE 2007). Hairy vetch is the most widely used winter annual legume in northern regions because of its high N content, winter hardiness, and high productivity (Lu et al. 2000), and crimson clover is considered one of the best cover crops for southern regions due to its fast matureness and large N addition to the following crops (SARE 2007). Grass cover crops include annual cereals (such as rye, wheat, barley, and oats), annual or perennial forage grasses (such as ryegrass), and warm-season grasses (such as sorghum–Sudan grass) (SARE 2007). Besides legumes and grasses, buckwheat and *Brassica* (such as mustard, rapeseed, and forage radish) can also be used as cover crops (SARE 2007).

The use of cover crops as a cropping strategy is not new. It was practiced by people in ancient Greece, Rome, and China as early as 3,000 years ago (Langdale et al. 1991). Cover crops were first used in the US in the 18th century and extensively expanded in the 19th century (Groff 2015), although by that time they were used mainly as green manures. The affordability and ease of use of synthetic fertilizer at the end of World War II, however, attracted farmers to utilize more synthetic fertilizer instead of cover crops to further improve crop yields. The use of cover crops in conventional agriculture has gradually become less common since then (Groff 2015).

Currently, the adoption rate of cover crops is low and varies by agricultural commodity type. According to Agricultural Resource Management Surveys (ARMS) in years of 2010, 2016, 2017, and 2018, the adoption rate ranged from just over 5 percent of acreage on corn-for-grain (2016) to 8.4 percent on soybeans (2018), around 13 percent on cotton (2015), and over 24 percent on corn-for-silage (2016) (Wallander et al. 2021), in stark contrast to the adoption rate of conservation tillage, for example, which is 67 percent of soybean acreage, according to USDA's 2018 ARMS Phase II, Soybean Production Practices and Costs Report.

Cover crops can protect and improve soil between periods of regular crop production (Schnepf and Cox 2006). Besides a variety of production, soil health, and environmental benefits such as increasing weed and pest suppression, reducing runoff of sediments and nutrients into waterways, and reducing soil erosion and compaction, cover crops can improve water infiltration, reduce water evaporation, and increase soil's water holding capacity (e.g., USDA NRCS 2018; Mitchell, Shrestha, and Irmak 2015; Blanco-Canqui et al. 2015; McDaniel, Tiemann, and Grandy 2014; Laloy and Bielders 2010; Dean and Weil 2009; Sainju, Singh, and Whitehead 2002; Sainju et al. 2006).

There are, however, well-recognized trade-offs and limitations in adopting cover crops as a conservation strategy (SARE 2017). In addition to the costs of soil preparation, seeds, and labor, there are challenges in implementation and management, such as the selection of cover crop species, planting and termination time—which may interfere with fall harvest or spring planting—and producing too much surface residue (CTIC 2015; Sackett

2013; Miller, Chin, and Zook 2012; Snapp et al. 2005). Moreover, there are concerns that water needed by cover crops may reduce the amount of water available to the following main crop (SARE 2017; Clark et al. 1997; Corak, Frye, and Smith 1991; Ebelhar, Frye, and Blevins 1984; Munawar et al. 1990).

Along with the above concerns, the effect on crop yield and risk is another important factor in farmers' adoption decision of cover crops, as yield and risk directly affect farmers' economic returns. The Iowa Farm and Rural Life Poll in 2015 (Arbuckle 2016) reveals that 74 percent of the farmers believe that economic factors have a moderate to very strong influence on their changes in management practices. The 2017 Cover Crop Survey conducted by SARE (2017) also shows that the fear of a lack of economic returns (54 percent of respondents), increasing production risk (48 percent of respondents), and potential yield reduction (44 percent) are among the major concerns for non-users. Another cover crop survey conducted in 2015 shows that the potential yield benefit to cash crops is an important factor in decision making, especially for non-adopters (CTIC 2015). Therefore, an analysis of the effects of management practices on yield and its risk is essential to find effective supporting programs to promote good management practice adoption.

The results of existing studies of yield effects of cover crops are mixed. A meta-analysis of the response of corn yield to cover crops by Miguez and Bollero (2005) concludes that legume cover crops increase corn yield by 37 percent. Similarly, Andraski and Bundy (2005) and Muñoz et al. (2014) find a positive effect of cover crop biomass on corn yields. Contrastingly, Reddy (2017) discovers lower soybean yield with cover crops compared with no cover crops. Nielsen et al. (2016) find that there was an average 10 percent reduction in wheat yield following a cover crop compared with following fallow, regardless of whether the cover crop was grown in a mixture or in a single-species planting; in addition, yield reductions were greater under drier conditions. In comparison, Acharya et al. (2019); Smith, Atwood, and Warren (2014); Hunter et al. (2019), and Acuña and Villamil (2014) locate no benefits of growing cover crops on subsequent crop yield. Note that all the findings are subject to certain conditions, such as soil types, other production practices (e.g., tillage), cover crop species, and precipitation.

Previous studies of yield effects of cover crops are mainly conducted in field experimental plots using agronomic models. A study based on a large number of fields with different agroecological characteristics and under varied weather is in need. In addition, increasing the frequency and severity of adverse events can expose farms to significant production uncertainty. Therefore, special attention is paid to downside risk exposure. In general, the downside risk is the risk associated with unfavorable events and located in the lower tail of the yield or return distribution (Kim et al. 2014). As pointed out by Hardaker et al. (2004), Kim et al. (2014), and OECD (2011), analyzing both the exposure to risk and levels of downside risk in agriculture is a key component in assessing welfare impacts. While cover cropping is recommended to farmers to deal with drought, its effects on farm yield risk and especially downside risk are not well documented. This study aims to fill the literature gap by analyzing the effects of cover crops on yield and its risk with varied weather, regional, and field characteristics.

In this study, we focus on US soybean production. The US is the world's second-largest soybean producer and exporter, accounting for 31 percent of world total production and 36 percent of world total exports, respectively. US farmers planted 87.2 million acres of soybeans in 2021, behind only corn. The growth and productivity of soybeans are adversely affected by various environmental stresses, among which drought stress is considered the most devastating event (Le et al. 2012; Shaheen et al. 2016). Drought stress, especially occurring at late vegetative stages, may cause significant soybean production losses of up to 40 percent (Specht, Hume, and Kumudin 1999; Le et al. 2012) by inhibiting increases in the soybean plant height and leaf area (Dong et al. 2019). Several studies find that cover crops improve soybean soil moisture (e.g., Acharya et al. 2019; Chu et al. 2017), although some do not (Barker et al. 2018).

The paper makes three contributions. First, it explores the factors that affect the adoption of cover cropping. We consider not only land characteristics and farmers' demographics and concerns but also droughts in previous years. This enables us to reveal whether farmers view cover cropping as an effective means of increasing resilience to drought. Second, the paper examines the effects of cover crops on yield variation and downside risk. We employ moments of yield distribution to evaluate the exposure to yield variation and downside risk. Disentangling the yield effects of adaptation is of paramount importance. It will reveal whether farmers who adopt cover crops are indeed getting benefits in terms of an increase in crop yield, a benefit crucial to broader adoption. Third, the paper utilizes a data set covering the majority of soybean fields in the US with significantly different soil types and weather conditions. Two interplays, one between soil types and the adaptation strategy (namely, cover cropping), and the other between weather and the cover cropping practice, are included in yield and its risk analysis. The interplay along with farmer demographics, farm characteristics, and input use allow us to examine the effectiveness of the managerial options for risk mitigation under varied soil and weather conditions, especially in the threat of drought.

The rest of the paper is organized as follows. The next section discusses the theoretical framework and empirical models, followed by a description of data and variables. Then estimation results are discussed, followed by conclusions at the end.

7.2   Theoretical Framework and Empirical Models

Consider a farmer who uses a vector of inputs $\mathbf{x}$ and drought adaptation strategies (e.g., cover cropping, mulching, or drought-resistant seeds) C to produce a single output Q through a technology described by a well-behaved (i.e., continuous and twice differentiable) production function $Q(\cdot)$. The farmer can choose to adopt ($C = 1$) or not ($C = 0$) a drought adaptation strategy. Use e to indicate random and uncontrollable factors reflecting production risk (e.g., drought effect) whose distribution is $F(e)$. The production technology can thus be represented by $Q = Q(\mathbf{x}, C, e)$. Use p to indicate the output price and w a vector of input prices. The net return is represented by $\pi = pQ(\mathbf{x}, C, e) - w\mathbf{x}$.

Let $U(\pi)$ be a von Neumann-Morgenstern utility function that represents farmers' preferences regarding income. To simplify the analysis, I assume that the only risk that farmers are facing is production risk and both output and input prices are given or nonrandom. Being risk averse, farmers are assumed to maximize expected utility $EU(\pi)$, where $EU(\pi) = \int U(\pi)dF(e)$ with E as the expectation operator. For the decision on the drought adaptation strategy C, for example, if $EU(\pi \mid C = 1) - EU(\pi \mid C = 0) > 0$, then the farmer would choose to adopt the drought adaptation strategy; otherwise, the farmer would choose not to do so. In addition, the greater the difference between the expected utilities, the higher the probability of adoption. Very often there is a requirement of investment and/or possible uncertainty in profit due either to a lack of the exact performance of the adaptation strategy or to the higher probability of erring in the use of the adaptation strategy (Koundouri, Nauges, and Tzouvelekas 2006). In those cases, the farmer may choose to delay the adoption to achieve more information (Koundouri, Nauges, and Tzouvelekas 2006). Consequently, the farmer will choose the adaptation strategy *iff* $EU(\pi \mid C = 1) - EU(\pi \mid C = 0) > V$, where $V \geq 0$ is the value of new information essential for the farmer to make adoption decision which depends on the investment, the uncertainty related to the use of the strategy, and the farmers' characteristics (Koundouri, Nauges, and Tzouvelekas 2006). Therefore, drought adaptation strategies that require less investment and have less uncertainty in profit will have a higher level of adoption. For example, if technical assistance and extension service are provided to farmers for adopting an adaptation strategy, then the uncertainty in profit will be lower. Consequently, it is more possible that farmers will choose to adopt the strategy. In addition, farmers' characteristics such as their education level or their concerns about the environment may also play a role in the adoption decision. The more concerned the farmer is about an environmental issue, the higher probability of adopting a practice that can address the concern.

With differentiability of $U(\pi)$, $EU(\pi)$ can be approximated by taking the expectation of an kth-order Taylor series expansion of $U(\pi)$ at the mean net return $E\pi$ where $E\pi = \int \pi dF(\epsilon) = u_1$ and is written as

(1)
$$EU(\pi) \approx E\left[\sum_{j=0}^{k} \left(\frac{1}{j!} \frac{\partial^j U}{\partial \pi^j}(u_1) \times (\pi - u_1)^j \right) \right]$$

$$= U(u_1) + \sum_{j=1}^{k} \left[\frac{1}{j!} \frac{\partial^j U}{\partial \pi^j}(u_1) \times E(\pi - u_1)^j \right]$$

$$= U(u_1) + \sum_{j=1}^{k} \left[\frac{1}{j!} \frac{\partial^j U}{\partial \pi^j}(u_1) \times E(\pi - u_1)^j \right].$$

Equation (1) shows that the expected utility depends on the mean net return u_1 and the jth ($j = 2, 3, \ldots, k$) central moment of net return, $u_j = E[(\pi - u_1)^j]$. When $j=2$, u_j is the second moment or the variance, and when $j = 3$, the third moment or the skewness of the net return. The skewness measures the asymmetry of the distribution around its mean, with a negative skewness implying a distribution skewed to the left; and a positive one implying a distribution skewed to the right. A lower skewness generates a greater exposure to downside risk.

By normalizing prices so that $p = 1$, a farm's net return can be expressed by $\pi = Q(\mathbf{x}, C, \mathbf{e}) - (\mathbf{wx} / p)$. The equation explicitly shows that the production function $Q(\mathbf{x}, C, \mathbf{e})$ provides all the relevant information for analyzing risk exposure on farms adopting drought adaptation strategies. To empirically investigate the impacts of cover crops on crop yield and yield risks, we start with the moment functions of crop yield.

7.2.1   Moment Representation of Production Function

Our empirical model is based on Antle's (1983) moment-based approach, which provides a flexible and convenient basis for evaluating exposure to production risk. As discussed in Antle (1983), a stochastic production function can be represented by a general parameterization of the moment functions. Using $\boldsymbol{\beta}_1$ to indicate a vector of technology parameters and as discussed above, the production technology can be represented by $Q = Q(x, C, e, \boldsymbol{\beta}_1)$. The production function is stochastic given the random error term e. Let the stochastic output Q have a cumulative distribution $F(e)$, then the first and the ith central moments of output Q can be represented, respectively, as

(2) $m_1(x,C,\boldsymbol{\beta}_1) = E[Q(x,C,e,\boldsymbol{\beta}_1)] = \int Q(x,C,e,\boldsymbol{\beta}_1)dF(e)$,

(3) $m_i(x,C,\boldsymbol{\beta}_i) = E[\{Q(x,C,e,\boldsymbol{\beta}_1) - m_1(x,C,\boldsymbol{\beta}_1)\}^i]$

$$= \int (Q(x,C,e,\boldsymbol{\beta}_1) - m_1(x,S,\boldsymbol{\beta}_1))^i dF(e) \; for \; i \geq 2.$$

Here, E is the expectation operator; m_1 is the mean and m_i is the ith moment of output (for $i \geq 2$); and $\boldsymbol{\beta}_i$ is a vector of parameters. The models in (1) and

(2) have the advantage of being flexible as there are no restrictions within or cross moments.

By rewriting equations (2) and (3), we get the following equations

$$(4) \qquad \varepsilon_1 = Q(x,C,\beta_1) - m_1(x,C,\beta_1),$$

$$(5) \qquad (\varepsilon_1)^i = m_i(x,C,\beta_i) + \varepsilon_i, i \geq 2.$$

Here, $E(\varepsilon_j)$ and $E(\varepsilon_j \varepsilon_{j'}) = 0$ ($j = 1, 2, \ldots, n$ and $j \neq j'$).

As discussed by Kendall and Stuart (1977) and shown in many empirical analyses (e.g., Day 1965; Di Falco and Chavas 2009; Di Falco and Veronesi 2014; Tack, Harri, and Coble 2012; and Anderson, Dillon, and Hardaker 1980), the first three moments including location (mean), dispersion (variance), and skewness (the third moment) of a given distribution can adequately approximate the distribution. We, therefore, choose the first three moments to represent the distribution of yield in our analysis. While the variance (m_2) is a traditional measure of risk, the skewness of the output measure (m_3) captures the tail asymmetry of a yield distribution around its mean. A negative (positive) skewness implies a distribution skewed to the left (right). A lower skewness presents a greater exposure to the downside risk of unexpected low yield, i.e., crop failure.

From equations (4) and (5), we have the mean, variance, and skewness of yield as the following,

$$(6) \qquad Q = m_1(x,C,\beta_1) + \varepsilon_1,$$

$$(7) \qquad (\varepsilon_1)^2 = m_2(x,C,\beta_i) + \varepsilon_2,$$

$$(8) \qquad (\varepsilon_1)^3 = m_3(x,C,\beta_i) + \varepsilon_3.$$

Here again, $E(\varepsilon_j) = 0$ and $E(\varepsilon_j \varepsilon_{j'}) = 0$ ($j = 1, 2, 3$ and $j \neq j'$). Empirically, if β_1^* is a consistent estimator of β_1 from a sample of observed outputs, then $\varepsilon_1^* = Q(x, C, \beta_1) - m_1(x, C, \beta_1^*)$ is a consistent estimator of $\varepsilon_1 = Q(x, C, \beta_1) - m_1(x, C, \beta_1)$. It also suggests that $(\varepsilon_i^*)^i = m_i(x, S, \beta_i^*) + \varepsilon_i, i \geq 2$ is a consistent estimator of $(\varepsilon_1)^i$. The models in (5), (6), and (7) have no restrictions within or cross moments and thus are flexible.

The adoption of cover cropping C is variance increasing, variance neutral, or variance decreasing if $(\partial m_2 / \partial_C) > 0, = 0$, or < 0, respectively. For a risk averse farmer, $(\partial m_2 / \partial_C) > 0$ meaning that the adoption of cover crops creates a greater risk in output is undesirable. Similarly, the adoption of cover cropping C is skewness increasing, skewness neutral, or skewness decreasing if $(\partial m_3 / \partial_C) > 0, = 0$, or < 0, respectively. And, for a risk-averse farmer, $(\partial m_3 / \partial_C) < 0$ meaning that the adoption of cover crops increasing the exposure to a lower output is undesirable.

Some factors that are known by farmers but unknown to economists may affect both yield and the cover cropping decision. Consequently, when empirically estimating the yield equation as shown in (6), what arises is a concern that the adoption of cover crops may be endogenous. The endogeneity may result in inconsistent and biased estimates. To address the potential endogeneity issue, we use a two-stage method, which is one of the most potent and versatile tools available to treat endogeneity (Antonakis et al. 2014).

We will first estimate the use of cover crops with an instrumental variable approach. We model the adoption of cover crops in a logit model as follows:

$$\text{Cover cropping: Cover cropping: } y_{cc}^* = \alpha' x + \gamma Z + \varepsilon$$

$$(9) \qquad y_{cc} = \begin{cases} 1 \text{ if } y_{cc}^* > 0 \\ 0 \text{ otherwise} \end{cases}.$$

Here y_{cc}^* is a latent continuous variable associated with the adoption of cover cropping; y_{cc} is the corresponding observed binary outcome with a value of 1 if cover cropping is adopted and 0 otherwise; α' is a transposed vector for parameters to be estimated; and Z is a vector of instrumental variables.

7.3   Data and Statistics

We apply our analysis to US soybean production. We construct the data from the USDA's 2018 ARMS Phase II, Soybean Production Practices and Costs Report, and Phase III Soybean Costs and Returns Report. The Phase II survey covers a cross-section of soybean fields in 19 states and collects information on production and management practices, input uses, and field characteristics. The Phase III report provides information on farm operators and financial characteristics. Farm-level survey data provide us a good opportunity to look more closely at farm activities and the motives behind them (Dong, Hennessy, and Jensen 2010; Dong, Hennessy, Jensen, and Volpe 2016).

As conventional and organic production are significantly different in production practices, we only use data from conventional soybean growers. We delete all observations with missing values, leaving a total of 1,177 observations. ARMS has a complex survey design and is a probability-based survey with unequal probability sampling (National Research Council 2007). To account for the survey design, we use the sampling weights (expansion factors) provided by USDA NASS to expand the sample to generate population estimates in the statistical analysis. With the survey weights applied to the sample observations, the weighted sample represents approximately 835,530 soybean fields in the United States.

7.3.1   Variables in Moment Equations

The dependent variables in equations (6) to (8) are the first three moments of the distribution of soybean yield per acre, respectively. Variable *phos-*

phorus measuring the use of phosphorus per acre is included along with its quadratic term. They are expected to increase the yield. Other independent variables include those on production practices, field characteristics, regional location, and weather/climate. To capture regional differences, indicator variables were constructed based on Farm Resource Regions (USDA ERS 2000), which are defined based on farm, soil, and climate characteristics rather than state boundaries. The regional dummy variables (*Eastern Uplands, Heartland, Mississippi Portal, Northern Crescent, Northern Great Plains, Prairie Gateway, and Southern Seaboard*) are equal to 1 if the field locates in the corresponding region and 0 otherwise. Two variables (*PlantLate* and *ReplantPct*) are used to capture the impacts of adverse factors negatively affecting soybean yields at the start of the planting season. Variable *PlantLate* is equal to 1 if the planting date fell in the last 15 percent percentile of the state, and 0 otherwise. Variable *ReplantPct* is the proportion of fields that was replanted. Both might be resulted from adverse weather and result in shorter growing seasons for soybeans, and thus are expected to have a negative effect on yield. Dummy variable *manure* has a value of 1 if manure was applied to the field and 0 otherwise. Several field characteristics may affect yields such as soil texture and slope of the field (Butcher et al. 2018; Arora et al. 2011; Shane and Barker 1986; Kaspar et al. 2004; Jiang and Thelen 2004; Kravchenko, Bullock, and Boast 2000; Linkemer, Board, and Musgrave 1998; Nelson and Meinhardt 2011) are also included in the yield moment functions. Soil texture is categorized into five types: loam, clay, sandy, mixed, and silty. Slope is categorized into two levels: nearly level and moderate/steep grade (even or variable). While the soybean growing season is from May to September, weather in both July and August is important for soybean yields (Westcott and Jewison 2013). The county-level US Drought Monitor (USDM) indicator jointly produced by the National Drought Mitigation Center (NDMC), the National Oceanic and Atmospheric Administration (NOAA), and the US Department of Agriculture is used. The USDM indicator has five categories: D0–D4, of which D0 indicates abnormally dry but not in drought, while D1–D4 indicates moderate drought to exceptional drought. The USDM indicator is based on inputs including the Palmer Drought Severity Index (PDSI), the Standardized Precipitation Index (SPI), satellite-based assessments of vegetation health, and various indicators of soil moisture as well as hydrologic data (NDMC 2021). If on a weekly average over 10 percent of a county area is categorized as D1 or above in either July or August of 2018, then an indicator variable *drought18* is equal to 1 and 0 otherwise. Moreover, modified growing degree days (mGDD) and overheating growing days (ODD) for soybeans during July and August of 2018 are also included to explicitly capture temperature or heat effects on the growth and development of soybean plants. GDD is one of the most important

Table 7.1   **Survey means and standard errors of variables in moment functions**

Variable and unit	Survey Mean	Std. Err
Yield (bushels/acre)	52.354	0.097
Heartland (0/1)	0.498	0.003
Northern Crescent (0/1)	0.141	0.003
Northern Great Plains (0/1)	0.066	0.001
Prairie Gateway (0/1)	0.054	0.001
Eastern Uplands (0/1)	0.053	0.002
Southern Seaboard (0/1)	0.105	0.002
Mississippi Portal (0/1)	0.083	0.001
PlantLate (0/1)	0.129	0.002
ReplantPct	0.041	0.002
Manure (0/1)	0.043	0.001
Phosphorus (lbs/acre)	25.550	0.229
drought18 (0/1)	0.175	0.002
Soil texture: loam (0/1)	0.369	0.003
Soil texture: clay (0/1)	0.170	0.003
Soil texture: sandy (0/1)	0.069	0.002
Soil texture: mixed (0/1)	0.369	0.005
Soil texture: silty (0/1)	0.019	0.001
Growing degree days (Celsius)	3304.675	4.396
Overheating degree days (days)	30.778	0.133
Cover crops	0.100	0.003

factors influencing the rate of development in soybean (Major et al. 1975; Pedersen and Licht 2014; Kessler, Archontoulis and Lich 2020). Daily mGDD is calculated as

$$mGDD = \max\left(\frac{\min(daily\,max\,temp, higher\,development\,threshold) + \max(daily\,min\,temp, lower\,development\,thresholds)}{2} - 50, 0\right).$$

GDD depends upon the minimum and maximum temperatures which affect the plant's growth. The higher and lower development thresholds are 86°F and 50°F, respectively, for soybeans. Daily mGDD in July and August is accumulated to get mGDD for the two months. ODD is the count of days in July and August with a temperature over 89.6°F and measures the heat stress for crops. Cover cropping is included in the function as a dummy variable. It is set equal to 1 if the cover cropping was adopted and 0 otherwise. In addition, interactions between cover cropping and drought status *drought18* are also included to capture the effects of the adaptation strategy on yield moments conditional on drought events. Survey population and sample summary statistics for variables used in the moment functions are reported in table 7.1.

7.3.2   Variables in Logit Model

The instrumental variables used in the cover crop equation include farmers' concerns about soil and water-related issues. Seven dummy variables are constructed, indicating concerns on water-driven erosion, wind-driven erosion, soil compaction, poor drainage, low organic matter, water quality, and other concerns, respectively, and taking a value of 1 if a farmer had such a concern and 0 otherwise. We believe that farmers having concerns about soil erosion or soil quality may have more intention to adopt soil conservation practices. The variable of land ownership is included as another instrumental variable. The variable takes a value of 1 if the operator owned the land and 0 otherwise. We expect that land ownership may increase the likelihood of fields adopting cover crops as landowners may care more about soil erosion and soil quality on their own land and thus have more motivation to adopt conservation practices.

Farm size and some other field characteristics such as field size and whether any part of the field was classified as "highly erodible," and whether the field contained a wetland are also included in the cover crop equation given their possible effect on farmers' cover crop decision making (Ding, Schoengold, and Tadesse 2009; Vitale et al. 2011; Wandel and Smithers 2000). Highly erodible land is any land that can erode at an excessive rate due to its soil properties. Farmers are required to farm such land in accordance with a conservation plan or system approved by NRCS (USDA Risk Management Agency 2015). We expect that field classified as "highly erodible" is more likely to adopt cover crops for their vulnerability to soil erosion. Both farm and field sizes are measured in acres. We expect that larger farms and larger fields may be less likely to adopt cover crops given the time and labor requirements of cover crop implementation and management. Moreover, farmers' age, education, years of experience in farming, and off-farm work are also included. We expect that older farmers or farmers with off-farm work may be less likely to adopt cover crops given the time and labor investment needed for cover crop implementation and management.

Several studies have found evidence that extreme weather affects farmers' adoption of practices. We use the variable *drought5yr* to indicate the number of years in which a D1 degree or above drought happened in at least 10 percent of the county areas during July and August in the last five years. Survey population and sample summary statistics for variables used in the logit model are reported in table 7.2. To avoid forbidden regression, all exogenous variables in yield moment equations are also included in the estimation of the cover crop equation. However, since many of them do not have realistic meanings (e.g., phosphorus use should not affect cover crop adoption), we only report several of their estimates, including regional dummy variables, soil texture, and slope of the field in the next section.

Table 7.2   Survey means and standard errors of variables in the logit model

Variable and Unit	Survey Mean	Std.Err
Cover crops (0/1)	0.100	0.003
Heartland (0/1)	0.498	0.003
Northern Great Plains (0/1)	0.141	0.003
Prairie Gateway (0/1)	0.066	0.001
Eastern Uplands (0/1)	0.054	0.001
Southern Seaboard (0/1)	0.053	0.002
Mississippi Portal (0/1)	0.105	0.002
Northern Crescent (0/1)	0.083	0.001
Fields having moderate or steeper slope (0/1)	0.556	0.004
Soil texture: loam (0/1)	0.369	0.003
Soil texture: clay (0/1)	0.170	0.003
Soil texture: sandy (0/1)	0.069	0.002
Soil texture: mixed (0/1)	0.369	0.005
Soil texture: silty (0/1)	0.019	0.001
Concern about water-driven erosion (0/1)	0.262	0.003
Concern about wind-driven erosion (0/1)	0.080	0.002
Concern about soil compaction (0/1)	0.261	0.004
Concern about poor drainage (0/1)	0.232	0.004
Concern about low organic matter (0/1)	0.109	0.003
Concern about water quality (0/1)	0.066	0.002
Other concerns (0/1)	0.025	0.001
Field classified as "highly erodible" (0/1)	0.180	0.003
Field contains wetland (0/1)	0.041	0.002
Age (years)	57.730	0.100
Years of experience (years)	33.326	0.121
College education (0/1)	0.251	0.002
Off-farm work (0/1)	0.185	0.004
Land ownership (0/1)	0.491	0.004
Farm size (acres)	1348.206	11.558
drought5yr (years)	0.764	0.007
Field size (acres)	50.015	0.328

7.4   Estimation Results

Utilizing a two-stage method, we estimate the logit model specified in equation (9) in the first stage to address the endogeneity issue. We explore the determinants of cover crop adoption with a focus on the effects of climate change, farmers' demographic information, and field characteristics. In the second stage, we estimate the moment equations of (6), (7), and (8).

7.4.1   Results of Logit Model for Cover Crop Adoption

The logit model estimation results are reported in table 7.3. The results show that there existed regional differences in the adoption of cover cropping. Compared to fields in Heartland, those in Northern Crescent, Northern Great Plains, Prairie Gateway, and Mississippi Portal were less likely

Table 7.3   **Estimates of parameters of the logit model**

Parameter	Estimation	Bootstrapped Std. Err.
Northern Crescent	−1.049***	0.157
Northern Great Plains	−3.100***	0.212
Prairie Gateway	−0.605***	0.205
Eastern Uplands	2.363***	0.129
Southern Seaboard	2.788***	0.172
Mississippi Portal	−1.187***	0.202
slope	0.520***	0.057
Soil texture: clay	0.711***	0.114
Soil texture: sandy	−0.249*	0.144
Soil texture: mixed	0.413***	0.087
Soil texture: silty	0.330	0.277
Concern about water-driven erosion	−0.413***	0.121
Concern about wind-driven erosion	0.532***	0.102
Concern about soil compaction	0.223*	0.129
Concern about poor drainage	−0.218**	0.105
Concern about low organic matter	−0.793***	0.121
Concern about water quality	0.901***	0.112
Other concerns	0.976***	0.142
Field "highly erodible"	0.545***	0.120
Field contains wetland	0.355	0.531
age	−0.024***	0.006
Years of experience	−0.005	0.005
College education	0.551***	0.078
Off-farm work	−0.258**	0.109
Land ownership	0.230*	0.080
Farm size	−9.280E-05***	3.160E-05
drought5yr	−0.089	0.068
Field size	−0.003***	0.001
Constant	−3.119***	0.667

Note: Statistical significance at 1%, 5%, and 10% are denoted as ***, **, and *, respectively.

to adopt cover crops while those in Eastern Uplands and Southern Seaboard were more likely to do so. The higher adoption rate of cover crops in Heartland, Eastern Uplands, and Southern Seaboard comparting to that in other regions can be attributed to several factors. In addition to federal programs such as the Environmental Quality Incentives Program (EQIP) and Conservation Stewardship Program (CSP), many states in Heartland, Eastern Uplands, and Southern Seaboard have implemented state incentive programs, which have been found positively correlated with the adoption of cover crops (e.g., Fleming 2017; Lichtenberg, Wang, and Newburn 2018; Wallander et al. 2021). The top seven state-funded cover-crop programs in terms of acreage in the US are all in the three regions (Wallander et al. 2021). In addition, tax credits, reduction on crop insurance premiums, and programs that rent out or loan equipment related to cover cropping in the three regions (Wallander et al. 2021) may also contribute to their higher adoption rates. Moreover, access to technical assistance and extension service may also cause variations in cover crop adoption across regions. Regional differences in the adoption of cover crops or other conservation practices have been found in other studies. Unger and Vigil (1998) find that the decision on adopting cover crops may vary from semiarid regions to humid/sub-humid regions depending on whether water is scarce or not. Given that western regions had a lower adoption rate than Heartland, Eastern Uplands, and Southern Seaboard, concerns about cover crops depleting soil water for the following main crops probably dominated in those drier western regions. In addition, as suggested in some other studies (e.g., Davey and Furtan 2008, Ding, Schoengold, and Tadesse 2009, and Claassen et al. 2018), a higher rate of adoption in Eastern Uplands and Southern Seaboard may reflect the concern about soil erosion given more rainfall in the regions.

The likelihood of cover crop adoption was also affected by farmers' concerns. Farmers who had concerns over wind-driven erosion, soil compaction, water quality, or other concerns were more likely to adopt cover crops than those who did not have such concerns. The result aligns itself with the benefits that cover crops are supposed to provide. In contrast, farmers who had concerns over water-driven erosion, poor drainage, or low organic matter were less likely to adopt than those without such concerns. Steele, Coale, and Hill (2012) did not observe consistent differences in total organic matter and labile organic matter between the winter cover crop and control soils in an experiment with 13 years of cover crop use. Our finding on the effect of concerns over low organic matters is consistent with Steele, Coale, and Hill's (2012) finding. While plenty of studies have found a positive effect of cover crops on soil organic matters and erosion by water (e.g., Shanks, Moore, and Sanders 1998; Ding et al. 2006; Dube, Chiduza, and Muchaonyerwa 2012), many studies have concluded that cover crops' effect on organic matter may vary with cover crop species, soil type, and other practices, such as tillage and rotation (e.g., Wulanningtyas et al. 2021; Abdollahi and Munkholm 2014; Dube, Chiduza, and Muchaonyerwa 2012; and Motta et al. 2007). Our results imply that cover crops might either have not practically worked well on improving water-driven erosion, poor drainage and organic matter conditional on commonly used management and practices in soybean production, or complexities of management and implementation of cover crops as well as their interactions with other practices have discouraged its adoption to address those concerns.

Field characteristics also affected the adoption of cover crops. As expected, if a field was classified as "highly erodible," it was more likely to adopt cover crops; fields with slopes were also more likely to adopt cover crops since they are more vulnerable to soil erosion. Whether a field contained wetland did not statistically significantly affect the adoption. Soil texture also affected farmers' decisions on cover crop adoption. Compared to loam soil, fields with clay or mixed soil were more likely to adopt cover crops and fields with sandy soil were less likely to do so. There was, however, no statistically significant difference between loam soil and silty soil in cover crop adoption. Larger farms and larger fields were less likely to adopt cover crops. This is expected since the larger the farm/field, the more labor and time are needed for the implementation and management of cover crops.

Farmers' demographic characteristics played an important role in practice adoption decisions. Older farmers or farmers working at least 50 percent off farms were less likely to adopt cover crops. This might be due to the labor and time requirements of cover crop implementation and management, as discussed previously. In addition, more educated farmers were more likely to adopt cover crops as they might better understand the importance of cover crops in the environment and agricultural sustainability. Consistent with expectation, farmers who had land ownership were more likely to adopt cover crops as they care more about their own land.

Drought in the last five years did not affect the likelihood of cover crop adoption. This might be due to the same reason as discussed above. Cover crops' drought mitigating effect may interact with other factors such as cover crop species, planting and termination time, tillage, rotation, soil type, etc. Proper combinations of the use of cover crops and other practices conditional on soil and weather conditions are required and many farmers might not have observed desired results of the use of cover crops as a drought adaptation strategy.

7.4.2   Results of Yield Moment Functions

Results of the first moment yield equation are presented in table 7.4. As expected, more mGDD increased the soybean yield while more ODD decreased the soybean yield. If the year had a drought in July or August, the mean yield decreased. Regional differences were shown in soybean yield. Fields in Heartland, which includes Illinois, Indiana, Iowa, and parts of Missouri, Nebraska, South Dakota, Minnesota, Ohio, and Kentucky, had the highest yield among all regions. Unsurprisingly, if soybeans were planted late or a bigger proportion of fields were replanted, the field had a lower yield, resulting from shortened vegetative and reproductive intervals. Inputs of fertilizer did help increase the yield. The more phosphorus was applied, the higher the yield. With the small parameter for the quadratic term, the effect of phosphorus on yield was close to linear. The application of manure also helped improve yield by adding more nutrients to the soil. Without cover crops, soybean yield in loam soil was higher than in clay, sandy, or mixed soil, but not significantly different from that in silty soil. This is consistent with the findings of previous studies such as Radočaj et al. (2020) and He, Luo, and Sun (2014).

Regional differences also showed in yield variance and skewness. Compared to Heartland, soybean yield in Southern Seaboard had lower varia-

Table 7.4 **Estimates of parameters of yield moment equations**

Parameter	First moment (Mean) equation		Second moment (variance) equation		Third moment (skewness) equation	
	Estimation	Bootstrapped Std. Err	Estimation	Bootstrapped Std. Err	Estimation	Bootstrapped Std. Err
Cover crops	−3.890	2.720	48.337***	9.092	2868.821***	341.341
Drought×cover crops	−20.160***	4.452	223.235***	21.604	16562.100***	1050.960
Cover crops×clay soil	6.869***	2.517	4.441	32.237	−12716.120***	1430.358
Cover crops×sandy soil	17.406***	2.509	−79.215***	17.073	−6299.486***	559.366
Cover crops×mixed soil	18.586***	1.655	−48.293***	18.088	−8944.720***	667.661
Cover crops×silty soil	14.110***	1.926	−154.608***	23.123	−4624.771***	793.497
Northern Crescent	−2.994***	0.314	−3.153	2.946	−263.873**	110.773
Northern Great Plains	−9.363***	0.472	19.329***	4.805	377.511***	146.329
Prairie Gateway	−4.436***	0.603	103.903***	7.312	−1081.599***	259.667
Eastern Uplands	−17.638***	1.147	39.311***	6.498	1982.181***	286.687
Southern Seaboard	−25.983***	0.819	−20.231***	4.326	932.423***	131.663
Mississippi Portal	−4.093***	0.668	153.902***	12.949	−2364.662***	500.023
PlantLate	−4.934***	0.256	−20.423***	−20.423***	−461.056***	77.673
ReplantPct	−6.639***	1.069	42.540***	11.625	4547.295***	499.821
Manure	1.956***	0.440	3.777	4.742	547.647***	184.636
Phosphorus	0.020***	0.004	−0.569***	0.062	−3.022	1.899
Phosphorus²	1.735E-04***	2.680E-05	0.003***	0.001	0.003	0.012
Drought18	−3.696***	0.401	30.540***	4.323	−856.462***	151.164
Soil texture: clay	−1.507***	0.389	7.641**	3.275	−788.829***	113.448
Soil texture: sandy	−10.515***	0.477	71.824***	5.500	2274.254***	211.552
Soil texture: mixed	−3.742***	0.243	−17.207***	2.521	446.028***	84.746
Soil texture: silty	0.485	0.429	1.400	10.029	−1525.396***	389.240
mGDD	0.013***	0.001	0.041***	0.007	−0.314	0.242
ODD	−0.184***	0.016	−0.650***	0.189	−5.393	6.670
Constant	23.424***	1.390	7.472	18.158	1100.763*	648.716
R^2	0.334		0.095		0.112	

Note: Statistical significance at 1%, 5%, and 10% are denoted as ***, **, and *, respectively.

tions, and it had higher variations in the Northern Great Plains, Prairie Gateway, Eastern Uplands, and Mississippi Portal. In addition, Northern Crescent, Prairie Gateway, and Mississippi Portal had higher downside risk; in contrast, Northern Great Plains, Eastern Upland, and Southern Seaboard had lower downside risk. The higher percentage of field replanted, the more variation and the lower downside risk in yield, probably resulting from the replacement of damaged plants, for example by frosts. If soybeans were planted late, their yield varied less but had a higher downside risk, probably due to a shorter growing season or higher probability of frost before harvest. At the mean application level, phosphorus inputs reduced the variation in yield, but manure application did not. In addition, more mGDD increased the variance while more ODD decreased the variance. Both did not change the downside risk. As expected, a drought that occurred in July or August increased the variance of yield and in the meantime increased the risk of crop failure. Yield moments also showed a heterogeneous effect on soil texture. Compared to loam soil, clay and sandy soils had higher variance; clay and silty soils had lower skewness or higher downside risks.

Given the interaction terms with soil texture and weather, marginal effects of cover crops were calculated and presented in table 7.5. Standard errors are calculated using the delta method. As shown in table 7.5, if there was no drought in July and August that year, then cover crops statistically significantly increased the yield of soybeans planted in sandy, silty, or mixed soil; but there was no statistically significant effect on soybean yield in loam and clay soils. This is consistent with many studies and experiments finding that cover crops help increase cash crop yield. In the meantime, cover crops increased yield variance in loam and clay soils, but decreased yield variance in sandy and silty soils and had no significant effect in mixed soil. In addition, planting cover crops reduced downside risk in loam soil while increased downside risk in all other types of soils. From the above, we can see that the effects of cover crops depend on soil types when there was no drought. And there was always a trade-off between the mean, the variation, and the downside risk of yield, i.e., there was no simultaneous positive effect of cover crops on the three moments of yield, which affected farmers' expected utility.

With droughts in July and August, cover crops reduced soybean yield in all soil types, although the effects in sandy and mixed soil were not statistically significant. This implies that cover crops consumed water in the soil for their own growth and reduced water available for the following cash crops. When drought occurred, the water supply worsened to a point where crop yield decreased. In addition, cover crops increased yield variance but reduced the risk of crop failure in all types of soils when drought occurred. The mixture of positive and negative effects of cover crops on yield moments is somewhat consistent with the finding in the first stage that previous droughts did not affect farmers' adoption of cover crops, implying a divided acceptance of cover crops as a drought adaptation strategy among farmers.

Table 7.5   **Marginal effects of cover crops under different conditions**

Soil type	Drought	yield equation		second moment (variance) equation		third moment (skewness) equation	
		Estimation	Std. Err	Estimation	Std. Err	Estimation	Std. Err
loam	No	−3.890	2.720	48.337***	9.092	2868.821***	341.341
	Yes	−24.049***	4.579	271.572***	26.449	19430.921***	1244.251
Clay	No	2.980	3.108	52.778*	31.177	−9847.299***	1426.819
	Yes	−17.180***	3.826	276.013***	31.774	6714.801***	1166.486
sandy	No	13.516***	2.349	−30.878***	13.859	−3430.665***	391.965
cover crops	Yes	−6.643	4.586	192.357***	23.436	13131.435***	1080.542
mixed	No	14.696***	2.375	0.045	13.002	−6075.899***	461.852
	Yes	−5.463	4.216	223.280***	17.548	10486.201***	701.112
silty	No	10.220***	1.839	−106.271***	21.328	−1755.950***	709.924
	Yes	−9.940***	4.234	116.964***	26.404	14806.150***	981.327

Note: Statistical significance at 1%, 5%, and 10% are denoted as ***, **, and *, respectively.

7.5   Conclusions and Discussions

We explored factors affecting farmers' adoption of cover crops by a logit model and examined the effects of cover crops on soybean yield and its risk by three moment functions. By incorporating two interplays between cover crops and soil type, and cover crops and drought, we were able to explore the varying effects of cover crops in drought and different soil types. While we found that the adoption of cover crops varied in regions and soil types and was affected by field properties and farmers' demographic characteristics and concerns, we did not find a significant effect of previous droughts on the adoption. The results from the moment functions of soybean yield confirmed what the results in the first stage suggest. When there was a drought, cover crops reduced yield and increased yield variance. However, cover crops also reduced the downside risk of crop failure in the meantime. The mixed effects of cover crops on yield and its risk associated with an occurrence of drought support the statistically insignificant effect of the previous drought on cover crop adoption, implying that farmers were divided in the acceptance of cover crops as a means to build resilience to drought. The mixed effect of cover crops also warrants a further study to calculate the certainty equivalent of net economic return of soybeans with cover crops, which requires information on cover crop seed, planting, and termination cost as well as additional or saved fertilizer and pesticide costs. The certainty equivalent of net economic returns of soybeans with cover crops may provide more information on economic impediments to farmer adoption.

The low adoption rate of cover crops may also be related to complex interactions between management and cultural practices including species selection, planting and termination date, rotation, and termination method. Achieving desired benefits requires significant training, learning, and adjustments in many aspects of the farming system (Wallander et al. 2021). As shown in the National Cover Crop Survey 2020 (Conservation Technology Information Center 2020), roughly 70 percent of respondents said that they typically used their own experience of trial and error for cover cropping. About 67 percent and 60 percent of the respondents considered the two approaches, i.e., local farm tours to see how cover crops worked and one-on-one technical assistance to select, plant, or manage cover crops, very helpful or moderately helpful, respectively, in encouraging them to try cover cropping. Therefore, programs that provide necessary training and showcase cover crop management to farmers could address an important lack of information.

Greater soil and environmental benefits can be achieved when cover crops are utilized in conjunction with other practices (Wallander et al. 2021), such as conservative tillage, irrigation, crop rotation, nutrient management, and adoption of drought tolerant seeds—which are currently available for maize. A broader range of research that finds proper combinations of cover crops and other practices conditional on soil types and weather are crucial for establishing practice guidance for farmers. Such guidance can help farmers achieve desired results by using cover crops as a drought adaptation strategy as well as a tool for improving soil and environmental benefits along with a suite of other conservation practices.

The recent increase in cover crop adoption has been accompanied by financial incentives. Given financial support, cover crop acres enrolled in the Environmental Quality Incentives Program (EQIP) increased from 312.6 thousand acres in 2009 to 2,443.1 thousand acres in 2020 (USDA Climate Hubs 2021). In 2018, about one-third of the acreage planted with a cover crop received a financial assistance payment for cover crop adoption from either federal, state, or other programs, ranging from $12 per acre to $92 per acre (Wallander et al. 2021). The USDA NRCS recently announced a program to promote the use of soil health practices, especially cover crops. The initiative sets a goal of doubling the number of corn and soybean acres using cover crops to 30 million acres by 2030 (USDA 2022). Given the mixed effects of cover crops on soybean yield and yield risks found in this study, financial incentives can help improve the certainty equivalent of net returns and thus encourage more risk-averse farmers to adopt cover crops. In addition, farmers have recently been paid to plant cover crops by large seeds, chemical, and food companies to generate carbon credits to offset their environmental footprints (Reuters 2022). The payments, however, are generally not as much as those from EQIP and CSP. In addition, the current carbon credit market lacks transparency and liquidity (Ag Decision Maker 2021). It is facing several challenges including setting up protocols to ensure the additionality and permanence of net greenhouse gas (GHG) reductions (Blaustein-Rejto 2021).

We conclude the paper by recognizing a key limitation of this study: it does not consider the long-term effects of sustainable practice adoption by using cross-sectional data. Cover crops have multiple benefits to soil and the environment. Cover crops can be used not only as a drought adaptation strategy but also to reduce soil erosion, enhance weed control, improve soil health, increase carbon storage, improve water quality through reduced nutrient and sediment runoff, and increase biological diversity. While solely comparing the cost of seed, seeding, and management to the impact on the yield of the following main crop may show a loss in the first few years, cover crops may possibly improve the efficiency and resiliency of the entire farm over time, resulting in a net benefit from the broad, holistic standpoint (Myers, Weber, and Tellatin, 2019). Myers, Weber, and Tellatin (2019), for example, show that the adoption of cover crops may have negative net returns in the first year, negligible net returns in three years, but about $18 net returns in five years. In addition, if cover crops are used to address more than one yield-limiting factor in a field such as for grazing, improving soil health, and weed impression, then the net return can be larger and faster.

This can be applied to many other sustainable practices. For example, the payoffs from investments in improving soil fertility and reducing soil erosion are cumulative and may take several years. And the subsequent improvement in soil fertility and reduction in soil erosion can reduce future expenses for crop nutrients, irrigation, and energy (Lee 2005; Tilman et al. 2001). If such long-term positive net returns can be demonstrated by more farmers who are supported financial assistance from federal and state programs to offset a portion of upfront investments—which have been proved very useful in increasing the adoption (Bowman and Lynch 2019)—then the adoption of cover crops, as well as other drought adaptation strategies, may surge.

References

Abdollahi, L., and L. Munkholm 2014. "Tillage System and Cover Crop Effects on Soil Quality: I. Chemical, Mechanical, and Biological Properties." *Soil Science Society of America Journal* 78 (1): 262–70.

Acharya, B., S. Dodla, L. Gaston, M. Darapuneni, J. Wang, S. Sepat, and H. Bohara. 2019. "Winter Cover Crops Effect on Soil Moisture and Soybean Growth and Yield under Different Tillage Systems." *Soil and Tillage Research* 195: 104430.

Acuña, J. C., and M. Villamil. 2014. "Short-Term Effects of Cover Crops and Compaction on Soil Properties And Soybean Production in Illinois. *Agronomy Journal* 106 (3): 860–70.

Ag Decision Maker. 2021. How to Grow and Sell Carbon Credits in US Agriculture. Accessed June 12, 2022. https://www.extension.iastate.edu/agdm/crops/pdf/a1-76.pdf.

Anderson, J. R., J. D. Dillion, and B. Hardaker. 1980. *Agricultural Decision Analysis*. Ames, Iowa: The Iowa State University Press.

Andraski, T. W., and L. G. Bundy. 2005. "Cover Crop Effects on Corn Yield Response to Nitrogen on an Irrigated Sandy Soil." *Agronomy Journal* 97 (4): 1239–1244.

Antle, J. 1983. "Testing the Stochastic Structure of Production: A Flexible Moment-Based Approach." *Journal of Business & Economic Statistics* 1 (3): 192–201.

Antonakis, J., S. Bendahan, P. Jacquart, and R. Lalive. 2014. "Causality and Endogeneity: Problems and Solutions." In *The Oxford Handbook of Leadership and Organizations*, edited by D. V. Day, 93–117. New York: Oxford University Press.

Arbuckle, J. 2016. Iowa Farm and Rural Life Poll: 2015 Summary Report. Accessed November 12, 2021. https://dr.lib.iastate.edu/server/api/core/bitstreams/b8e999e7-f588-4dfb-9ed6-9ff93781f37a/content.

Arora, V. K., C. Singh, A. Sidhu, and S. Thind. 2011. "Irrigation, Tillage and Mulching Effects on Soybean Yield and Water Productivity in Relation to Soil Texture." *Agricultural Water Management* 98 (4): 563–68.

Barker, J. B., D. M. Heeren, K. Koehler-Cole, C. A. Shapiro, H. Blanco-Canqui, R. W. Elmore, C. A. Proctor, S. Irmak, C. A. Francis, T. M. Shaver, and A. T. Mohammed. 2018. "Cover Crops Have Negligible Impact on Soil Water in Nebraska Maize-Soybean Rotation." *Agronomy Journal* 110: 1–1.

Blanco-Canqui, H., T. Shaver, J. Lindquist, C. Shapiro, R. Elmore, C. Francis, and G. Hergert. 2015. "Cover Crops and Ecosystem Services: Insights from Studies in Temperate Soils." *Agronomy Journal* 107 (6): 2449–2474.

Blaustein-Rejto, D. 2021. "Dishing the Dirt on Ag Carbon Credits." Accessed June 13, 2022. https://agfundernews.com/carbon-credits-in-ag-dishing-the-dirt.

Bowman, M., and L. Lynch. 2019. "Government Programs that Support Farmer Adoption of Soil Health Practices: A Focus on Maryland's Agricultural Water Quality Cost-Share Program." *Choices*. Quarter 2.

Butcher, K., A. F. Wick, T. DeSutter, A. Chatterjee, and J. Harmon. 2018. "Corn and Soybean Yield Response to Salinity Influenced by Soil Texture." *Agronomy Journal* 110: 1243–1253.

Chu, M., S. Jagadamma, F. Walker, N. Eash, M. Buschermohle, and L. Duncan. 2017. "Effect of multispecies cover crop mixture on soil properties and crop yield." *Agricultural & Environmental Letters* 2 (1): 170030.

Claassen, R., M. Bowman, J. McFadden, D. Smith, and S. Wallander. 2018. "Tillage Intensity and Conservation Cropping in the United States." Economic Information Bulletin No. 197. Economics Research Services, USDA.

Clark, A., A. Decker, J. Meisinger, and M. McIntosh. 1997. "Kill Date of Vetch, Rye, and a Vetch–Rye Mixture: I. Cover Crop and Corn Nitrogen." *Agronomy Journal* 89 (3): 427–34.

Conservation Technology Information Center (CTIC). 2015. Report of the 2014–2015 Cover Crop Survey. Joint publication of the CTIC and the North Central Region Sustainable Agriculture Research and Education Program. Accessed January 15, 2022. West Lafayette, Indiana.

———.2020. National Cover Crop Survey. https://www.ctic.org/files/CoverCrop Survey 2020 24mb.pdf.

Corak, S. J., W. W. Frye, and M. S. Smith. 1991. "Legume Mulch and Nitrogen Fertilizer Effects on Soil Water and Corn Production." *Soil Science Society of America Journal* 55 (5): 1395–1400.

Davey, K. A., and W. H. Furtan. 2008. "Factors That Affect the Adoption Decision of Conservation Tillage in the Prairie Region of Canada." *Canadian Journal of Agricultural Economics* 56 (3): 257–75.

Dean, J. E., and R. Weil. 2009. "Brassica Cover Crops for Nitrogen Retention in the Mid-Atlantic Coastal Plain." *Journal of Environmental Quality* 38 (2): 520–28.

Day, R. H. 1965. "Probability Distributions of Field Crops." *Journal of Farm Economics* 47: 713–41.

Di Falco, S., and J. Chavas. 2009. "On Crop Biodiversity, Risk Exposure, and Food Security in the Highlands of Ethiopia." *American Journal of Agricultural Economics* 91 (3): 599–611.

Di Falco, S., and M. Veronesi. 2014. "Managing Environmental Risk in Presence of Climate Change: The Role of Adaptation in the Nile Basin of Ethiopia." *Environmental and Resource Economics* 57 (4): 553–77.

Ding, G., X. Liu, S. Herbert, J. Novak, D. Amarasiriwardena, and B. Xing. 2006. "Effect of Cover Crop Management on Soil Organic Matter." *Geoderma* 130 (3–4): 229–39.

Ding, Y., K. Schoengold, and T. Tadesse. 2009. "The Impact of Weather Extremes on Agricultural Production Methods: Does Drought Increase Adoption of Conservation Tillage Practices?" *Journal of Agricultural and Resource Economics* 34: 395–411.

Dong, F., D. Hennessy, and H. Jensen. 2010. "Contract and Exit Decisions in Finisher Hog Production." *American Journal of Agricultural Economics* 92 (3): 667–84.

Dong, F., D. Hennessy, H. Jensen, and R. Volpe. 2016. "Technical Efficiency, Herd Size, and Exit Intentions in US Dairy Farms." *Agricultural Economics* 47 (5): 533–45.

Dong, S., Y. Jiang, Y. Dong, L. Wang, W. Wang, Z. Ma, C. Yan, C. Ma, and L. Liu. 2019. "A Study on Soybean Responses to Drought Stress and Rehydration." *Saudi Journal of Biological Sciences* 26 (8): 2006–2017.

Dube, E., C. Chiduza, and P. Muchaonyerwa. 2012. "Conservation Agriculture Effects on Soil Organic Matter on a Haplic Cambisol after Four Years of Maize–Oat and Maize–Grazing Vetch Rotations in South Africa." *Soil and Tillage Research* 123: 21–28.

Ebelhar, S. A., W. W. Frye, and R. L. Blevins. 1984. "Nitrogen from Legume Cover Crops for No-Tillage Corn 1." *Agronomy Journal* 76 (1): 51–55.

Fleming, P. 2017. "Agricultural Cost Sharing and Water Quality in the Chesapeake Bay: Estimating Indirect Effects of Environmental Payments." *American Journal of Agricultural Economics* 99 (5): 1208–1227.

Groff, S. 2015. "The Past, Present, and Future of the Cover Crop Industry." *Journal of Soil and Water Conservation* 70 (6):130A–133A.

Hardaker, J. B., J. Richardson, G. Lien, and K. Schumann. 2004. "Stochastic Efficiency Analysis with Risk Aversion Bounds: A Simplified Approach." *Australian Journal of Agricultural and Resource Economics* 48 (2): 253–70.

He, W., X. Luo, and G. Sun. 2014. "The Trend of GIS-Based Suitable Planting Areas for Chinese Soybean under the Future Climate Scenario." In *Ecosystem Assessment and Fuzzy Systems Management*, edited by B. Y. Cao, S. Q. Ma, and H. H. Cao, 325–38. Cham, Switzerland: Springer International Publishing.

Hunter, M., M. Schipanski, M. Burgess, J. LaChance, B. Bradley, M. Barbercheck, J. Kaye, and D. Mortensen. 2019. "Cover Crop Mixture Effects on Maize, Soybean, and Wheat Yield in Rotation." *Agricultural and Environmental Letters* 4 (1).

Jiang, P., and K. Thelen. 2004. "Effect of Soil and Topographic Properties on Crop Yield in a North-Central Corn–Soybean Cropping System." *Agronomy Journal* 96: 252–58.

Karl, T. 1986. "The Sensitivity of the Palmer Drought Severity Index and Palmer's Z-Index to their Calibration Coefficients Including Potential Evapotranspiration." *Journal of Climate and Applied Meteorology* 25 (1): 77–86.

Kaspar, T. C., D. J. Pulido, T. Fenton, T. Colvin, D. Karlen, D. Jaynes, and D. Meek. 2004. "Relationship of Corn and Soybean Yield to Soil and Terrain Properties." *Agronomy Journal* 96: 700–709.

Kendall, M. G., and A. Stuart. 1977. *The Advanced Theory of Statistics*, Vol. 1, 168. New York: Macmillan.

Kessler, A., S. Archontoulis, and M. Licht. 2020. "Soybean Yield and Crop Stage Response to Planting Date and Cultivar Maturity in Iowa, USA." *Agronomy Journal* 112: 382–94.

Kim, K., J. P. Chavas, B. Barham, and J. Foltz. 2014. "Rice, Irrigation and Downside Risk: A Quantile Analysis of Risk Exposure and Mitigation on Korean Farms." *European Review of Agricultural Economics* 41 (5): 775–815.

Koundouri, P., C. Nauges, and V. Tzouvelekas. 2006. "Technology adoption under production uncertainty: theory and application to irrigation technology." *American Journal of Agricultural Economics* 88 (3): 657–670.

Kravchenko, A. N., D. Bullock, and C. Boast. 2000. "Joint Multifractal Analysis of Crop Yield and Terrain Slope." *Agronomy Journal* 92: 1279–1290.

Laloy, E., and C. Bielders. 2010. "Effect of Intercropping Period Management on Runoff and Erosion in a Maize Cropping System." *Journal of Environmental Quality* 39 (3): 1001–1008.

Langdale, G. W., R. L. Blevins, D. L. Karlen, D. K. McCool, M. A. Nearing, E. L. Skidmore, A. W. Thomas, D. D. Tyler, and J. R. Williams. 1991. "Cover Crop

Effects on Soil Erosion by Wind and Water." In *Cover Crops for Clean Water*, edited by W. L. Hargrove, 15–21. Ankeny, IA: SWCS.

Le, D. T., R. Nishiyama, Y. Watanabe, M. Tanaka, M. Seki, K. Yamaguchi-Shinozaki, K. Shinozaki, L. Tran. 2012. "Differential Gene Expression in Soybean Leaf Tissues at Late Developmental Stages under Drought Stress Revealed by Genome-Wide Transcriptome Analysis." *PLoS One* 7: e49522.

Lee, D. 2005. "Agricultural Sustainability and Technology Adoption: Issues and Policies for Developing Countries." *American Journal of Agricultural Economics* 87: 1325–1334.

Lichtenberg, E., H. Wang, and D. Newburn. 2018. "Uptake and Additionality in a Green Payment Program: A Panel Data Study of the Maryland Cover Crop Program." Paper presented at the annual meeting of the Agricultural and Applied Economics Association, August 5–7, Washington, DC.

Linkemer, G., J. Board, and M. Musgrave. 1998. "Waterlogging Effects on Growth and Yield Components in Late-Planted Soybean." *Crop Science* 38: 1576–1584.

Lu, Y. C., K. B. Watkins, J. R. Teasdale, and A. A. Abdul-Baki. 2000. "Cover crops in sustainable food production." *Food Reviews International* 16 (2): 121–157.

Major, D. J., D. Johnson, J. Tanner, and I. Anderson. 1975. "Effects of Day Length and Temperature on Soybean Development." *Crop Science* 15: 174–79.

McDaniel, M. D., L. Tiemann, and A. Grandy. 2014. "Does Agricultural Crop Diversity Enhance Soil Microbial Biomass and Organic Matter Dynamics? A Meta-analysis." *Ecological Applications* 24 (3): 560–70.

Myers, R., A. Weber, and S. Tellatin. 2019. "Cover Crop Economics: Opportunities to Improve Your Bottom Line in Row Crops." SARE Technical Bulletin. Accessed June 12, 2022. https://www.sare.org/wp-content/uploads/Cover-Crop-Economics.pdf.

Miguez, F. E. and G. A. Bollero. 2005. "Review of Corn Yield Response under Winter Cover Cropping Systems Using Meta-analytic Methods." *Crop Science* 45 (6): 2318–2329.

Miller, L., J. Chin, and K. Zook. 2012. "Policy Opportunities to Increase Cover Crop Adoption on North Carolina Farms." Masters Project, Duke University, Durham, NC.

Mitchell J. P., A. Shrestha, and S. Irmak. 2015. "Trade-offs between Winter Cover Crop Production and Soil Water Depletion in the San Joaquin Valley, California." *Journal of Soil and Water Conservation* 70 (6):430–40,

Motta, A. C., D. Wayne Reeves, C. Burmester, and Y. Feng. 2007. "Conservation Tillage, Rotations, and Cover Crop Affecting Soil Quality in the Tennessee Valley: Particulate Organic Matter, Organic Matter, and Microbial Biomass." *Communications in Soil Science and Plant Analysis* 38 (19–20): 2831–2847.

Munawar, A., R. Blevins, W. Frye, and M. Saul. 1990. "Tillage and Cover Crop Management for Soil Water Conservation." *Agronomy Journal* 82 (4): 773–77.

Muñoz, J. D., J. Steibel, S. Snapp, and A. Kravchenko. 2014. "Cover Crop Effect on Corn Growth and Yield as Influenced by Topography." *Agriculture, Ecosystems & Environment* 189: 229–39.

National Drought Mitigation Center (NDMC). 2021. "What Is the USDM." Accessed November 11, 2021. https://droughtmonitor.unl.edu/About/WhatistheUSDM.aspx.

National Research Council, 2007. *Understanding American Agriculture: Challenges for the Agricultural Resource Management Survey*. Washington, DC: The National Academies Press.

Nelson, K. A., and C. Meinhardt. 2011. "Soybean Yield Response to Pyraclostrobin and Drainage Water Management." *Agronomy Journal* 103: 1359–1365.

Nielsen, D. C., D. Lyon, R. Higgins, G. Hergert, J. Holman, and M. Vigil. 2016. "Cover Crop Effect on Subsequent Wheat Yield in the Central Great Plains." *Agronomy Journal* 108 (1): 243–56.

OECD. 2011. *Managing Risk in Agriculture: Policy Assessment and Design*. Paris: Organization for Economic Cooperation and Development.

Paz, J., W. D. Batchelor, G. L. Tylka, and R. G. Hartzler. 2001. "A Modeling Approach to Quantify the Effects of Spatial Soybean Yield Limiting Factors." *Transactions of the ASAE* 44 (5): 1329–1334.

Pedersen, P., and M. Licht. 2014. "Soybean Growth and Development." PM 19948. Ames, IA: Iowa State University Extension and Outreach.

Radočaj, D., M. Jurišić, V. Zebec, and I. Plaščak. 2020. "Delineation of Soil Texture Suitability Zones for Soybean Cultivation: A Case Study in Continental Croatia." *Agronomy* 10 (6): 823.

Reddy, K. 2001. "Effects of Cereal and Legume Cover Crop Residues on Weeds, Yield, and Net Return in Soybean (Glycine max)." *Weed Technology* 15 (4): 660–68.

Reddy, P. P. 2017. "Cover/Green Manure Cropping." In *Agro-ecological Approaches to Pest Management for Sustainable Agriculture*, 91–107. Springer, Singapore.

Reuters. 2022. "Off-Season 'Cover' Crops Expand as U.S. Growers Eye Low-Carbon Future." Accessed January 5, 2022. https://www.reuters.com/markets /commodities/farming-climate-off-season-cover-crops-expand-us-growers-eye -low-carbon-future-2022-01-04/.

Robertson, G. P., T. Bruulsema, R. Gehl, D. Kanter, D. Mauzerall, C. Rotz, and C. Williams. 2013. "Nitrogen–Climate Interactions in US Agriculture." *Biogeochemistry* 114 (1): 41–70.

Sackett, J. L. 2013. "An NCR-SARE Cover Crop Project: Farmer-Cooperator Motivation and Agronomic Practices." *Journal of the NACAA* 6 (2).

Sainju, U., B. Singh, and W. Whitehead. 2002. "Long-term effects of tillage, cover crops, and nitrogen fertilization on organic carbon and nitrogen concentrations in sandy loam soils in Georgia, USA." *Soil & Tillage Research* 63: 167–179.

Sainju, U., B. Singh, F. Whitehead, and S. Wang. 2006. "Carbon supply and storage in tilled and nontilled soils as influenced by cover crops and nitrogen fertilization." *Journal of Environmental Quality*: 1507–17.

Schnepf, M., and C. Cox. 2006. "Environmental benefits of conservation on cropland: the status of our knowledge." In *Environmental Benefits of Conservation on Cropland: The Status of Our Knowledge*, edited by M. Schnepf and C. Cox. Ankeny, IA: Soil and Water Conservation Society.

Shaheen, T., M. Rahman, M. Shahid Riaz, Y. Zafar, and M. Rahman. 2016. "Soybean Production and Drought Stress." In *Abiotic and Biotic Stresses in Soybean Production*, edited by M. Miransari. San Diego, CA: Academic Press.

Shane, W. W., and K. Barker. 1986. "Effects of Temperature, Plant Age, Soil Texture, and Meloidogyne incognita on Early Growth of Soybean." *Journal of Nematology* 18 (3): 320–26.

Shanks L., D. Moore, and C. Sanders. 1998. "Soil erosion." In *Cover Cropping in Vineyards. A Grower's Handbook*, edited by C. A. Ingels, R. L. Bugg, G. T. McGourty, and L. P. Christensen, 80–85. Oakland, CA: University of California. Publication 3338.

Smith, R. G., L. Atwood, and N. Warren. 2014. "Increased Productivity of a Cover Crop Mixture Is Not Associated with Enhanced Agroecosystem Services." *PloS One* 9 (5): p.e97351.

Snapp, S. S., S. Swinton, R. Labarta, D. Mutch, J. Black, R. Leep, J. Nyiraneza, and K. O'Neil. 2005. "Evaluating Cover Crops for Benefits, Costs and Performance within Cropping System Niches." *Agronomy Journal* 97 (1): 322–32.

Specht, J. E., D. Hume, and S. Kumudin. 1999. "Soybean Yield Potential—A Genetic and Physiological Perspective." *Crop Science* 39: 1560–1570.

Steele, M., F. J. Coale, R. L. Hill. 2012. "Winter Annual Cover Crop Impacts on No-Till Soil Physical Properties and Organic Matter." *Soil Science Society of America. Journal* 76: 2164–2173.

SARE (Sustainable Agriculture Research and Education Program). 2007. *Managing Cover Crops Profitably*, 3rd edition. Accessed on March 2, 2023. https://www.sare.org/Learning-Center/Books/Managing-Cover-Crops-Profitably-3rd-Edition.

———. 2017. "Cover Crops. What Is Sustainable Agriculture?" UC Division of Agriculture and Natural Resources. Accessed January 5, 2022.https://sarep.ucdavis.edu/sustainable-ag/cover-crops.

Tack, J., A. Harri, and K. Coble. 2012. "More Than Mean Effects: Modeling the Effect of Climate on the Higher Order Moments of Crop Yields." *American Journal of Agricultural Economics* 94 (5): 1037–1054.

Tilman, D. K., J. Fargione, B. Wolff, C. D'Antonio, A. Dobson, R. Howarth, D. Schindler, W. Schlesinger, D. Simberloff, and D. Swackhamer. 2001. "Forecasting Agriculturally Driven Global Environmental Change." *Science* 292: 281–84.

Unger, P. W., and M. F. Vigil. 1998. "Cover Crop Effects on Soil Water Relationships." *Journal of Soil and Water Conservation* 53 (3): 200–207.

USDA. 2022. "USDA Offers Expanded Conservation Program Opportunities to Support Climate Smart Agriculture in 2022." Accessed January 15, 2022. https://www.usda.gov/media/press-releases/2022/01/10/usda-offers-expanded-conservation-program-opportunities-support.

USDA Climate Hubs. 2021. "Drought Resistant Practices." Accessed October 20, 2021. https://www.climatehubs.usda.gov/hubs/northeast/topic/drought-resistant-practices.

USDA ERS. 2000. "Farm Resource Regions." Agricultural Information Bulletin Number 760. Accessed on September 5, 2021. https://www.ers.usda.gov/webdocs/publications/42298/32489_aib-760_002.pdf?v=42487#:~:text=The%20Farm%20Resource%20Regions%20are%20derived%20from%20four,Land%20Resource%20Regions%2C%20and%20NASS%20Crop%20Reporting%20Districts.

USDA NRCS. 2018. Technical Note: Iowa Agronomy Technical Note 38 (Cover Crop Management). Accessed November 1, 2021. https://efotg.sc.egov.usda.gov/references/public/IA/Cover_Crop_Management_38_AGR_TN_2018_08.pdf.

———. 2019. NRCS Cover Crop Termination Guidelines Version 4.

USDA Risk Management Agency. 2015. "Conservation Compliance—Highly Erodible Land and Wetlands. Risk Management Agency Fact Sheet." Accessed January 15, 2022. https://www.rma.usda.gov/en/Fact-Sheets/National-Fact-Sheets/Conservation-Compliance-Highly-Erodible-Land-and-Wetlands#:~:text=Highly%20erodible%20land%20is%20any%20land%20that%20can,total%20field%20acreage%20that%20contains%20highly%20erodible%20soils.

Vitale, J. D., C. Godsey, J. Edwards, and R. Taylor. 2011. "The Adoption of Conservation Tillage Practices in Oklahoma: Findings from a Producer Survey." *Journal of Soil and Water Conservation* 66 (4): 250–64.

Wallander, S., D. Smith, M. Bowman, and R. Claassen. 2021. *Cover Crop Trends, Programs, and Practices in the United States*. Economic Information Bulletin number 222, U.S. Department of Agriculture, Economic Research Service.

Wandel, J., and J. Smithers. 2000. "Factors Affecting the Adoption of Conservation Tillage on Clay Soils in Southwestern Ontario, Canada." *American Journal of Alternative Agriculture* 15 (4): 181–88.

Westcott, P., and M. Jewison. 2013. "Weather Effects on Expected Corn and Soybean Yields." FDS-13g-01. Economic Research Service/USDA.

Wulanningtyas, H., Y. Gong, P. Li, N. Sakagami, J. Nishiwaki, and M. Komatsuzaki. 2021. "A Cover Crop and No-Tillage System for Enhancing Soil Health by Increasing Soil Organic Matter in Soybean Cultivation." *Soil and Tillage Research* 205.

Climate Change and Downstream Water Quality in Agricultural Production
The Case of Nutrient Runoff to the Gulf of Mexico

Levan Elbakidze, Yuelu Xu, Philip W. Gassman,
Jeffrey G. Arnold, and Haw Yen

8.1 Introduction

The Mississippi River basin (MRB) spans more than 3.2 million square kilometers, is dominated by agricultural land use, and is the largest drainage basin in the US. Approximately 70 percent of US cropland is in the MRB (Kumar and Merwade 2011; Marshall et al. 2018). Agricultural production in the MRB relies on intensive nitrogen (N) fertilizer use with a well-documented negative externality in the form of hypoxia in the Gulf of Mexico.

Hypoxia in the gulf has been a public concern for decades due to the detrimental consequences for the aquatic ecosystems (US EPA 2019). N runoff to the gulf and the consequent eutrophication of coastal waters promotes algal bloom. Decomposing algae depletes the marine ecosystem of dissolved

Levan Elbakidze is an associate professor in the Division of Resource Economics and Management, Davis College of Agriculture, Natural Resources and Design, and a faculty research associate at the Regional Research Institute and at the Center for Innovation in Gas Research and Utilization at West Virginia University.

Yuelu Xu is a postdoctoral fellow in the Division of Resource Economics and Management, Davis College of Agriculture, Natural Resources and Design at West Virginia University.

Philip W. Gassman is an associate scientist at the Center for Agricultural and Rural Development at Iowa State University.

Jeffrey G. Arnold is an agricultural engineer at the Grassland Soil and Water Research Laboratory of the Agricultural Research Service of the US Department of Agriculture (USDA-ARS).

Haw Yen is a senior scientist at Bayer Crop Science, and an Affiliate Assistant Professor at the College of Forestry, Wildlife, and Environment at Auburn University.

For acknowledgments, sources of research support, and disclosure of the authors' material financial relationships, if any, please see https://www.nber.org/books-and-chapters/american -agriculture-water-resources-and-climate-change/climate-change-and-downstream-water -quality-agricultural-production-case-nutrient-runoff-gulf-mexico.

oxygen, which is critical for sustaining aquatic ecosystems. Oxygen depletion results in hypoxic or "dead" zones as marine life either dies or migrates to other areas. In 2001, the EPA established the Gulf of Mexico Hypoxia Task Force to reduce the size of the hypoxic zone to 5,000 km^2 by 2035 (US EPA 2014). In 2021, the hypoxic zone in the gulf still reached 16,405 km^2, significantly exceeding the EPA goal (US EPA 2021a).

Climate change, with higher temperatures, more variable rainfall, and elevated CO_2 concentrations, can alter crop yields and agricultural production. Previous literature documents mixed expected impacts of climate change on crop yields in the MRB. Panagopoulos et al. (2014) simulated corn and soybean yields in the Upper Mississippi River basin (UMRB, a subbasin of the MRB) using the Soil and Water Assessment Tool (SWAT) for the baseline climate (1981–2000) and seven future (2046–2065) GCM climate projections under four agricultural management scenarios. Predicted corn and soybean yields modestly decline relative to the baseline climate conditions under all future climates and agricultural management scenarios. Panagopoulos et al. (2015) reported similar results for the Ohio-Tennessee River basin (OTRB, a subbasin of the MRB), with predicted corn and soybean yields in all examined future climates and agricultural management practices declining relative to the corresponding baseline scenarios. Chen et al. (2019) modeled the effects of climate change on crop yields in the northern High Plains of Texas (partially located within the MRB) using SWAT. They found that the median irrigated corn and sorghum yields would decrease by 3–22 percent and 6–42 percent, respectively, relative to the historical values. Median non-irrigated sorghum yield would decrease by up to 10 percent.

The changes in crop yields in the MRB may influence agricultural input and land use with associated implications for environmental outcomes in the Gulf of Mexico. On the one hand, the use of N fertilizer may intensify to compensate for losses in crop yields. This may increase N runoff from the MRB and exacerbate hypoxia in the Gulf of Mexico. On the other hand, lower yields may reduce profitability of crop production and may result in decreased crop acreage, which could decrease N runoff to the Gulf of Mexico. The net effect of climate change–driven changes in crop yields on N runoff to the Gulf of Mexico is thus unclear and should be examined empirically.

The MRB is the largest basin in the US and includes several large subbasins with different agricultural practices and contributions to the Gulf N runoff. For example, UMRB and OTRB are major N contributors to the Gulf (Kling et al. 2014; White et al. 2014). In the Corn Belt, highly fertile soils, relatively level land, hot days and nights, and well-distributed precipitation during the growing season provide ideal conditions for crop production (Wu, Qu, and Hao 2015). These factors have led to prevalent corn-soybean rotation with high fertilizer use and tile drainage systems.

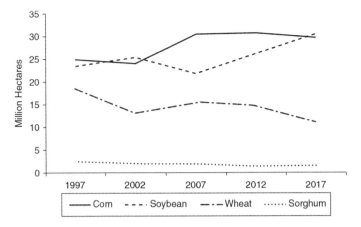

Fig. 8.1   **Harvested acreage within the MRB over time (ha)**

The Missouri and Arkansas-Red-White River basin includes both rainfed and irrigated crop production. In Nebraska, western Kansas, Oklahoma, and north Texas, groundwater from Ogallala aquifer is a major source of irrigation for agricultural production (Xu et al. 2022). Some of the climate projection scenarios suggest that regions with rainfed agriculture will be wetter and regions relying on irrigation will be drier (NCAR 2022a). These spatially heterogeneous changes, and the corresponding adaptations, are important to examine in terms of implications for environmental outcomes.

The MRB contains 962,342 square kilometers of cropland. Corn, soybean, and wheat are dominant crops, which account for 34.6 percent, 23.1 percent, and 18.0 percent of cropland, respectively (Marshall et al. 2018). Figure 8.1 presents the harvested acreages of major crops planted in the MRB from 1997 to 2017 (USDA NASS 2019). Corn and soybean acreages increased substantially over time mainly due to the increasing demand for feedstock sources in bioenergy production and feed for both domestic and overseas livestock operations (USDA ERS 2022). Meanwhile, wheat and sorghum acreages have decreased. Correspondingly, irrigated corn and soybean acreages grew significantly from 1997 to 2017, while irrigated wheat and sorghum acreages declined (figure 8.2).

There are several farmer adaptation options to climate-driven changes in crop yields. For example, technological developments, government and insurance programs, alternative farm production practices like new irrigation systems, and more drought tolerant crops can mitigate some of the climate impacts on agriculture (Smit and Skinner 2002). While these options are important for a comprehensive examination, in this study, we offer a partial analysis of farmers' response to climate-driven changes in crop yields. We examine adaptation at the extensive (planting decisions for existing crops)

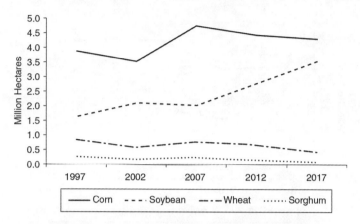

Fig. 8.2   Harvested irrigated acreage within the MRB over time (ha)

and intensive (per ha nitrogen use and irrigation) margins, *ceteris paribus*. This analysis offers an initial assessment of the relationship between N run-off and adaptation in agricultural production to climate change. Future studies should consider a wider set of adaptation alternatives including new crop varieties and production technologies.

While there is extensive literature on the impacts of agricultural production on N loading in surface water, few studies have evaluated this problem in the context of climate change. Bosch et al. (2018) and Xu et al. (2019) evaluated the effects of climate change on the costs of achieving water quality goals in an experimental watershed in Pennsylvania using an economic model and the SWAT-Variable Source Area model with climate predictions. Both studies showed that estimated costs of meeting water quality goals increase in future climates relative to the historical baseline. However, N fertilizer use in these studies is exogenously determined, which limits N use flexibility in response to variations in crop yields in future climate scenarios.

We contribute to previous literature by examining the effects of climate change on N runoff to the Gulf of Mexico with endogenous land and N use decisions. Our approach includes a behavioral crop production response to changes in productivity and evaluates N runoff accordingly. Our focus is on N and land use with associated impacts on N runoff to the gulf, as a response to crop yield changes in future climate scenarios. Our primary purpose is to draw attention to the implications of adaptation to climate change in agricultural production for N use and downstream water quality. This aspect of climate change and associated adaptation has not received much attention in scientific literature. It is important to note that the objective of this study is not to predict the changes in N runoff to the gulf under a changing climate, as the modeling exercise is based on several important assumptions and limitations that we discuss in the conclusions section. Instead, our goal is to provide a first, partial assessment of the sensitivity of gulf N runoff to the changes in crop yields and corresponding adaptation in crop production for some mid-century (2050–2068) climate change scenarios. The results of this study should encourage additional analysis of changes in N runoff as an externality from agricultural production adaptation to climate change.

8.2   Theoretical Framework

This section presents a theoretical economic framework and simplified analytical results illustrating the impact of climate-driven changes in crop yields on fertilizer use. A parsimonious welfare maximization model with a representative commodity market is considered as:

(1)
$$\max_{x,n_1,n_2,w_1} \pi = \int_0^x p(t)dt - C_n * (n_1 + n_2) - C_w * w_1$$

subject to

(2)
$$\alpha_1 * f(n_1,w_1) + \alpha_2 * g(n_2) \geq x,$$

where x is crop consumption $p(t)$ is the inverse commodity demand function. C_n and C_w are unit costs for fertilizer and water, respectively. Crop production takes place in irrigated region 1 and rainfed region 2. $f(n_1, w_1)$ is production function in region 1 requiring nitrogen (n_1) and water (w_1) as inputs, with $f' > 0$, and $f'' < 0$. $g(n_2)$ is production function in region 2 requiring only nitrogen (n_2), with $g' > 0$, and $g'' < 0$. For example, corn production in Illinois is mostly rainfed, while irrigated corn is prevalent in Kansas and Nebraska. α_1 and α_2 are the yield multipliers in future climates, with $\alpha > 1$ indicating an increase in crop yield and $0 < \alpha < 1$ indicating a reduction in crop yield. Equation (2) restricts crop consumption to not exceed production.

The appendix provides the Lagrangian and the first-order conditions, which are used to form the Hessian matrix. The determinant of the Hessian matrix is:

$$|H| = \alpha_1^2 \alpha_2 \lambda^2 \begin{bmatrix} 2\alpha_1 f_{n_1} f_{n_1w_1} f_{w_1} g_{n_2n_2} p_x - \alpha_1 f_{n_1}^2 f_{w_1w_1} g_{n_2n_2} p_x \\ + f_{n_1w_1}^2 (\lambda g_{n_2n_2} + \alpha_2 p_x g_{n_2}^2) \\ - f_{n_1n_1}(\lambda f_{w_1w_1} g_{n_2n_2} + \alpha_2 f_{w_1w_1} p_x g_{n_2}^2 + \alpha_1 f_{w_1}^2 g_{n_2n_2} p_x) \end{bmatrix}$$

Comparative statics for changes in variables of interest with respect to the change in α_1 are obtained using Cramer's rule:

(3)
$$\frac{\partial n_1}{\partial \alpha_1} = \frac{-\alpha_1 \alpha_2 \lambda^2 (f_{n_1w_1} f_{w_1} - f_{n_1} f_{w_1w_1})(\alpha_2 p_x g_{n_2}^2 + g_{n_2n_2}(\lambda + \alpha_1 p_x f(n_1,w_1)))}{|H|}$$

(4) $\dfrac{\partial n_2}{\partial \alpha_1} = \dfrac{-\alpha_1^2 \alpha_2 \lambda^2 g_{n_2} p_x [-2 f_{n_1} f_{n_1 w_1} f_{w_1} + f_{n_1 n_1} f_{w_1}^2 + f(n_1, w_1)(f_{n_1 w_1}^2 - f_{n_1 n_1} f_{w_1 w_1})]}{|H|}$

(5) $\dfrac{\partial w_1}{\partial \alpha_1} = \dfrac{-\lambda^2 \alpha_1 \alpha_2 (f_{n_1 w_1} f_{n_1} - f_{w_1} f_{n_1 n_1})(\alpha_2 p_x g_{n_2}^2 + g_{n_2 n_2}(\lambda + \alpha_1 p_x f(n_1, w_1)))}{|H|}$

The denominator $|H|$ in equations (3), (4) and (5) is positive according to the maximization requirements. Therefore, the sign of equation (3), which shows the effects of changes in crop yields in region 1 on the N use in region 1, depends on the signs of the numerator. The direction of the derivative is indeterminate and depends on the slope of the demand curve, production function, change in yield, and price of the commodity. The sign of equation (4), indicating the effects of changes in crop yields in region 1 on N use in region 2, is also ambiguous and depends on the relative magnitudes of commodity price, yield and yield changes with respect to irrigation and fertilizer, and slope of the demand curve. Similar results can be observed for productivity changes in region 2 (α_2) and are provided in the appendix. Since nutrient runoff to the gulf depends on per ha use of N and on acreage decisions, the combined effect of changes in productivity (α) on N runoff is ambiguous.

The sign of equation (5), which shows the effects of changes in crop yields in region 1 on water use in region 1, is also ambiguous. The direction of the change in water use in region 1 under climate change depends on the production function, the price of the commodity, and magnitudes of changes in both crop yields. Similar results hold for the effect of region to yield changes (α_2) on water use in region 1 (see appendix).

The simplified analytical model provides a theoretical insight for the effect of altered crop yields on input use as a form of adaptation to climate change. The result shows theoretical foundations for the need to consider the behavioral response to climate change alongside biophysical parameters in assessing the impacts of changes in production environment on production decisions that generate externalities for downstream water quality. Economic factors including prices and demand, and biophysical production parameters determine the first-order conditions. Therefore, rigorous assessments of changes in N runoff from agricultural production in response to climate change should combine biophysical and economic modeling systems that account for adaptation in production activities. For the sake of parsimony, the theoretical analysis only considers two regions and a representative commodity rather than a set of crops, which is important to consider empirically as relocation of crop production will alter spatial N use distribution and runoff to the gulf. In the empirical analysis, we use a spatially explicit model with four N intensive crops that combines biophysical and economic components to examine changes in N runoff.

8.3   Methods and Data

We use the Integrated Hydro-Economic Agricultural Land use (IHEAL) model (Xu et al. 2022) to empirically assess the effects of climate change–driven crop yield variation on N runoff to the Gulf of Mexico. IHEAL is an integrated hydro-economic agricultural land use model, which combines a national price endogenous partial equilibrium commodity market formulation for select crops and a process-based SWAT. Corn, soybean, wheat, and sorghum are included in the model as individual commodities because these crops are the most fertilizer-intensive crops planted in the US (USDA NASS 2020; Marshall et al. 2015; Steiner et al. 2021). Production of all other commodities is combined to account for county-scale agricultural land use. The model includes county-scale crop planting, fertilizer use, and irrigation decisions. Production activities generate national commodity supply estimates that are combined with corresponding national commodity demand functions to produce equilibrium prices, quantities, and producer and consumer surplus estimates. The model endogenously determines annual county crop planting acreage, N use, and irrigation based on constrained consumer and producer welfare maximization in the select crop markets.

The IHEAL model maximizes consumer and producer welfare in the US subject to commodity specific supply-demand balance, including exports and imports, production technology constraints, irrigated acreage constraints, and land allocation constraints that represent a convex combination of historically observed and synthetic county crop acreages. Historical and synthetic crop acreage proportions at the county scale are used to constrain planting decisions, so that model solutions reflect agronomic, managerial and technologic requirements for crop rotation. Synthetic acreages are obtained using own and cross-price elasticities and own and cross acreage price elasticities following Chen and Önal (2012). Elasticity estimates are obtained using fixed effect Arellano-Bond estimator and county production and price data from 2005 to 2019.

Hydrologic and Water Quality System (HAWQS) is used to obtain long-run crop yields and N runoff to the gulf for the baseline time period (2000–2018) (HAWQS 2020). HAWQS also provides future (2050–2068) crop yields for five different Coupled Model Intercomparison Project Phase 5 (CMIP5) climate models, including ACCESS1.3, MIROC5, IPSL-CM5A-LR, MIROC-ESM-CHEM and CCSM4[1]. Table 8.1 presents the list of climate models used in this study. The performance of the selected climate models is discussed in Harding, Snyder, and Liess (2013). Figure 8.3 presents average crop yields across all counties within the MRB under baseline (historical) and future

---

1. The climate models in our study were selected based on the availability in HAWQS, and inclusion in Harding, Snyder, and Liess (2013) assessment.

Table 8.1   List of climate models used in this study

Model	Institution	Resolution
Access1.3	CSIRO-BOM (Australia)	1.875*1.25
CCSM	NCAR (USA)	0.9*1.25
IPSL-CM5A-LR	IPSL (France)	1.875*3.75
MIROC-ESM-CHEM	MIROC (Japan)	2.8*2.8
MIROC5	MIROC (Japan)	2.8*2.8

Source: Harding, Snyder, and Liess (2013).

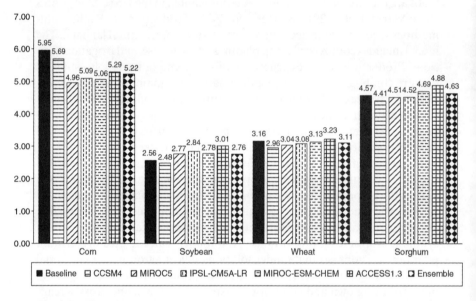

Fig. 8.3   The mean of crop yields under historical and future climates over all counties within the MRB (t/ha)

climate scenarios. The "Ensemble" scenario is the mean across all climate change models. The impacts of climate change on corn yields are negative in all climate scenarios relative to the baseline, which is consistent with previous literature (Panagopoulos et al. 2014, 2015; Chen et al. 2019). The impacts on soybean, wheat, and sorghum yields are mixed across climate models.

The IHEAL model includes crop production activities in 2,788 counties in the contiguous US where at least one of the crops included in this model was planted in at least one year from 2005 to 2019. These counties include 1,620 that are located within the MRB and 1,168 outside. Per ha crop yields in the counties located within the MRB are expressed as functions of N use and irrigation using SWAT parameter outputs from HAWQS. Per ha crop yields in counties outside of the MRB are fixed based on the USDA data and do not vary with irrigation and N use. To account for the aggregate impact of climate change on yields outside the MRB, we discount corn, soybean, and sorghum yields by 1.6 percent, 2.7 percent, and 6 percent, respectively, and increase wheat yields by 7 percent relative to their corresponding baseline values (Basche et al. 2016; Karimi et al. 2017; Chen et al. 2019). County planted acreages within and outside of the MRB are endogenously estimated.

The parametric model data include crop demand elasticities, market prices, county-specific historical crop acreage, historical county maximum irrigated acreage, and input costs, including energy, fertilizer, water, and other production costs. The crop demand elasticities are obtained from previous literature (Westcott and Hoffman 1999; Piggott and Wohlgenant 2002; Ishida and Jaime 2015). The crop market prices and historical crop acreage are collected from USDA NASS (USDA NASS 2020). The county maximum observed irrigated acreages are obtained from US Geological Survey data (Dieter et al. 2018; USGS 2018). The upper bounds on county scale irrigated acreage restrict model solutions from irrigating lands that have never been irrigated due to water, water right, and/or capital limitations. Energy input, fertilizer, water and other production costs are obtained from USDA ERS (USDA ERS 2019). IHEAL combines county production activities, including crop planting acreage, irrigation, fertilizer use and leaching with the watershed SWAT delivery ratios to estimate annual N runoff from crop production to the Gulf of Mexico (White et al. 2014).

8.4   Results and Discussion

Section 8.4 is organized as follows. We first present the validation and baseline results. Next, we discuss aggregate MRB results for crop production and N runoff with adjusted crop yields within the MRB under future climate scenarios. Then, we evaluate crop production and N runoff to the MRB under altered precipitation within the MRB and crop yields outside the MRB in future climates. Finally, we present the corresponding spatial results for the changes in N use and delivery to the Gulf of Mexico relative to the baseline values.

8.4.1   Validation and Baseline Results

The purpose of this section is twofold. One is to validate the model solutions in terms of replicating observed market data. The other is to obtain baseline estimates of N runoff to the gulf, to be used as benchmarks for subsequent climate scenario analyses.

For model validation purposes, the model is solved using observed county historical crop mix data. We present the 2018 observed values and the corresponding key baseline model solutions, including crop production, crop prices, the amount of N delivered to the Gulf of Mexico, irrigated crop

Table 8.2   Validation and baseline results

	Validation results (historical crop mix)	Observed in 2018[a][b]	Baseline results (historical and synthetic crop mix)
LAND USE (MILLION HECTARES) FOR THE CONTIGUOUS UNITED STATES			
Corn	39.6	36.0	38.2
Soybean	39.1	36.1	37.6
Winter wheat	14.5	13.2	12.4
Sorghum	2.4	2.3	2.2
PRICES ($/METRIC TON)			
Corn Price	140.6	142	147.7
Soybean Price	312.6	314	335.4
Wheat Price	182.3	190	216.0
Sorghum Price	119.0	117	133.5

	Validation results (historical crop mix)	Values from literature	Baseline results (historical and synthetic crop mix)
Total irrigated acreage (million ha)	3.92 (MRB)	7.49 (MRB)[c]	3.96 (MRB)
Total water use (million acre-feet)	4.52 (MRB)	83.40 (U.S.)[a]	4.57 (MRB)
N applied within the MRB (1000 metric ton)	6,835 (MRB)	12,610 (U.S.)[d]	6,798 (MRB)
N delivered to the Gulf of Mexico from fertilizer application (metric ton)	370,140 (MRB)	796,000 (MRB)[e][f]	369,190 (MRB)

[a] *Source*: USDA NASS (2019).

[b] Baseline model data, including prices and quantities for commodity demands are from 2018. Hence, we compare the baseline results with data observed in 2018.

[c] Total irrigated acreage of corn, soybean wheat and sorghum in the MRB in 2018 were 7,489,765 ha (USDA NASS 2019).

[d] The sum of county-level farm N fertilizer use (Falcone 2021).

[e] Source: White et al. 2014.

[f] N fertilizer use in crop production accounts for 68% of N delivered to the Gulf of Mexico from agriculture. The rest of N exported to the gulf from agriculture comes from confined animal operations and legume crops (USGS 2017).

acreage, and the irrigation water used for corn, soybean, sorghum, and wheat within the MRB as part of model validation (table 8.2). The model overestimates cumulative crop acreage for corn, soybean, wheat, and sorghum by 10.0 percent, 8.3 percent, 9.9 percent and 4.4 percent, respectively, relative to the acreages observed in 2018. All estimated crop prices are close to the observed values in 2018, with all deviations less than 3 percent.

Baseline water use, N use, and N delivery to the Gulf of Mexico are also presented in table 8.2. The estimated irrigated acreage of corn, soybean, wheat, and sorghum within the MRB is 3.92 million ha, representing 65.93 percent of irrigated acreage for these crops in the US in 2018. The annual water use within the MRB is 4.52 million acre-feet, which accounts for 5.42 percent[2] of the total observed irrigation water use in the US. Annual N use within the MRB for corn, soybean, wheat, and sorghum is 6,835 thousand metric tons, which is 54.20 percent of the total N use in the US. The corresponding N delivered to the Gulf of Mexico from fertilizer use in corn, soybean, wheat, and sorghum fields is 370,140 metric tons, accounting for 46.5 percent of the total N delivered to the Gulf of Mexico from the agricultural sector in the MRB (White et al. 2014). These solutions provide a firm footing and benchmark for the subsequent analysis of N runoff scenarios.

We use the historical and synthetic crop mix data to generate baseline model results as a reference point for comparison to the solutions from the climate change scenarios (column 3, table 8.2). Synthetic crop acreages allow for greater model flexibility than the model that uses only historical crop mix. The added flexibility is advantageous for the scenarios with constraints or parameter values that fall outside of historically observed settings. We use these baseline results as benchmarks, rather than the results in column 1, for greater consistency between long-run equilibrium results of scenarios with and without added restrictions. The baseline N runoff to the Gulf of Mexico is 369,190 metric tons.

8.4.2   Results for Future Climate Scenarios

This section presents the results from the IHEAL model with predicted changes in crop yields within the MRB for 2050–2068. Table 8.3 shows aggregate MRB results for crop acreage and production, irrigated acreage, water use, N fertilizer use, and corresponding runoff to the Gulf of Mexico under baseline and future climates. Results from five climate models, including ACCESS1.3, MIROC5, IPSL-CM5A-LR, MIROC-ESM-CHEM and CCSM4, are presented. Among these models, CCSM4 and IPSL-CM5A-LR scenarios produce the lowest and highest impacts on N runoff to the gulf. We focus our discussion of results on these models as these provide the upper and lower bounds for N runoff impacts. In addition, we also provide the results from the ensemble climate scenario where future crop yields are averages across five climate prediction models. We refer to this model as the "Ensemble Mean" in the following discussion.

Table 8.3 indicates that the impact of climate change on crop acreages and production within the MRB is mixed. Relative to the baseline with no climate change, corn acreage declines by 0.3 percent in CCSM4, and increases by 2.5 percent and 2.8 percent in the Ensemble Mean and IPSL-CM5A-LR, respectively. However, corn production decreases consistently in all models. Soybean acreage (production) decreases (increases) in future climates

---

2. This value does not include other irrigation intensive crops like rice and alfalfa grown in the MRB.

Table 8.3 Results under future climates

	Baseline	Ensemble Mean	CCSM4	ACCESS1.3	IPSL-CM5A-LR	MIROC-ESM-CHEM	MIROC5
Corn acreage within the MRB (million ha)	31.6	32.5	31.5	32.8	32.4	32.8	32.5
Corn production within the MRB (million metric ton)	320.3	294.4	308.4	307.6	280.4	280.1	276.8
Soybean acreage within the MRB (million ha)	29.1	28.3	29.2	27.3	27.8	28.1	28
Soybean production within the MRB (million metric ton)	98.4	103.3	94	111.9	104.1	102	101.7
Wheat acreage within the MRB (million ha)	9.4	9.1	9.2	8.8	9.2	9.4	8.8
Wheat production within the MRB (million metric ton)	21.9	23.0	20.9	25.5	21.7	24.8	22.6
Sorghum acreage within the MRB (million ha)	1.8	1.7	1.7	1.7	1.5	1.6	1.6
Sorghum production within the MRB (million metric ton)	7.6	7.3	7	8.4	5.8	6.5	6.5
Irrigated Acreage within the MRB (ha)	3,955,607	3,979,146	3,934,678	3,953,137	3,919,521	3,922,389	3,916,433
Total water use within the MRB (million acre-feet)	4.57	4.11	4.5	4.16	4.62	4.69	4.07
N applied within the MRB (1000 metric ton)	6,798	6,930	6,747	6,931	6,948	7,006	6,874
N delivered to the Gulf of Mexico from fertilizer application (metric ton)	369,190	372,410	370,650	370,990	375,010	373,310	372,940
Consumer and producer surplus for four commodities (billion $)	204.8	202.1	201.3	207.7	199.8	199.2	198.6
Consumer and producer surplus with a 45% N runoff reduction from MRB relative to the baseline (billion $)	197.0	194.9	193.2	201.4	192.1	192.3	191.1

by 4.5 percent (5.8 percent) and 2.7 percent (5.0 percent) in the Ensemble Mean and IPSL-CM5A-LR, respectively. In the CCSM climate, soybean acreage increases by 0.3 percent and production decreases by 4.4 percent, respectively. Wheat acreage in future climates consistently declines relative to the baseline result. Changes in wheat production within the MRB are −4.6 percent, −0.9 percent and 5.0 percent under CCSM4, IPSL-CM5A-LR and the Ensemble Mean, respectively. Sorghum acreage and production decline in all models. Sorghum acreage (production) drops by 5.6 percent (8.3 percent), 16.7 percent (24.0 percent) and 5.6 percent (4.3 percent) in CCSM4, IPSL-CM5A-LR and the Ensemble Mean climates, respectively.

Changes in N use relative to the baseline are −0.8 percent, 2.2 percent and 1.9 percent in CCSM4, IPSL-CM5A-LR and the Ensemble Mean climate scenarios, respectively. Although changes in N use within the MRB are mixed across models, N delivered to the Gulf of Mexico consistently increases across all models (table 8.3). Annual N runoff to the Gulf of Mexico increases compared to the baseline by 0.4 percent (CCSM4), 2.2 percent (IPSL-CM5A-LR), and 0.9 percent (Ensemble Mean). Although aggregate N use decreases in some models, N-intensive crop production shifts spatially to areas with high edge-of-field N leakage and gulf runoff potential. As a result, cumulative N runoff to the gulf increases in all models.

We also examine the implications of reducing N runoff to the gulf by 45 percent following EPA Hypoxia Task Force goal (Robertson and Saad 2013) for consumer and producer surplus in each of the considered climate scenarios. We estimate the opportunity cost of reducing N runoff in terms of foregone consumer and producer surplus in the four considered commodity markets as N runoff externality is restricted. The last two rows of table 8.3 show consumer and producer surplus values with and without the constraint limiting N runoff to the gulf by 45 percent. The change in consumer and producer surplus estimates due to the N runoff constraint represents the opportunity cost of internalizing the N runoff externality (Xu et al. 2022). In the baseline scenario without climate change, consumer and producer surplus in the four commodity markets declines by $7.8 billion. This estimate varies between $6.3 and $8.1 billion depending on climate scenario. Hence, the opportunity cost of reducing the externality by 45 percent can increase by 3 percent (8.1/7.8) or decrease by 20 percent (6.3/7.8) depending on climate prediction models.

8.4.3   N Runoff with Altered Precipitation in the MRB and Crop Yields Outside the MRB

Next, we extend the preceding analysis by accounting for the effects of likely changes in precipitation within the MRB and changes in crop yields outside the MRB. We use predicted precipitation for future climate scenarios as a proxy for water availability in counties with irrigated agriculture within the MRB. We obtain 2050–2068 annual precipitation projections from

GFDL-ESM2M-RegCM4, HadGEM2-ES-RegCM4, and MPI-ESM-LR-RegCM4 models provided by the National Center for Atmospheric Research (NCAR) (NCAR 2022b).[3] We use these data to obtain mean annual precipitation across three models. Predicted changes in precipitation are combined with the baseline IHEAL water use solutions to generate the county-scale water availability constraints for future climate change scenarios.[4]

In this analysis, we also make an effort to account for the likely change in crop yields outside the MRB. Unfortunately, we do not have data on county specific effects of climate change on crop yields outside the MRB. Although land use outside the MRB is not critical for the purposes of this study, it is important to account for yield changes outside the MRB because of implications for national commodity supply and price. Therefore, we use the result from previous literature to adjust crop yields outside the MRB uniformly (Basche et al. 2016; Karimi et al. 2017; Chen et al. 2019). In particular, we assume that corn, soybean, wheat, and sorghum yields outside of MRB will change by −1.6 percent, −2.7 percent, 7.0 percent, and −6.0 percent, respectively. We apply these adjustments to all models in table 8.4.

Table 8.4 presents the aggregate MRB results from five climate models and the Ensemble Mean, including crop acreage and production, irrigated acreage, water use, N use, and N delivery to the Gulf of Mexico. Values in parentheses are percentage changes relative to the baseline scenario in table 8.3 (no climate change). We mainly discuss the Ensemble Mean model in this section. Ensemble Mean changes in corn, soybean, and wheat acreages and production are consistent with the corresponding results in table 8.3 in terms of signs and magnitudes. Ensemble Mean sorghum acreage within the MRB is the same in tables 8.3 and 8.4. However, unlike table 8.3, production increases in table 8.4.

Changes in irrigated acreage and water use relative to the baseline scenario are consistent across Ensemble Mean solutions in tables 8.3 and 8.4. However, Ensemble Mean irrigated acreage increases while water use declines within the MRB in table 8.4 relative to table 8.3. Two reasons explain this change. First, future precipitation is predicted to decline in counties located in southern Kansas, eastern New Mexico, northern Texas, and Oklahoma, where agricultural production heavily relies on irrigation and precipitation. Water availability in these MRB counties decreases in table 8.4 relative to table 8.3, which leads to a reduction in total water use. Second, decrease in crop yields outside the MRB in table 8.4 relative to table 8.3 results in real-

---

3. RegCM4 (the Regional Climate Model version 4) is widely used to downscale global climate models for regional climate projections in the US (Mei, Wang, and Gu 2013; Ashfaq et al. 2016). Our selection of global climate models for precipitation projection data is based on the availability of downscaled data in the NCAR database.

4. Ensemble precipitation change is used for all climate model scenarios. A preferred approach would be to use precipitation change corresponding to each climate model used in IHEAL. Unfortunately, the precipitation prediction data for ACCESS1.3, MIROC5, IPSL-CM5A-LR, MIROC-ESM-CHEM, and CCSM4 models are not available from the NCAR database.

Table 8.4   Results with changes in water availability and crop yields adjusted outside the MRB under future climates

	Ensemble Mean	CCSM4	ACCESS1.3	IPSL-CM5A-LR	MIROC-ESM-CHEM	MIROC5
Corn acreage within the MRB (million ha)	32.6 (3.2%)	31.5 (−0.3%)	32.8 (3.8%)	32.5 (2.8%)	32.9 (4.1%)	32.6 (3.2%)
Corn production within the MRB (million metric ton)	294.4 (−8.1%)	308.6 (−3.7%)	307.6 (−4.0%)	280.8 (−12.3%)	280.2 (−12.5%)	277.1 (−13.5%)
Soybean acreage within the MRB (million ha)	28.4 (−2.4%)	29.2 (0.3%)	27.4 (−5.8%)	27.8 (−4.5%)	28.1 (−3.4%)	28.1 (−3.4%)
Soybean production within the MRB (million metric ton)	103.6 (5.3%)	94.1 (−4.4%)	112.2 (14.0%)	104.2 (5.9%)	102.2 (3.9%)	101.9 (3.6%)
Wheat acreage within the MRB (million ha)	8.9 (−5.3%)	8.8 (−6.4%)	8.6 (−8.5%)	8.8 (−6.4%)	8.9 (−5.3%)	8.6 (−8.5%)
Wheat production within the MRB (million metric ton)	22.4 (2.3%)	20.0 (−8.7%)	24.8 (13.2%)	20.9 (−4.6%)	23.6 (7.8%)	22.1 (0.9%)
Sorghum acreage within the MRB (million ha)	1.7 (−5.6%)	1.7 (−5.6%)	1.7 (−5.6%)	1.6 (−11.1%)	1.6 (−11.1%)	1.6 (−11.1%)
Sorghum production within the MRB (million metric ton)	7.7 (0.9%)	7.4 (−3.0%)	8.4 (10.1%)	6.5 (−14.8%)	6.7 (−12.2%)	6.8 (−10.9%)
Irrigated Acreage within the MRB (ha)	3,990,864 (0.9%)	3,949,977 (−0.1%)	3,933,342 (−0.6%)	3,937,504 (−0.5%)	3,927,531 (−0.7%)	3,922,191 (−0.8%)
Total water use within the MRB (million acre-feet)	3.91 (−14.4%)	4.45 (−2.6%)	3.90 (14.7%)	4.41 (−3.5%)	4.37 (−4.4%)	3.80 (−16.8%)
N applied within the MRB (1000 metric ton)	6,915 (1.7%)	6,720 (−1.1%)	6,912 (1.7%)	6,927 (1.9%)	6,971 (2.5%)	6,871 (1.1%)
N delivered to the Gulf of Mexico from fertilizer application (metric ton)	372,900 (1.0%)	370,880 (0.5%)	371,420 (0.6%)	375,170 (1.6%)	373,480 (1.2%)	373,050 (1.0%)
Consumer and producer surplus for four commodities (billion $)	201.9	201.1	207.5	199.6	199.0	198.4
Consumer and producer surplus with a 45% N runoff reduction from MRB relative to the baseline (billion $)	194.5	192.8	201.1	191.7	191.9	190.7

location of some of the acreage from outside to inside the MRB. Hence, after adjusting water availability within the MRB and yields outside the MRB, acreage with irrigation increases, but total water use within the MRB declines in table 8.4 relative to table 8.3.

The Ensemble Mean N fertilizer use within the MRB is 30,000 metric tons lower in table 8.4 than in table 8.3. However, N runoff to the Gulf of Mexico is 490 metric tons greater in table 8.4 than in table 8.3. Two factors contribute to this divergence between N use and runoff in the Gulf of Mexico. First, within the MRB, corn, soybean, and sorghum acreages increase by 0.05, 0.11 and 0.04 million ha, respectively, while wheat acreage decreases by 0.22 million ha. Cumulatively, the acreage of these crops decreases in table 8.4 relative to table 8.3, which leads to the modest decline in N use. Second, the increased corn, soybean, and sorghum acreages occur in regions with both higher productivity and higher N runoff potential. As a result, N runoff to the Gulf of Mexico increases from crop production within the MRB. We explore the spatial distribution of N use and associated runoff to the gulf in the next section.

Table 8.4 also shows estimates for consumer and producer surplus changes in the four commodity markets across climate scenarios and for the corresponding 45 percent N runoff reduction scenarios. Estimates for consumer and producer surplus do not change significantly relative to the corresponding estimates in table 8.3. All estimates of consumer and producer surplus without the N runoff reduction policy decline by less than 1 percent relative to table 8.3. Similar to the results in table 8.3, the opportunity cost of reducing N runoff by 45 percent varies between $6.4 and $8.3 billion.

8.4.4   Spatial Distribution of N Use and Delivery to the Gulf of Mexico

The aggregate results show that in future climate scenarios, N delivery to the Gulf of Mexico from N fertilizer use within the MRB increases relative to the baseline. However, spatial heterogeneity is observed in terms of use and runoff contribution. In this section, the spatial distribution of N use (figure 8.4) and the corresponding runoff (figure 8.5) to the Gulf of Mexico is discussed, using the Ensemble Mean solutions in table 8.4.

N use declines in Oklahoma, South Dakota, and Texas, where corn yields in HAWQ-SWAT Ensemble Mean climate model decline by 10.8 percent, 13.3 percent and 3.2 percent, respectively. In these states, lower corn yields and greater demand for irrigation increase production costs, which leads to corn production shifting to other regions. Hence, N use in these regions declines (figure 8.4). However, N use increases in some areas of Colorado, western Kansas, Iowa, Illinois, Indiana, Minnesota, North Dakota, and Wisconsin. Although corn yields in these states also decrease, the higher marginal productivity of N fertilizer in these regions leads to more corn acreage and greater N use.

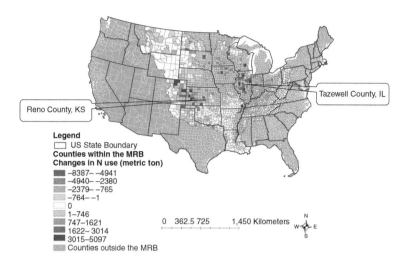

Fig. 8.4   Spatial distribution of N use in the Ensemble Mean of table 8.4

The largest increase in N use, from 11,903 to 17,000 metric tons per year, is observed in Tazewell County, Illinois. This growth in N use is due to the increase in corn and wheat acreages by 13,973 and 1,430 ha, respectively. Although corn yield in this county is predicted to decline by 8.5 percent, acreage increases as other counties suffer even greater yield losses and reduce corn production. The largest annual N use decrease from 10,087 to 1,700 metric tons is in Reno County, Kansas. This decrease is due to lower corn and wheat production as yields of these crops decline by 12.9 percent and 5.3 percent, respectively. In addition, precipitation in this county also declines by 0.1 percent.

Figure 8.5 presents county-specific changes in N delivery to the gulf for the Ensemble Mean analysis relative to the baseline results. Agricultural production in the UMRB and OTRB delivers most of the N runoff to the Gulf of Mexico that originates in the MRB (Kling et al. 2014). These regions are currently targeted by the EPA's Hypoxia Task Force goals to reduce N runoff. The figure shows that N runoff from the UMRB may increase with climate change, while runoff from the OTRB may decrease relative to the baseline. States located in the UMRB, including Iowa, Illinois, and Indiana, increase N delivery to the Gulf of Mexico relative to the baseline by 3,733 metric tons, a 1.4 percent increase. Increased N runoff from these states accounts for 99.3 percent of the predicted growth in N runoff to the gulf. On the other hand, N runoff from Ohio, Tennessee, and Kentucky (states located in OTRB) declines by 629 metric tons, a 2 percent reduction relative to the baseline runoff from these states.

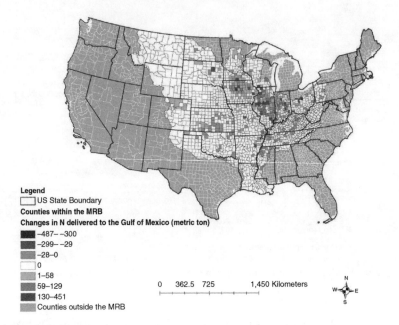

Legend
☐ US State Boundary
Counties within the MRB
Changes in N delivered to the Gulf of Mexico (metric ton)
■ −487−−300
■ −299−−29
■ −28−0
☐ 0
■ 1−58
■ 59−129
■ 130−451
■ Counties outside the MRB

0   362.5   725   1,450 Kilometers

Fig. 8.5   Spatial distribution of N delivered to the Gulf of Mexico in the Ensemble Mean of table 8.4

8.5   Conclusion

This paper examines some of the effects of climate change on downstream water quality externality from agricultural production. Specifically, we investigate how climate-driven changes in crop yields affect agricultural production in the MRB and the corresponding water quality outcomes in the Gulf of Mexico. Our purpose is to illustrate, rather than predict, the potential impact of climate change on agricultural production externality in the form of N runoff to the gulf. This dimension of the nexus between climate change and water resource sustainability has not received much attention in scientific literature. In this respect, our goal is to provide the first examination of its kind and spur additional research in this direction using integrated models with economic and biophysical components. The integrated approach is necessary because the behavioral response to environmental change is an important element of climate adaptation and can significantly affect downstream water quality.

This study differs from Metaxoglou and Smith in this volume in at least three important ways. First, we do not consider N legacy effects although it is an important part of hypoxia in the Gulf of Mexico. Second, the IHEAL model includes N runoff from only four crops and excludes other crops and sectors including livestock and industrial production. Third, this study models N loads, while Metaxoglou and Smith investigate N concentrations. These differences imply that the results from the two studies cannot be directly compared.

We obtain three main findings. First, climate-driven changes in crop yields affect agricultural production decisions in the MRB at intensive and extensive margins. Crop acreage and per acre N use are affected by changes in production conditions. These changes increase the overall N delivery to the Gulf of Mexico from agricultural production, *ceteris paribus*. The estimated increase in N runoff to the gulf is in the range of 0.5–1.6 percent (1,690–5,980 metric tons) relative to the baseline. These impacts are not substantial in terms of magnitude relative to current runoff. However, the corresponding marginal damages to aquatic ecosystems can be significant. Future studies should examine and evaluate the impacts of incremental increases in N runoff on gulf aquatic ecosystems under climate change. Second, the changes in production, including N use, are spatially heterogeneous. In some counties, N use will intensify, while in others, N use will decrease. Third, spatial heterogeneity also applies at a larger spatial scale. As major contributors to the N runoff from agricultural production to the gulf, the UMRB and OTRB are prioritized by the EPA's Hypoxia Task Force for reducing N runoff. In climate scenarios examined in this study, N runoff is expected to increase from the UMRB and decrease from the OTRB.

We also examine the sensitivity of the opportunity costs to reduce N runoff to the gulf by 45 percent across climate scenarios. The results show that without climate change, the opportunity cost is $7.8 billion while with climate change this estimate varies between $6.4 and $8.1 billion. Our N runoff reduction scenario is akin to a performance-based policy where internalizing the N runoff externality reduces N runoff by 45 percent. Although not directly addressed in this study, an example of a performance-based policy is tradeable pollution permit system that imposes an exogenous upper bound on environmental impact. With frictionless trade in the permits market, cost-effective distribution of production and mitigation efforts can be achieved under various emissions caps (Montgomery 1972; Cropper and Oates 1992). Cap and trade policies are operationally and politically challenging to implement even if technologically feasible. Nevertheless, while a detailed examination of tradable permit-based runoff mitigation is beyond the scope of this study, our results are informative in terms of providing an estimate for the opportunity cost of such a policy in the four commodity markets and in terms of examining the sensitivity of the estimated costs across several climate models.

Several limitations of this study should be mentioned for future research. First, climate change can affect not only crop yields but also water balance. In some regions, changes in climate can influence soil water properties and surface and groundwater interactions (Scibek et al. 2007; Saha et al. 2017; Guevara-Ochoa, Medina-Sierra, and Vives 2020). In this study, we do not account for ground versus surface water availability explicitly. Instead, precipitation changes, as predicted by the climate models included in this study and reported in the NCAR database, are used to examine the impact of changes in water availability. The explicit delineation between ground and surface water irrigation, and the associated impacts of climate change, will improve the accuracy of our estimates.

Second, the modeling exercise does not account for potential changes in the edge-of-field N runoff and N delivery ratios from cropland to the gulf in future climate scenarios. This may over or underestimate N loading in the Gulf of Mexico. Unfortunately, estimates of climate impact on spatial and temporal attributes of N delivery ratios to the gulf have not been produced yet.

Third, crop yield changes under future climates outside the MRB are assumed to be uniform across all counties. The assumed uniformity in yield change outside the MRB precludes the analysis of impacts on N runoff outside the MRB but is less critical for the purpose of this paper. We use these uniform yield changes outside the MRB to account for the potential effect on national commodity supply and prices which can influence production decisions within the MRB and associated N runoff. More detailed modeling of yield changes in areas outside the MRB may improve the accuracy of our estimates and enable analysis of N impacts outside of the MRB.

Fourth, we do not explicitly account for the effect of precipitation change in non-irrigated regions. Instead, we assume that precipitation affects water availability only in the areas with non-zero irrigation, as observed in the past data because irrigation water availability depends at least in part on precipitation. In addition, we do not explicitly account for irrigation infrastructure that links precipitation and irrigation water supply. For non-irrigated regions, we do not have estimates for the effect of precipitation or irrigation on crop yields. This is an important caveat that should be addressed in future studies. A decline in precipitation in rainfed crop production regions may prompt investment in irrigation infrastructure, which we do not include in the current study. Conversely, we also do not account for potential increase in precipitation or flooding effects in non-irrigated regions that can influence production decisions and N delivery ratios.

Fifth, the IHEAL model corresponds to the social planner's problem with perfect information. Crop production, land and input use (N and water) are obtained based on social welfare maximization. This framework is consistent with Potential Pareto Optimality criteria but does not explicitly consider implications for strict Pareto Optimality (Griffin 1995). Nevertheless, in terms of long-run equilibrium outcomes, the model provides useful insights for illustrating the potential impacts of agricultural production on downstream water quality. Such models have been extensively used for various policy-relevant analyses (Havlik et al. 2011; Chen et al. 2014; Xu et al. 2022).

Despite the limitations, the study provides a useful initial evaluation of the impacts of agricultural production adaptation to climate change on downstream water quality. Our purpose in this study is not to predict the water quality outcomes. Instead, our purpose is to draw attention to a previously unaddressed climate related issue, which is the externality of agricultural production adaptation to climate change in terms of nutrient runoff and downstream water quality. The initial estimates in this study show that N runoff can increase by 0.5 percent–1.6 percent (1,690–5,980 metric tons), and reducing N runoff by 45 percent will be from 18.0 percent less to 6.4 percent more costly depending on climate change scenario relative to the baseline. We do not claim to have addressed this issue comprehensively, but the results suggest that future studies should examine the nutrient runoff externalities from agricultural production adaptation to climate change in greater detail.

Appendix

(S1)
$$\max_{x,n_1,n_2,w_1} \pi = \int_0^x p(t)dt - C_n * (n_1 + n_2) - C_w * w_1$$

subject to

(S2)
$$\alpha_1 * f(n_1,w_1) + \alpha_2 * g(n_2) \geq x$$

Lagrangian and corresponding first-order conditions are as follows:

(S3)
$$L = \int_0^x p(t)dt - C_n * (n_1 + n_2) - C_w * w_1$$
$$+ \lambda(\alpha_1 * f(n_1,w_1) + \alpha_2 * g(n_2) - x)$$

(S4)
$$[x]\frac{\partial L}{\partial x} = p(x) - \lambda = 0$$

$$[n_1]\frac{\partial L}{\partial n_1} = -C_n + \lambda\alpha_1 f_{n_1} = 0$$

$$[n_2]\frac{\partial L}{\partial n_2} = -C_n + \lambda\alpha_2 g_{n_2} = 0$$

$$[w_1]\frac{\partial L}{\partial w_1} = -C_w + \lambda\alpha_1 f_{w_1} = 0$$

$$[\lambda]\frac{\partial L}{\partial \lambda} = \alpha_1 * f(n_1,w_1) + \alpha_2 * g(n_2) - x = 0$$

Total differentiation of the first-order conditions with respect to α_1 gives:

(S5) $\quad [x] \; p_x \dfrac{\partial x}{\partial \alpha_1} - \dfrac{\partial \lambda}{\partial \alpha_1} = 0$

$$[n_1] \; \lambda\alpha_1 f_{n_1 n_1} \frac{\partial n_1}{\partial \alpha_1} + \lambda\alpha_1 f_{n_1 w_1} \frac{\partial w_1}{\partial \alpha_1} + \alpha_1 f_{n_1} \frac{\partial \lambda}{\partial \alpha_1} = -\lambda f_{n_1}$$

$$[n_2] \; \lambda\alpha_2 g_{n_2 n_2} \frac{\partial n_2}{\partial \alpha_1} + \alpha_2 g_{n_2} \frac{\partial \lambda}{\partial \alpha_1} = 0$$

$$[w_1] \; \lambda\alpha_1 f_{w_1 n_1} \frac{\partial n_1}{\partial \alpha_1} + \lambda\alpha_1 f_{w_1 w_1} \frac{\partial w_1}{\partial \alpha_1} + \alpha_1 f_{w_1} \frac{\partial \lambda}{\partial \alpha_1} = -\lambda f_{w_1}$$

$$[\lambda] \; \alpha_1 f_{n_1} \frac{\partial n_1}{\partial \alpha_1} + \alpha_1 f_{w_1} \frac{\partial w_1}{\partial \alpha_1} + \alpha_2 g_{n_2} \frac{\partial n_2}{\partial \alpha_1} - \frac{\partial x}{\partial \alpha_1} = -f(n_1, w_1)$$

The second-order conditions can be expressed in terms of the Bordered Hessian representation as AH = B, where $A = [\partial x / \partial \alpha_1, \partial n_1 / \partial \alpha_1, \partial n_2 / \partial \alpha_1, \partial w_1 / \partial \alpha_1, \partial \lambda / \partial \alpha_1]$ is the vector of derivatives of all endogenous variables w.r.t τ. H is the Hessian matrix shown below, and $B = [0, -\lambda f_{n_1}, 0, -\lambda f_{w_1}, -f(n_1, w_1)]$.

(S6) $\quad H = \begin{bmatrix} p_x & 0 & 0 & 0 & -1 \\ 0 & \lambda\alpha_1 f_{n_1 n_1} & 0 & \lambda\alpha_1 f_{w_1 n_1} & \alpha_1 f_{n_1} \\ 0 & 0 & \lambda\alpha_2 g_{n_2 n_2} & 0 & \alpha_2 g_{n_2} \\ 0 & \lambda\alpha_1 f_{n_1 w_1} & 0 & \lambda\alpha_1 f_{w_1 w_1} & \alpha_1 f_{w_1} \\ -1 & \alpha_1 f_{n_1} & \alpha_2 g_{n_2} & \alpha_1 f_{w_1} & 0 \end{bmatrix}$

(S7) $\quad |H| = \alpha_1^2 \alpha_2 \lambda^2 \begin{bmatrix} 2\alpha_1 f_{n_1} f_{n_1 w_1} f_{w_1} g_{n_2 n_2} p_x - \alpha_1 f_{n_1}^2 f_{w_1 w_1} g_{n_2 n_2} p_x + f_{n_1 w_1}^2 (\lambda g_{n_2 n_2} + \alpha_2 p_x g_{n_2}^2) \\ - f_{n_1 n_1} (\lambda f_{w_1 w_1} g_{n_2 n_2} + \alpha_2 f_{w_1 w_1} p_x g_{n_2}^2 + \alpha_1 f_{w_1}^2 g_{n_2 n_2} p_x) \end{bmatrix}$

(S8) $\quad \dfrac{\partial n_1}{\partial \alpha_1} = \dfrac{|H_{n_1}|}{|H|} = \dfrac{-\alpha_1 \alpha_2 \lambda^2 (f_{n_1 w_1} f_{w_1} - f_{n_1} f_{w_1 w_1})(\alpha_2 p_x g_{n_2}^2 + g_{n_2 n_2}(\lambda + \alpha_1 p_x f(n_1, w_1)))}{\alpha_1^2 \alpha_2 \lambda^2 \begin{bmatrix} 2\alpha_1 f_{n_1} f_{n_1 w_1} f_{w_1} g_{n_2 n_2} p_x - \alpha_1 f_{n_1}^2 f_{w_1 w_1} g_{n_2 n_2} p_x \\ + f_{n_1 w_1}^2 (\lambda g_{n_2 n_2} + \alpha_2 p_x g_{n_2}^2) \\ - f_{n_1 n_1} (\lambda f_{w_1 w_1} g_{n_2 n_2} + \alpha_2 f_{w_1 w_1} p_x g_{n_2}^2 + \alpha_1 f_{w_1}^2 g_{n_2 n_2} p_x) \end{bmatrix}}$

$\qquad = \dfrac{-(f_{n_1 w_1} f_{w_1} - f_{n_1} f_{w_1 w_1})(\alpha_2 p_x g_{n_2}^2 + g_{n_2 n_2}(\lambda + \alpha_1 p_x f(n_1, w_1)))}{\alpha_1 \begin{bmatrix} 2\alpha_1 f_{n_1} f_{n_1 w_1} f_{w_1} g_{n_2 n_2} p_x - \alpha_1 f_{n_1}^2 f_{w_1 w_1} g_{n_2 n_2} p_x \\ + f_{n_1 w_1}^2 (\lambda g_{n_2 n_2} + \alpha_2 p_x g_{n_2}^2) \\ - f_{n_1 n_1} (\lambda f_{w_1 w_1} g_{n_2 n_2} + \alpha_2 f_{w_1 w_1} p_x g_{n_2}^2 + \alpha_1 f_{w_1}^2 g_{n_2 n_2} p_x) \end{bmatrix}}$

$$(S9) \quad \frac{\partial n_2}{\partial \alpha_1} = \frac{|H_{n_2}|}{|H|} = \frac{-\alpha_1^2 \alpha_2 \lambda^2 g_{n_2} p_x \left[-2 f_{n_1} f_{n_1 w_1} f_{w_1} + f_{n_1 n_1} f_{w_1}^2 + f(n_1, w_1)(f_{n_1 w_1}^2 - f_{n_1 n_1} f_{w_1 w_1})\right]}{\alpha_1^2 \alpha_2 \lambda^2 \begin{bmatrix} 2\alpha_1 f_{n_1} f_{n_1 w_1} f_{w_1} g_{n_2 n_2} p_x - \alpha_1 f_{n_1}^2 f_{w_1 w_1} g_{n_2 n_2} p_x \\ + f_{n_1 w_1}^2 (\lambda g_{n_2 n_2} + \alpha_2 p_x g_{n_2}^2) \\ - f_{n_1 n_1}(\lambda f_{w_1 w_1} g_{n_2 n_2} + \alpha_2 f_{w_1 w_1} p_x g_{n_2}^2 + \alpha_1 f_{w_1}^2 g_{n_2 n_2} p_x) \end{bmatrix}}$$

$$= \frac{-g_{n_2} p_x \left[-2 f_{n_1} f_{n_1 w_1} f_{w_1} + f_{n_1 n_1} f_{w_1}^2 + f(n_1, w_1)(f_{n_1 w_1}^2 - f_{n_1 n_1} f_{w_1 w_1})\right]}{2\alpha_1 f_{n_1} f_{n_1 w_1} f_{w_1} g_{n_2 n_2} p_x - \alpha_1 f_{n_1}^2 f_{w_1 w_1} g_{n_2 n_2} p_x + f_{n_1 w_1}^2 (\lambda g_{n_2 n_2} + \alpha_2 p_x g_{n_2}^2)}$$
$$- f_{n_1 n_1}(\lambda f_{w_1 w_1} g_{n_2 n_2} + \alpha_2 f_{w_1 w_1} p_x g_{n_2}^2 + \alpha_1 f_{w_1}^2 g_{n_2 n_2} p_x)$$

$$(S10) \quad \frac{\partial w_1}{\partial \alpha_1} = \frac{|H_{w_1}|}{|H|} = \frac{-\lambda^2 \alpha_1 \alpha_2 (f_{n_1 w_1} f_{n_1} - f_{w_1} f_{n_1 n_1})(\alpha_2 p_x g_{n_2}^2 + g_{n_2 n_2}(\lambda + \alpha_1 p_x f(n_1, w_1)))}{\alpha_1^2 \alpha_2 \lambda^2 \begin{bmatrix} 2\alpha_1 f_{n_1} f_{n_1 w_1} f_{w_1} g_{n_2 n_2} p_x - \alpha_1 f_{n_1}^2 f_{w_1 w_1} g_{n_2 n_2} p_x \\ + f_{n_1 w_1}^2 (\lambda g_{n_2 n_2} + \alpha_2 p_x g_{n_2}^2) \\ - f_{n_1 n_1}(\lambda f_{w_1 w_1} g_{n_2 n_2} + \alpha_2 f_{w_1 w_1} p_x g_{n_2}^2 + \alpha_1 f_{w_1}^2 g_{n_2 n_2} p_x) \end{bmatrix}}$$

$$= \frac{-(f_{n_1 w_1} f_{n_1} - f_{w_1} f_{n_1 n_1})(\alpha_2 p_x g_{n_2}^2 + g_{n_2 n_2}(\lambda + \alpha_1 p_x f(n_1, w_1)))}{\alpha_1 \begin{bmatrix} 2\alpha_1 f_{n_1} f_{n_1 w_1} f_{w_1} g_{n_2 n_2} p_x - \alpha_1 f_{n_1}^2 f_{w_1 w_1} g_{n_2 n_2} p_x \\ + f_{n_1 w_1}^2 (\lambda g_{n_2 n_2} + \alpha_2 p_x g_{n_2}^2) \\ - f_{n_1 n_1}(\lambda f_{w_1 w_1} g_{n_2 n_2} + \alpha_2 f_{w_1 w_1} p_x g_{n_2}^2 + \alpha_1 f_{w_1}^2 g_{n_2 n_2} p_x) \end{bmatrix}}$$

Total differentiation of the first-order conditions with respect to α_2 gives:

$$(S11) \quad [x]\, p_x \frac{\partial x}{\partial \alpha_2} - \frac{\partial \lambda}{\partial \alpha_2} = 0$$

$$[n_1]\, \lambda \alpha_1 f_{n_1 n_1} \frac{\partial n_1}{\partial \alpha_2} + \lambda \alpha_1 f_{n_1 w_1} \frac{\partial w_1}{\partial \alpha_2} + \alpha_1 f_{n_1} \frac{\partial \lambda}{\partial \alpha_2} = 0$$

$$[n_2]\, \lambda \alpha_2 g_{n_2 n_2} \frac{\partial n_2}{\partial \alpha_2} + \alpha_2 g_{n_2} \frac{\partial \lambda}{\partial \alpha_2} = -\lambda g_{n_2}$$

$$[w_1]\, \lambda \alpha_1 f_{w_1 n_1} \frac{\partial n_1}{\partial \alpha_2} + \lambda \alpha_1 f_{w_1 w_1} \frac{\partial w_1}{\partial \alpha_2} + \alpha_1 f_{w_1} \frac{\partial \lambda}{\partial \alpha_2} = 0$$

$$[\lambda]\, \alpha_1 f_{n_1} \frac{\partial n_1}{\partial \alpha_2} + \alpha_1 f_{w_1} \frac{\partial w_1}{\partial \alpha_2} + \alpha_2 g_{n_2} \frac{\partial n_2}{\partial \alpha_2} - \frac{\partial x_1}{\partial \alpha_2} = -g(n_2)$$

The second-order conditions can be expressed in terms of the Bordered Hessian representation as $AH = B$, where $A = [\partial x / \partial \alpha_2, \partial n_1 / \partial \alpha_2, \partial n_2 / \partial \alpha_2, \partial w_1 / \partial \alpha_2, \partial \lambda / \partial \alpha_2]$ is the vector of derivatives of all endogenous variables w.r.t τ. H is the Hessian matrix shown below, and $B = [0, 0, -\lambda g_{n_2}, 0, -g(n_2)]$.

$$(S12) \qquad H = \begin{bmatrix} p_{x_1} & 0 & 0 & 0 & -1 \\ 0 & \lambda\alpha_1 f_{n_1 n_1} & 0 & \lambda\alpha_1 f_{w_1 n_1} & \alpha_1 f_{n_1} \\ 0 & 0 & \lambda\alpha_2 g_{n_2 n_2} & 0 & \alpha_2 g_{n_2} \\ 0 & \lambda\alpha_1 f_{n_1 w_1} & 0 & \lambda\alpha_1 f_{w_1 w_1} & \alpha_1 f_{w_1} \\ -1 & \alpha_1 f_{n_1} & \alpha_2 g_{n_2} & \alpha_1 f_{w_1} & 0 \end{bmatrix}$$

$$(S13) \quad |H| = \alpha_1^2 \alpha_2 \lambda^2 \begin{bmatrix} 2\alpha_1 f_{n_1} f_{n_1 w_1} f_{w_1} g_{n_2 n_2} p_x - \alpha_1 f_{n_1}^2 f_{w_1 w_1} g_{n_2 n_2} p_x + f_{n_1 w_1}^2 (\lambda g_{n_2 n_2} + \alpha_2 p_x g_2^2) \\ -f_{n_1 n_1} (\lambda f_{w_1 w_1} g_{n_2 n_2} + \alpha_2 f_{w_1 w_1} p_x g_2^2 + \alpha_1 f_{w_1}^2 g_{n_2 n_2} p_x) \end{bmatrix}$$

$$(S14) \quad \frac{\partial n_1}{\partial \alpha_2} = \frac{|H_{n_1}|}{|H|} = \frac{\alpha_1^2 \alpha_2 \lambda^2 p_x (f_{n_1 w_1} f_{w_1} - f_{n_1} f_{w_1 w_1})}{\alpha_1^2 \alpha_2 \lambda^2 \begin{bmatrix} 2\alpha_1 f_{n_1} f_{n_1 w_1} f_{w_1} g_{n_2 n_2} p_x - \alpha_1 f_{n_1}^2 f_{w_1 w_1} g_{n_2 n_2} p_x \\ + f_{n_1 w_1}^2 (\lambda g_{n_2 n_2} + \alpha_2 p_x g_2^2) \\ - f_{n_1 n_1} (\lambda f_{w_1 w_1} g_{n_2 n_2} + \alpha_2 f_{w_1 w_1} p_x g_2^2 + \alpha_1 f_{w_1}^2 g_{n_2 n_2} p_x) \end{bmatrix}}$$

$$= \frac{p_x (f_{n_1 w_1} f_{w_1} - f_{n_1} f_{w_1 w_1})}{\begin{bmatrix} 2\alpha_1 f_{n_1} f_{n_1 w_1} f_{w_1} g_{n_2 n_2} p_x - \alpha_1 f_{n_1}^2 f_{w_1 w_1} g_{n_2 n_2} p_x \\ + f_{n_1 w_1}^2 (\lambda g_{n_2 n_2} + \alpha_2 p_x g_2^2) \\ - f_{n_1 n_1} (\lambda f_{w_1 w_1} g_{n_2 n_2} + \alpha_2 f_{w_1 w_1} p_x g_2^2 + \alpha_1 f_{w_1}^2 g_{n_2 n_2} p_x) \end{bmatrix}}$$

$$(S15) \quad \frac{\partial n_2}{\partial \alpha_2} = \frac{|H_{n_2}|}{|H|} = \frac{\alpha_1^2 g_{n_2} \lambda^2 \begin{bmatrix} -2\alpha_1 f_{n_1} f_{n_1 w_1} f_{w_1} p_x + \alpha_1 f_{n_1}^2 f_{w_1 w_1} g_{n_2 n_2} p_x \\ - f_{n_1 w_1}^2 (\lambda + \alpha_2 p_x g(n_2)) \\ + f_{n_1 n_1} (\lambda f_{w_1 w_1} + \alpha_1 f_{w_1}^2 p_x + \alpha_2 f_{w_1 w_1} g_{n_2} p_x) \end{bmatrix}}{\alpha_1^2 \alpha_2 \lambda^2 \begin{bmatrix} 2\alpha_1 f_{n_1} f_{n_1 w_1} f_{w_1} g_{n_2 n_2} p_x - \alpha_1 f_{n_1}^2 f_{w_1 w_1} g_{n_2 n_2} p_x \\ + f_{n_1 w_1}^2 (\lambda g_{n_2 n_2} + \alpha_2 p_x g_2^2) \\ - f_{n_1 n_1} (\lambda f_{w_1 w_1} g_{n_2 n_2} + \alpha_2 f_{w_1 w_1} p_x g_2^2 + \alpha_1 f_{w_1}^2 g_{n_2 n_2} p_x) \end{bmatrix}}$$

$$= \frac{g_{n_2} \begin{bmatrix} -2\alpha_1 f_{n_1} f_{n_1 w_1} f_{w_1} p_x + \alpha_1 f_{n_1}^2 f_{w_1 w_1} g_{n_2 n_2} p_x \\ - f_{n_1 w_1}^2 (\lambda + \alpha_2 p_x g(n_2)) \\ + f_{n_1 n_1} (\lambda f_{w_1 w_1} + \alpha_1 f_{w_1}^2 p_x + \alpha_2 f_{w_1 w_1} g_{n_2} p_x) \end{bmatrix}}{\alpha_2 \begin{bmatrix} 2\alpha_1 f_{n_1} f_{n_1 w_1} f_{w_1} g_{n_2 n_2} p_x - \alpha_1 f_{n_1}^2 f_{w_1 w_1} g_{n_2 n_2} p_x \\ + f_{n_1 w_1}^2 (\lambda g_{n_2 n_2} + \alpha_2 p_x g_2^2) \\ - f_{n_1 n_1} (\lambda f_{w_1 w_1} g_{n_2 n_2} + \alpha_2 f_{w_1 w_1} p_x g_2^2 + \alpha_1 f_{w_1}^2 g_{n_2 n_2} p_x) \end{bmatrix}}$$

$$(S16) \quad \frac{\partial w_1}{\partial \alpha_2} = \frac{|H_{w_1}|}{|H|} = \frac{\alpha_1^2 \alpha_2 \lambda^2 p_x (f_{n_1 w_1} f_{n_1} - f_{w_1} f_{n_1 n_1})(g_{n_2}^2 - g(n_2) g_{n_2 n_2})}{\alpha_1^2 \alpha_2 \lambda^2 \begin{bmatrix} 2\alpha_1 f_{n_1} f_{n_1 w_1} f_{w_1} g_{n_2 n_2} p_x - \alpha_1 f_{n_1}^2 f_{w_1 w_1} g_{n_2 n_2} p_x \\ + f_{n_1 w_1}^2 (\lambda g_{n_2 n_2} + \alpha_2 p_x g_{n_2}^2) \\ - f_{n_1 n_1} (\lambda f_{w_1 w_1} g_{n_2 n_2} + \alpha_2 f_{w_1 w_1} p_x g_{n_2}^2 + \alpha_1 f_{w_1}^2 g_{n_2 n_2} p_x) \end{bmatrix}}$$

$$= \frac{p_x (f_{n_1 w_1} f_{n_1} - f_{w_1} f_{n_1 n_1})(g_{n_2}^2 - g(n_2) g_{n_2 n_2})}{\begin{bmatrix} 2\alpha_1 f_{n_1} f_{n_1 w_1} f_{w_1} g_{n_2 n_2} p_x - \alpha_1 f_{n_1}^2 f_{w_1 w_1} g_{n_2 n_2} p_x \\ + f_{n_1 w_1}^2 (\lambda g_{n_2 n_2} + \alpha_2 p_x g_{n_2}^2) \\ - f_{n_1 n_1} (\lambda f_{w_1 w_1} g_{n_2 n_2} + \alpha_2 f_{w_1 w_1} p_x g_{n_2}^2 + \alpha_1 f_{w_1}^2 g_{n_2 n_2} p_x) \end{bmatrix}}$$

References

Ashfaq, M., D. Rastogi, R. Mei, S. C. Kao, S. Gangrade, B. S. Naz, and D. Touma. 2016. "High-Resolution Ensemble Projections of Near-Term Regional Climate over the Continental United States." *Journal of Geophysical Research: Atmospheres* 121 (17): 9943–963.

Basche, A. D., S. V. Archontoulis, T. C. Kaspar, D. B. Jaynes, T. B. Parkin, and F. E. Miguez. 2016. "Simulating Long-Term Impacts of Cover Crops and Climate Change on Crop Production and Environmental Outcomes in the Midwestern United States." *Agriculture, Ecosystems & Environment* 218: 95–106.

Bosch, D. J. M. B. Wagena, A. C. Ross, A. S. Collick, and Z. M. Easton. 2018. "Meeting Water Quality Goals under Climate Change in Chesapeake Bay Watershed, USA." *JAWRA Journal of the American Water Resources Association* 54 (6): 1239–1257.

Chen, X., and H. Önal. 2012. "Modeling Agricultural Supply Response Using Mathematical Programming and Crop Mixes." *American Journal of Agricultural Economics* 94 (3): 674–86.

Chen, X., H. Huang, M. Khanna, and H. Önal. 2014. "Alternative Transportation Fuel Standards: Welfare Effects and Climate Benefits." *Journal of Environmental Economics and Management* 67 (3): 241–57.

Chen, Y., G. W. Marek, T. H. Marek, J. E. Moorhead, K. R. Heflin, D. K. Brauer, P. H. Gowda, and R. Srinivasan. 2019. "Simulating the Impacts of Climate Change on Hydrology and Crop Production in the Northern High Plains of Texas Using an Improved SWAT Model." *Agricultural Water Management* 221: 13–24.

Cropper, M. L., and W. E. Oates. 1992. "Environmental Economics: A Survey." *Journal of Economic Literature* 30 (2): 675–740.

Dieter, C. A., K. S. Linsey, R. R. Caldwell, M. A. Harris, T. I. Ivahnenko, J. K. Lovelace, M. A. Maupin, and N. L. Barber. 2018. "Estimated Use of Water in the United States County-Level Data for 2015 (ver. 2.0, June 2018)." US Geological Survey data release, 10, p.F7TB15V5.

Falcone, J. A. 2021. "Estimates of county-level nitrogen and phosphorus from fertilizer and manure from 1950 through 2017 in the conterminous United States."

U.S. Geological Survey Open-File Report 2020–1153. https://doi.org/10.3133/ofr20201153.

Griffin, R. C. 1995. "On the Meaning of Economic Efficiency in Policy Analysis." *Land Economics*: 1–15.

Guevara-Ochoa, C., A. Medina-Sierra, and L. Vives. 2020. "Spatio-temporal Effect of Climate Change on Water Balance and Interactions between Groundwater and Surface Water in Plains." *Science of the Total Environment* 722: 137886.

Harding, K. J., P. K. Snyder, and S. Liess. 2013. "Use of Dynamical Downscaling to Improve the Simulation of Central US Warm Season Precipitation in CMIP5 Models." *Journal of Geophysical Research: Atmospheres* 118 (22): 12–522.

HAWQS. 2020. "HAWQS System and Data to Model the Lower 48 Conterminous U.S. Using the SWAT Model." Texas Data Repository Dataverse, V1. https://doi.org/10.18738/T8/XN3TE0.

Havlík, P., U. A. Schneider, E. Schmid, H. Böttcher, S. Fritz, R. Skalský, K. Aoki, S. De Cara, G. Kindermann, F. Kraxner, and S. Leduc. 2011. "Global Land-Use Implications of First and Second Generation Biofuel Targets." *Energy Policy* 39 (10): 5690–5702.

Ishida, K., and M. Jaime. 2015. "A Partial Equilibrium of the Sorghum Markets in US, Mexico, and Japan." No. 330–2016–13894, pp. 1–1.

Karimi, T., C. O. Stöckle, S. S. Higgins, R. L. Nelson, and D. Huggins. 2017. "Projected Dryland Cropping System Shifts in the Pacific Northwest in Response to Climate Change." *Frontiers in Ecology and Evolution* 5: 20.

Kling, C. L., Y. Panagopoulos, S. S. Rabotyagov, A. M. Valcu, P. W. Gassman, T. Campbell, M. J. White, J. G. Arnold, R. Srinivasan, M. K. Jha, and J. J. Richardson. 2014. "LUMINATE: Linking Agricultural Land Use, Local Water Quality and Gulf of Mexico Hypoxia." *European Review of Agricultural Economics* 41 (3): 431–59.

Kumar, S., and V. Merwade. 2011. "Evaluation of NARR and CLM3. 5 Outputs for Surface Water and Energy Budgets in the Mississippi River basin." *Journal of Geophysical Research: Atmospheres* 116(D8).

Marshall, K. K., S. M. Riche, R. M. Seeley, and P. C. Westcott. 2015. "Effects of Recent Energy Price Reductions on US Agriculture." United States Department of Agriculture, Economic Research Service.

Marshall, Elizabeth, Marcel Aillery, Marc Ribaudo, Nigel Key, Stacy Sneeringer, LeRoy Hansen, Scott Malcolm, and Anne Riddle. 2018. "Reducing Nutrient Losses From Cropland in the Mississippi/Atchafalaya River Basin: Cost Efficiency and Regional Distribution," ERR-258. U.S. Department of Agriculture, Economic Research Service.

Montgomery, W.D. 1972. "Markets in Licenses and Efficient Pollution Control Programs." *Journal of Economic Theory* 5 (3): 395–418.

Mei, R., G. Wang, and H. Gu. 2013. "Summer Land–Atmosphere Coupling Strength over the United States: Results from the Regional Climate Model RegCM4–CLM3. 5." *Journal of Hydrometeorology* 14 (3): 946–62.

National Center for Atmospheric Research (NCAR). 2022a. North American Regional Climate Change Assessment Program. Accessed March 10, 2022. https://www.narccap.ucar.edu/results/index.html#climate-change.

NCAR. 2022b. Climate Data Gateway at NCAR. Accessed March 10, 2022. https://www.earthsystemgrid.org/.

Panagopoulos, Y., P. W. Gassman, R. W. Arritt, D. E. Herzmann, T. D. Campbell, M. K. Jha, C. L. Kling, R. Srinivasan, M. White, and J. G. Arnold. 2014. "Surface Water Quality and Cropping Systems Sustainability under a Changing Climate in the Upper Mississippi River basin." *Journal of Soil and Water Conservation* 69 (6): 483–94.

Panagopoulos, Y., P. W. Gassman, R. W. Arritt, D. E. Herzmann, T. D. Campbell, A. Valcu, M. K. Jha, C. L. Kling, R. Srinivasan, M. White, and J. G. Arnold. 2015. "Impacts of Climate Change on Hydrology, Water Quality and Crop Productivity in the Ohio-Tennessee River Basin." *International Journal of Agricultural and Biological Engineering* 8 (3): 36–53.

Piggott, N. E., and M. K. Wohlgenant, 2002. "Price Elasticities, Joint Products, and International Trade." *Australian Journal of Agricultural and Resource Economics* 46 (4): 487–500.

Robertson, D. M., and D. A. Saad. 2013. "SPARROW Models Used to Understand Nutrient Sources in the Mississippi/Atchafalaya River Basin." *Journal of Environmental Quality* 42 (5): 1422–1440.

Saha, G. C., J. Li, R. W. Thring, F. Hirshfield, and S. S. Paul. 2017. "Temporal Dynamics of Groundwater-Surface Water Interaction under the Effects of Climate Change: A Case Study in the Kiskatinaw River Watershed, Canada." *Journal of Hydrology* 551: 440–52.

Scibek, J., D. M. Allen, A. J. Cannon, and P. H. Whitfield. 2007. "Groundwater–Surface Water Interaction under Scenarios of Climate Change Using a High-Resolution Transient Groundwater Model." *Journal of Hydrology* 333 (2–4): 165–81.

Smit, B., and M. W. Skinner. 2002. "Adaptation Options in Agriculture to Climate Change: A Typology." *Mitigation and Adaptation Strategies for Global Change* 7 (1): 85–114.

Steiner, J. L., D. L. Devlin, S. Perkins, J. P. Aguilar, B. Golden, E. A. Santos, and M. Unruh, 2021. "Policy, Technology, and Management Options for Water Conservation in the Ogallala Aquifer in Kansas, USA." *Water* 13 (23): 3406.

U.S. Environmental Protection Agency (US EPA). 2014. Mississippi River Gulf of Mexico Watershed Nutrient Task Force New Goal Framework. (Accessed December 2021). https://www.epa.gov/sites/production/files/2015-07/documents/htf-goals-framework-2015.pdf.

US EPA. 2019. Hypoxia 101. Accessed Dec 2021. https://www.epa.gov/ms-htf/hypoxia-101.

US EPA. 2021a. Northern Gulf of Mexico Hypoxic Zone. Accessed December 2021. https://www.epa.gov/ms-htf/northern-gulf-mexico-hypoxic-zone.

US EPA. 2021b. Hypoxia Task Force Nutrient Reduction Strategies. Accessed December 2021. https://www.epa.gov/ms-htf/hypoxia-task-force-nutrient-reduction-strategies.

USDA ERS (United States Department of Agriculture Economic Research Service). 2019. Fertilizer Use and Price. Accessed October 30, 2019. https://www.ers.usda.gov/data-products/fertilizer-use-and-price.

USDA ERS. 2021. Irrigation & Water Use. Accessed December 2021. https://www.ers.usda.gov/topics/farm-practices-management/irrigation-water-use/.

USDA ERS. 2022. Irrigated cropping patterns in the United States have evolved significantly since 1964. Accessed July 2022. https://www.ers.usda.gov/data-products/chart-gallery/gallery/chart-detail/?chartId=103568.

U.S. Department of Agriculture National Agricultural Statistics Service (USDA NASS). 2019. 2017 Census of Agriculture. Accessed December 2021. www.nass.usda.gov/AgCensus.

USDA NASS. 2020. U.S. & All States County Data–Crops. Accessed May 13, 2021. Washington, DC. http://www.nass.usda.gov/.

U.S. Geological Survey (USGS). 2017. "The Challenge of Tracking Nutrient Pollution 2,300 Miles." USGS.

U.S. Geological Survey (USGS). 2018. "Water-Use Data Available from USGS." Accessed January 1, 2022. https://water.usgs.gov/watuse/data/index.html.

Westcott, P. C. and L. A. Hoffman, 1999. "Price Determination for Corn and Wheat: The Role of Market Factors and Government Programs." No. 1488–2016–123383.

White, M. J., C. Santhi, N. Kannan, J. G. Arnold, D. Harmel, L. Norfleet, P. Allen, M. DiLuzio, X. Wang, J. Atwood, and E. Haney. 2014. "Nutrient Delivery from the Mississippi River to the Gulf of Mexico and Effects of Cropland Conservation." *Journal of Soil and Water Conservation* 69 (1): 26–40.

Wu, D., J. J. Qu, and X. Hao. 2015. "Agricultural Drought Monitoring Using MODIS-Based Drought Indices over the USA Corn Belt." *International Journal of Remote Sensing* 36 (21): 5403–5425.

Xu, Y., D. J. Bosch, M. B. Wagena, A. S. Collick, and Z. M. Easton. 2019. "Meeting Water Quality Goals by Spatial Targeting of Best Management Practices under Climate Change." *Environmental Management* 63 (2): 173–84.

Xu, Y., L. Elbakidze, H. Yen, J. G. Arnold, P. W. Gassman, J. Hubbart, and M. P. Strager. 2022. "Integrated Assessment of Nitrogen Runoff to the Gulf of Mexico." *Resource and Energy Economics* 67: 101279.

Nutrient Pollution and US Agriculture
Causal Effects, Integrated Assessment, and Implications of Climate Change

Konstantinos Metaxoglou and Aaron Smith

9.1 Introduction

Nutrient pollution is one of the country's most widespread, costly, and challenging environmental problems. It is caused by excess nitrogen and phosphorus in the air and water. Although nutrients such as nitrogen and phosphorous are chemical elements that plants and animals need to grow, when too much nitrogen and phosphorus enter the environment, usually from a wide range of human activities, the air and water can become severely polluted.

Some of the largest sources of nutrient pollution include commercial fertilizers, animal manure, sewage treatment plant discharge, storm water runoff, cars, and power plants. In the Mississippi River basin (MRB), which spans 31 states and drains 40 percent of the contiguous US (CONUS) into the Gulf of Mexico (GoM), nutrients from row crops, large farms, and concentrated animal feeding operations account for most of the nutrient pollution. Fertilizer runoff from agricultural crops has been estimated to

Konstantinos Metaxoglou is an associate professor of economics at Carleton University.

Aaron Smith is the DeLoach Professor of Agricultural Economics at the University of California, Davis.

We thank Joe Shapiro for helping us navigate through the USGS data in the very early stages of the paper, Sergey Robotyagov, and Cathy Kling for sharing results of previous work, Jeremy Proville of the Environmental Defense Fund for sharing the results of an in-progress report, and seminar participants at Oregon State and UC Berkeley for comments. We received feedback from Ariel Dinar, Gary Libecap, and Lynne Lewis that helped us to significantly improve the original draft. Any remaining errors are ours. For acknowledgments, sources of research support, and disclosure of the authors' material financial relationships, if any, please see https://www.nber.org/books-and-chapters/american-agriculture-water-resources-and-climate-change/nutrient-pollution-and-us-agriculture-causal-effects-integrated-assessment-and-implications-climate.

contribute somewhere between 50 percent (CENR 2000) and 76 percent (David, Drinkwater, and McIsaac 2010) of the annual and spring nitrogen riverine export from the MRB to the GoM fueling a hypoxic ("dead") zone, with oxygen levels that are too low for fish and other marine life to survive. The GoM hypoxic zone is the second largest in the world behind the dead zone in the Arabian Sea with a peak areal extent equal to that of New Jersey (8,776 square miles) recorded in the summer of 2017.

According to the EPA (2016), 46 percent (about 546,000 miles) of US streams and rivers are in poor condition in terms of their phosphorous levels, and 41 percent (about 495,000 miles) are in poor condition in terms of their nitrogen levels based on sampling results from almost 2,000 sites benchmarked against conditions represented by a set of least-disturbed sites. Excessive nitrogen and phosphorus in water and the air can cause health problems, damage land and water, and take a heavy toll on the economy.[1] Reducing the areal extent of the hypoxic zone to a five-year running average of 5,000 square kilometers, a target set in the Action Plan of the GoM Hypoxia Task Force, comes at an estimated price tag of $2.7 billion per year (Rabotyagov et al. 2014b).

In this chapter, we focus on water pollution and its relationship to US agriculture. We use regression analysis to establish a causal link between farmers' decisions about crop acreage and nutrient pollution that is detrimental to surface water quality. In particular, we estimate the causal effects of corn acreage on nitrogen concentration in water bodies using panel fixed-effect (FE) regressions and what we call "(c)ounty-centric" analysis. We make few and transparent assumptions that allow us to the assess the robustness of our findings to various factors. In contrast, most prior estimates of effects similar to the ones estimated in this chapter are based on agronomic and hydrologic models.

To perform our c-centric analysis, we combine annual county-level data on acres planted and nitrogen pollution. Data on acres planted are readily available from the US Department of Agriculture (USDA). We compile data on nitrogen pollution using US Geological Survey (USGS) monitoring sites within a 50-mile radius from the county centroids. Based on our preferred estimate of the elasticity of nitrogen concentration (mg/L) with respect to corn acreage of about 0.1, an increase in corn acres planted equal to 1 within-county standard deviation implies a 3.3 percent increase in the level of nitrogen concentration. At the average nitrogen concentration of about 2.5 mg/L and the average streamflow of the Mississippi River in the GoM in our sample, this effect entails close to 50,800 additional metric tons of nitrogen in the GoM. Using the median potential damages of nitrogen

---

1. See CENR (2000), EPA (2007), and, more recently, Olmstead (2010), GOMNTF (2013) and Rabotyagov et al. (2014a). Several papers assess the cost of nitrogen pollution employing a variety of methodologies; see Dodds et al. (2009), Compton et al. (2011), Birch et al. (2011), Rabotyagov et al. (2014b), and Sobota et al. (2015), among others.

due to declines in fisheries and estuarine/marine life of \$15.84 per kilogram (\$2008) from Sobota et al. (2015), the implied annual external cost is about \$800 million. The magnitude of the estimated effects depends on the amount of annual precipitation but not on extreme heat despite its well-documented negative impact on crop growth and, hence, nutrient uptake.

We also explore the implications of climate change for nitrogen pollution using the NASA Earth Exchange Global Daily Downscaled Projections (NEX-GDDP-CMIP6) data set to obtain out-of-sample projections for precipitation and temperature, which we translate into projections of corn acreage marginal effects on nitrogen pollution. The NEX-GDDP-CMIP6 data set is comprised of global downscaled climate scenarios derived from the General Circulation Model runs conducted under the Coupled Model Intercomparison Project Phase 6 and across two of the four "Tier 1" greenhouse gas emissions scenarios known as shared socioeconomic pathways (SSPs), namely, SSP2–4.5 and SSP5–8.5. Abstracting from the impact that climate change may have an acreage, yields, nitrogen fertilizer use, legacy nitrogen, runoff, and streamflow, all of which may contribute to nitrogen pollution, the out-of-sample precipitation and temperature projections imply similar effects of corn acreage on nitrogen concentration as in our estimation sample. This finding arises because the climate models project relatively small changes in precipitation and because our estimated effects of corn acreage on nitrogen concentration do not vary a lot with temperature.

The focus of this chapter is different from the chapter by Elbakidze et al. (2022) in this volume. Elbakidze et al. study the effects of changes in nitrogen fertilizer use by US farmers on surface water quality due to climate change. Investigating the effect of climate-driven productivity changes on water quality in the GoM using an integrated hydro-economic agricultural land use model (IHEAL), they find that land and nitrogen use adaptation in agricultural production to climate change increases nitrogen loads to the GoM by 0.4–1.58 percent. As we discuss later in the chapter, our findings are consistent with new research in environmental science arguing that there is a large amount of nitrogen stored in subsurface soil and groundwater contributing to the so-called legacy nitrogen, which may increase loadings in rivers and streams with a long delay. The work by Elbakidze et al. does not address legacy nitrogen. Elbakidze et al. account for farmers' adaptation to climate change in their analysis while our reduced-form econometric analysis does not.

The remainder of the paper is organized as follows. Section 9.2 provides a background on nutrient pollution emphasizing the role of agriculture and shedding light on the impacts of climate change. Section 9.3 is a simple theoretical backdrop for section 9.4, where we describe the empirical approach for estimating the causal effects of interest. Subsequently, having discussed the data and provided some descriptive analysis in section 9.5, we present the results from our regressions in section 9.6. We next explore the implications of climate change for nitrogen pollution in section 9.7. We finally conclude.

9.2   Background on Nutrient Pollution

Preamble. Nitrogen inputs to the ecosystem from both anthropogenic and natural sources are transported via atmospheric, surface flow, drain flow, and groundwater pathways. Nitrate-nitrogen concentrations in the Mississippi River, which drains most US cropland, increased dramatically in the second half of the last century, especially between the early 1960s and the mid-1980s, largely coinciding with the surge in commercial fertilizer use for row crops in the MRB states (e.g., see Capel et al. 2018). The corn-and-soybeans cropping system that dominates the Corn Belt is an inherently "leaky" system—some nitrogen loss to subsurface drainage water is inevitable (McLellan et al. 2015). In fact, the majority of agricultural nitrogen loss occurs via subsurface drainage water, either as seepage through soils and shallow geologic units or in engineered drainage structures such as drainage tiles and ditches.

Aside from oscillations in streamflow, artificial drainage and other changes to the hydrology of the Midwest (e.g., dams and reservoirs), atmospheric deposition of nitrates within the MRB, non-point discharges from urban and suburban areas, and point discharges, particularly from domestic wastewater treatment systems and feedlots, all contribute to the nutrients that reach the GoM (Goolsby et al. 1999). Between 1980 and 2016, close to 1.5 million metric tons of nitrogen (about 63 percent in the form of nitrate) per year were discharged, on average, to the GoM. From 1968 to 2016, the average annual Mississippi streamflow was close to 21,500 cubic meters per second.[2] During this time, there was a strong positive relationship between the streamflow of the Mississippi and nitrogen flux in the GoM.

Dairy, beef, hog, poultry, and aquaculture systems can also cause significant discharges of nutrients to streams and rivers. Untreated wastewater from these systems generally has very high concentrations of nitrogen, most often as ammonia-nitrogen, although high concentrations of nitrate-nitrogen are also possible. Urban and suburban areas have significant runoff from lawns, parking lots, rooftops, roads, highways, and other impervious sources. The major point sources of direct discharges of nutrients, particularly nitrogen-nitrogen, appear to be domestic wastewater treatment plants. Fossil-fuel combustion in car engines and electric generating plants also contributes to airborne nitrates that return to the earth's surface with rain, snow, and fog (wet deposition) or as gases and particulate (dry deposition).

---

2. We refer to the average flow and total Mississippi-Atchafalya nitrogen flux (sum of NO_3+NO_2, TKN, and NH_3) AMLE estimates using data in this link: http://toxics.usgs.gov/hypoxia/mississippi/flux_ests/delivery/index.html.

Table 9.1   Nitrogen pollution damages and abatement costs

Source	Damages	Details
	A. Damages	
Taylor and Heal (2021)	$583	U.S., per ton of nitrogen
Sobota et al. (2015)	$15,840	U.S., per ton of nitrogen
Van Grinsven et al. (2013)	$13,338–$53,351	E.U., per ton of nitrogen
Compton et al. (2011)	$56,000	GoM fisheries decline, per ton of nitrogen
Compton et al. (2011)	$6,380	CB recreational use, per ton of nitrogen
Blottnitz et al. (2006)	$300	E.U., per ton of nitrogen
Dodds et al. (2009)	$2.2 billion	U.S., freshwater eutrophication, annually
Kudela et al. (2015)	$4 billion	U.S., algal blooms, annually
UCS (2020)	$0.552–$2.4 billion	GoM fisheries & marine habitat, annually
Anderson et al. (2000)	$449 million	U.S., algal blooms, annually

Source	Abatement costs	Geographic scope
	B. Abatement costs	
Xu et al. (2021)	$6 billion	Mississippi River Basin
Tallis et al. (2019)	$2.6 billion	Mississippi River Basin
Marshall et al. (2018)	$1.9–$3.3 billion	Mississippi River Basin
McLellan et al. (2016)	$1.48 billion	Mississippi River Basin
Whittaker et al. (2015)	$9.25 billion	Mississippi River Basin
Rabotyagov et al. (2014a)	$2.6 billion	Mississippi River Basin
USEPA (2001)	<$1–$4.3 billion	US, national
Ribaudo et al. (2001)	$0.1–$7.91 billion	Mississippi River Basin
Doering et al. (1999)	−$0.1–$17.95 billion	Mississippi River Basin

Note: In Van Grinsven et al. (2013a), the reported cost of €25–100 billion per year implies a cost of €4.11–16.43 per lb of nitrogen using $0.6 \times 4.6 = 2.6$ million tons of nitrogen attributed to agricultural sources. At an exchange rate of $1.5/ in 2008, we have a cost of $6.05–$24.20 per lb of nitrogen in 2008. We report the cost per ton of nitrogen. In the case of USEPA (2001), the costs are per year for the development of TMDLs. Table IV-1 in USEPA (2001) shows the leading causes of water impairment (nutrients account for 11.5%) and leading sources (agriculture accounts for 24.6%). See table 6.1 in Doering et al. (1999), where the numbers are reported as net social benefits. See table 2 in Ribaudo et al. (2001), where the numbers are reported as net social benefits too. We use "CB" to refer to the Chesapeake Bay, "GoM" to refer to the Gulf of Mexico. For additional details, see section 9.2 in the main text and section A.1 of the online appendix (http://www.nber.org/data-appendix/c14692/appendix.pdf).

This nitrogen then enters streams and rivers and/or is retained in terrestrial systems in the same pathways as nitrate-nitrogen fertilizer.

Damages and abatement costs of nitrogen pollution. In table 9.1, we summarize studies related to damages and abatement costs associated with nitrogen pollution noting that the estimation of the economic value of the damages associated with nutrient pollution can be particularly challenging.[3] The social cost of pollution in the context of water quality has

3. EPA (2015) provides estimates of external costs associated with nutrient pollution impacts on tourism and recreation, commercial fishing, property values, human health, as well as drinking water treatment costs, mitigation costs, and restoration costs.

received less attention than the social cost of carbon in the context of climate change. Quantifying the social cost of nitrogen is challenging due to multiple loss pathways associated with damages to water quality, air quality, and climate change that occur over heterogeneous spatial and temporal scales (Gourevitch, Keeler, and Ricketts 2018). The diversity of nitrogen loss pathways and endpoints at which damages occur makes it challenging to construct a single cost metric. The impacts are largely driven by the location where the nitrogen is emitted and applied, the transport and transformation of nitrogen into different forms, and the expected damages along the flow path (Keeler et al. 2018).

Nitrogen pollution and agriculture. Using too little nitrogen for a highly responsive crop such as corn entails lower yields, poorer grain quality, and reduced profits. When too much nitrogen is applied, crop yields and quality are not affected, but profit can be reduced somewhat and negative environmental consequences are very likely. Thus, many farmers choose to err on the liberal side in terms of nitrogen application rates. This extra nitrogen is often called "insurance" nitrogen; see Mitsch et al. (1999) and CENR (2000), among others. Overall, nitrogen use efficiency (uptake) and the "4Rs" in nutrient management—right source, rate, time, and place for plant nutrient application based on local agronomic recommendations—in order to minimize nitrogen losses to the environment are of paramount importance for addressing nitrogen pollution.

The prevention of nutrient pollution, particularly in the form of nitrate-nitrogen, is possible through a number of general approaches and specific techniques, ranging from modification of agricultural practices to the construction and restoration of riparian zones and wetlands as buffer systems between agricultural lands and waterways.[4] To provide some examples, on-site control of agricultural drainage is possible via adoption of one or a combination of the following: nitrogen fertilizer application rates, management of manure spreading, timing of nitrogen application, the use of nitrification inhibitors, the change of plowing (tillage) methods, and increasing drainage tile spacing. Wetlands and riparian buffers can be effective means of off-site control.

Policy responses to nutrient pollution. As of this writing, the major federal response to nutrient pollution from agriculture continues to be through research, education, outreach, and voluntary technical and financial incentives. A number of USDA agencies provide support through education, outreach, and research, while federal funds are provided through conservation programs to help agricultural producers, who participate voluntarily, to adopt best management practices in crop production to achieve nutri-

---

4. EPA (2007), Ribaudo et al. (2011), NRCS (2017b), and Capel et al. (2018), among others, offer a very informative discussion on controlling nitrogen pollution from agricultural sources.

ent pollution reduction. At a very high level, the USDA programs are distinguished between land-retirement and working-land programs with the spending on conservation programs having increased substantially since the 2002 Farm Security and Rural Investment Act.[5] In the case of the land-retirement programs, landowners receive payments in exchange for taking land out of active agricultural production and putting the land into perennial grasses, trees, or wetland restoration. Landowners or producers participating in working-land programs receive payments to cover part or all of the costs of making changes in conservation practices and management decisions on their land that remains in agricultural production.

In one of the most comprehensive assessments of conservation practices by US farmers, the USDA Conservation Effects Assessment Project (CEAP) national nitrogen loss report (NRCS 2017b) found that 29 percent of nitrogen applied as commercial fertilizer or manure was lost from the fields through various pathways based on survey data for 2003–2006. The mean of the average annual estimates of total nitrogen loss was 34 lb per cultivated cropland acre per year. The amount varied considerably, however, among cultivated cropland acres. Total nitrogen losses were highest for acres receiving manure (56 lb per acre per year). Based on simulations performed using the APEX model in the report, the use of conservation practices during 2003–2006 reduced total nitrogen loss (all loss pathways) by 14.9 lb per acre per year, on average, representing a 30 percent reduction.

9.3   A Simple Theoretical Framework

We estimate the *reduced-form* effect of an increase in corn acreage on nitrogen pollution via OLS regressions. We focus on this relationship in part because corn acreage is the driving force behind the amount of nitrogen fertilizer used. In addition, acreage is much better measured than fertilizer use. We observe nitrogen fertilizer sales by county, but we do not know in which county or year that fertilizer was applied to a field. In contrast, we observe annual acreage by county.[6]

---

5. We refer to this link, https://www.fsa.usda.gov/programs-and-services/conservation-programs/index, and Capel et al. (2018) for a succinct and very informative discussion of the various USDA conservation programs.

6. Paudel and Crago (2020) use the nitrogen fertilizer sales data to estimate the effect of fertilizer on nitrogen pollution. They obtain an elasticity of nitrogen pollution with respect to nitrogen fertilizer of about 0.15 for the US. We find an elasticity of nitrogen pollution with respect to corn acres of a very similar magnitude. Adding the assumption of no substitution between nitrogen fertilizer and other inputs to the assumption of a fixed amount of nitrogen fertilizer per corn acre allows us to link the price elasticity of the demand for fertilizer (η_{fert}) to the price elasticity demand for corn (η_{corn}) via $\eta_{fert} = (p_{fert} / p_{corn})\eta_{corn}$. In terms of notation, p_{fert} and p_{corn} are the prices of nitrogen fertilizer and corn, respectively. As we discuss later in the paper, fertilizer costs account for about 20 percent of the value of corn production during the period we study, which coupled with a reasonable value of η_{corn} of about −0.3, also supported empirically in subsequent section, imply $\eta_{fert} = -0.3 \times 0.2 = -0.06$. Hence, the demand for nitrogen fertilizer is highly inelastic.

Our empirical analysis, which focuses on the relationship between corn acreage and nitrogen pollution, is motivated by the following. Farmers decide how to allocate acreage to various crops including corn, which is the most fertilizer intensive and is the crop we focus on. Soybeans, the other commonly planted crop in the US Corn Belt, require little nitrogen fertilizer. Farmers apply about 150 lb of nitrogen fertilizer per planted acre of corn and 5 lb per planted acre of soybeans. About 70 percent of soybean acres receive no nitrogen fertilizer.[7] Crop production requires various inputs such as labor, capital, fuel, seeds, fertilizers, and chemicals. Farmers' planting decisions are based on the expected post-harvest crop price and expected costs. Weather conditions, especially precipitation and temperature, during the growing season determine plant growth and eventually yields. Pre-planting weather conditions may also affect planting decisions.

As farmers plant more corn acres, they use more nitrogen fertilizer, generally following agronomic recommendations. The shape of the crop production function implies that fertilizer application in excess of agronomic recommendations does not reduce yields, which provides an insurance motivation to use extra fertilizer, as we discussed earlier. A combination of factors in and out of the farmers' control, including weather, determine the crop nitrogen uptake, and, hence, the amount of surplus (excess) nitrogen that will not be used by the plants and will remain in the soil. This surplus nitrogen will eventually find its way to lakes, rivers, and streams, contributing to nutrient pollution. The amount of surplus nitrogen that enters waterways is determined in part by the weather. Wetter conditions affect acreage, nutrient runoff, and streamflow, all of which can contribute to nutrient pollution. All else equal, more rainfall means more nutrients carried through the soil and along the surface into waterways. Thus, we expect increases in corn acreage to increase nitrogen concentration, especially in wet years. Similarly, extreme heat, which has a well-documented negative impact on crop growth (e.g., Jägermeyr et al. 2021, among others) may limit nutrient uptake and contribute to runoff. On the one hand, it is plausible that farmers may compensate for the loss in yields by fertilizing more. On the other hand, as discussed in the chapter by Elbakidze et al. (2022), lower yields may reduce the profitability of crop production and may result in decreased crop acreage, which could reduce nitrogen runoff.

In general, more rainfall due to a warmer and wetter atmosphere is increasing nitrogen pollution exacerbating algae growth and expanding dead zones in coastal areas.[8] Evidence suggests that several projected outcomes of global climate change will act to increase the prevalence and negative

---

7. Based on the USDA ERS Fertilizer Use and Price data for 2018 (US average).

8. In the US Gulf Coast, the frequency and severity of hurricanes, which have been linked to climate change, can also play an important role in the areal extent the hypoxic zone formed every summer.

impacts of dead zones.[9] Warmer waters hold less oxygen than cooler water, thus making it easier for dead zones to form. Warmer waters also increase metabolism of marine creatures, thereby increasing their need for oxygen. Additionally, warmer temperatures and increased runoff of fresh water will increase stratification of the water column, thus further promoting the formation of dead zones. Increased runoff will also increase nutrient inputs into coastal water bodies. On the other hand, projections of more intense tropical storms and lower runoff would act to decrease stratification and thus make dead zones less likely to form or less pronounced if they do form.[10]

Diaz and Rosenberg (2008) assembled a database of over 400 dead zones worldwide showing that their number is increasing exponentially over time. To characterize the severity of climate change that these ecosystems are likely to experience over the coming century, Diaz and Rosenberg also explored the future annual temperature anomalies predicted to occur for each of these systems. The majority of dead zones are in regions predicted to experience over 2°C warming by the end of this century. Sinha, Michalak, and Balaji (2017) show that precipitation changes due to climate changes alone will increase by 19 percent the riverine total nitrogen loading within the CONUS by the end of the century for their business-as-usual scenario. The impacts are particularly large in the Northeast (28 percent), the upper MRB (24 percent), and the Great Lakes Basin (21 percent). According to the authors, precipitation changes alone will lead to an 18 percent increase in nitrogen loads in the MRB, which would require a 30 percent reduction in nitrogen inputs. The target of a 20 percent load reduction set by the GoM Hypoxia Task Force in 2015 would require a 62 percent reduction in nitrogen inputs taking into account the confounding effect of precipitation.[11]

9.4   Empirical Approach

We estimate panel fixed-effect (FE) OLS regressions of the form:

$$(1) \qquad y_{it} = \delta_i + \beta_1 a_{it} + \beta_2 a_{it} p_{it} + \mathbf{z}'_{it}\gamma + g_i(t) + \varepsilon_{it},$$

9. Our discussion borrows heavily from the discussion on "Dead Zones and Climate Change" available in the VIMS website here: https://www.vims.edu/research/topics/dead_zones /climate_change/index.php.

10. According to Diaz and Rosenberg (2008), tropical storms and hurricanes influence the duration, distribution, and size of the GoM dead zone in a complex way. In 2005, four hurricanes (Cindy, Dennis, Katrina, and Rita) disrupted stratification and aerated bottom waters. After the first two storms, stratification was reestablished and hypoxia reoccurred, but the total area was a fourth less than predicted from spring nitrogen flux. The other two hurricanes occurred later in the season and dissipated hypoxia for the year.

11. In February 2015, the states and federal agencies that comprise the Mississippi River/ GoM Watershed Nutrient Task Force (Hypoxia Task Force, or HTF) announced that the HTF would retain its goal of reducing the areal extent of the GoM hypoxic zone to less than 5,000 km^2, but that it will take until 2035 to do so. The HTF agreed on an interim target of a 20 percent nutrient load reduction in the Gulf of Mexico by the year 2025 as a milestone toward achieving the final goal in 2035.

where i denotes the cross-sectional unit (county) and t denotes the time (year) in what we call the (c)ounty-centric (henceforth, *c-centric*) analysis. The dependent variable y_{it} is nitrogen concentration in milligrams per liter (mg/L), a_{it} denotes corn acres planted, p_{it} denotes precipitation, and $\mathbf{z}_{it}$ is a vector of weather-related control variables. The weather-related controls include precipitation, squared precipitation, moderate-heat, and extreme-heat degree days. We use $g_i(t)$ to denote alternative functions of time (e.g., time trend, year FE, etc.). Finally, ε_{it} is the error term.

For our c-centric analysis, y_{it} is the average nitrogen concentration recorded at USGS monitoring sites within a 50 mile-radius from the county centroids, and a_{it} are corn acres planted in county i at time t. As part of a series of robustness checks to our results, we estimate (1) using average nitrogen concentration recorded at sites within larger (100- and 200-mile) radii, as well as accounting for streamflow using only sites downstream of the county centroids.

Our specifications aim to capture the most salient factors that are both in the control and out of the control of US farmers and that influence the nitrogen concentration of waters draining cropland, some of which we have already discussed. Aside from weather, factors outside farmers' control include hydrologic conditions, terrain properties of the cropland (e.g., slope and elevation), and soil properties (e.g., depth, texture, mineralogy, capacity to support crop growth, and susceptibility to erosion). Factors in farmers' control include agricultural management practices used to boost profits, such as cropping systems, rate of and timing of nitrogen application, use and type of drainage and tillage systems, deployment of programs aiming to combat nutrient pollution by the US Environmental Protection Agency (EPA), and conservation programs administered by the USDA, among others.

Precipitation and temperature generally affect the farmers' decision making during the spring planting season (e.g., when and what to plant, and how much to fertilize). Miao, Khanna, and Huang (2015) include monthly precipitation in March to May to control for the effect of pre-planting weather conditions on corn acreage in the US. They argue that a wet spring can make it difficult for corn to be planted on time, and, hence, corn acreage may be switched to soybean acreage. During the growing season, which is somewhere between March and September for most of the US, both temperature and precipitation have an effect on crop growth and, hence, on the plants' nutrient uptake. In the absence of robust crop growth rates, nutrients that are not absorbed by the plants can be carried over to streams, rivers, and lakes, depending on soil characteristics and precipitation.

Nitrogen concentrations in a basin like the MRB, which drains most of the cropland where corn is grown and is characterized by an abundant supply of nitrogen in the soil, tend to peak in the late winter and spring when streamflow is highest, and lowest in the late summer and fall when streamflow is low. This strong positive relationship between concentration and streamflow has been well documented in the Midwest; see Goolsby et al.(1999) and the references cited. Importantly, the same strong positive relationship implies that nitrogen pollution is predominantly due to non-point sources. Nitrogen concentrations generally decrease in the summer and fall as streamflow and agricultural drainage decrease. Assimilation of nitrate by agricultural crops on the land and aquatic plants in streams also helps decrease nitrogen concentrations in streams during the summer. Moreover, in-stream denitrification rates also increase during the summer due to increased temperatures and longer residence times of water in the streams. Hence, temperature and precipitation are correlated with both acres planted and nitrogen concentration.

The fixed effects δ_i aim to capture time invariant spatial attributes such as soil properties and texture, and water infiltration rates that affect both the farmers' planting decisions and levels of nitrogen in the water due to, say, transport and attenuation. For example, soil texture—the proportions of sand, clay, and silt—influences the ease with which the soil can be worked, the amount of water and air the soil holds, and the rate at which the water can enter and move through the soil. Fine-grained (clayey) solid can hold more water than coarse-grained (sandy) soils.

Finally, $g_i(t)$ allows us to model in a flexible way trends in fertilization rates, as well as land management practices, such as tillage, and subsurface tile drainage, for which data with good spatial and time coverage are not available. They also allow us to account for farmers' participation in conservation programs administered by the USDA and other unobservables that may exhibit spatially differentiated trends and affect both the corn acreage and nitrogen concentration.

In the robustness checks discussed later in the chapter, we consider a long list of additional controls to capture factors that may be correlated with both corn acres planted and nitrogen concentration as discussed above to alleviate concerns for potentially biased estimates. We also explore alternative ways to measure nitrogen concentration including distance, streamflow, and time of the year, as well as spatial and temporal variation in the effects of corn acreage on nitrogen concentration.

9.5   Data

9.5.1   Data Sources

Water quality. The data on nitrogen concentration are from the Water Quality Portal (WQP). The WQP is a cooperative service sponsored by the USGS, the EPA, and the National Water Quality Monitoring Council. It serves data collected by over 400 state, federal, tribal, and local agencies with more than more than 297 million water quality records.

We accessed WQP data on sites and sample results (physical/chemical metadata) associated with the parameter code 00600, which is described as "total nitrogen [nitrate + nitrite + ammonia + organic-N], water, unfiltered, milligrams per liter" without imposing any other of the additional filters available in the portal in December 2019. At the time we accessed the WQP data, there were close to 754,000 observations in the sample results data and 41,800 observations in the site data.[12]

The site data contain information regarding the site's location such as longitude and latitude, county, and the eight-digit hydrologic unit (HUC8). The site data also contain information on the agency operating the site (e.g., "USGS-IL") and the site type (e.g., "stream," "facility," "lake," "well," etc.) The sample results data contain a long list of variables related to water quality measures, such as the date, time, and method of the water sample collection. Linking the site to the sample results data is straightforward using the site location identifier field, which is present in both data sets.

We measure nitrogen pollution using concentration in milligrams per liter (mg/L). We limit the data to those for sites in the CONUS and for which we track "surface water" and "groundwater" concentration in the sample results data. For the interested reader, some additional information regarding the WQP data used in the paper is available in sections A.2–A.4 of the online appendix.

Crops. Annual county-level data on corn acres planted are available from the National Agricultural Statistics Service (NASS) of the USDA.[13] Following Schlenker and Roberts (2009) and Annan and Schlenker (2015), among others, in a long stream of literature in agricultural economics, and to focus on rainfed agriculture, we limit our sample to counties east of the 100th meridian and exclude Florida. This is the part of the country that accounts for more than 95 percent of the corn produced during the time relevant for our analysis; as part of our robustness checks, we expand the geographic scope of our analysis to the CONUS.

Weather. We use updated temperature and precipitation data from Schlenker and Roberts (2009), which are available for each county during the growing season for 1970–2017 and are based on PRISM gridded weather data. The data from Schlenker and Roberts have been used extensively in the literature on the effects of climate change on US agriculture and are discussed in great detail elsewhere (Roberts, Schlenker, and Eyer 2012).

---

12. The WQP data can be accessed in this link, https://www.waterqualitydata.us/ using web service calls. A parameter code is a five-digit number used in the National Water Information System (NWIS) to uniquely identify a water quality characteristic.

13. Table A1 in the online appendix (http://www.nber.org/data-appendix/c14692/appendix .pdf) shows corn production by state for 1970–2017.

Following this stream of the literature, we use precipitation, the square of precipitation, cumulative degree days (DDs) between 10°C and 29°C (moderate heat), and cumulative degree days above 29°C (extreme heat). In what follows, the precipitation is measured in meters, the moderate heat is measured in 1,000 DDs, and the extreme heat is measured in 100 DDs.

Hydrologic Units. We use the USDA Natural Resources Conservation Service (NRCS) watershed boundary data set (WBD) to identify hydrologic units of different size.[14] We use two-digit hydrologic unit codes (HUC2s) to explore spatial variation in our estimated acreage effects in the panel FE regressions and to construct spatial FEs in robustness checks that pertain to cross-section regressions. We use four-digit hydrologic unit codes (HUC4s) to cluster the standard errors in our regressions. We use HUC8s in an analysis based on an alternative data aggregation scheme, as part of our robustness checks.

National hydrography data set plus V21. As in Keiser and Shapiro (2018), we use the NHD Plus flowline network to follow water pollution upstream and downstream. In particular, we use the National Seamless Geodatabase built on NHD Plus to identify monitoring sites downstream of counties of interest.

9.5.2   Data Overview and Descriptive Statistics

For our baseline estimates, we use data for counties east of the 100th meridian (EAST-100) excluding Florida for 1970–2017. We use the latitude and longitude of the county centroids to identify the relevant EAST-100 counties which we obtain from the CENSUS TIGER shape files. As we discussed earlier, we calculate nitrogen concentration using USGS monitoring sites within a 50-mile radius from the county centroids.

Table 9.2 shows basic summary statistics for nitrogen concentration, our measure of pollution, and corn acres planted. These are the dependent and main explanatory variables of interest in our regression models. The table also shows summary statistics for precipitation (total annual and total by month), as well as for moderate and extreme heat by month. Precipitation

---

14. The GBD files for hydrologic units of different size are available in the following link: https://nrcs.app.box.com/v/gateway/folder/18546994164. The US is divided into successively smaller hydrologic units which are classified into four levels: regions, subregions, accounting units, and cataloging units. The hydrologic units are arranged or nested within each other from the largest geographic areas (regions) to the smallest geographic areas (cataloging units). Each hydrologic unit is identified by a unique hydrologic unit code (HUC) consisting of two to eight digits based on the four levels of classification in the hydrologic unit system. It is common to refer to hydrologic units as watersheds, and what we describe here as hydrologic accounting is also described as watershed delineation. The word *watershed* is sometimes used interchangeably with *drainage basin* or *catchment*.

Table 9.2   **Summary statistics**

Variable	Panel	obs	Years	Mean	s.d. B	s.d. W	Median
nitrogen	2,232	64,121	28.7	2.451	1.645	1.663	1.683
acres planted	2,232	64,121	28.7	0.038	0.048	0.011	0.015
precipitation annual	2,232	64,121	28.7	1.088	0.259	0.174	1.070
precipitation jan	2,232	64,121	28.7	0.073	0.041	0.039	0.060
precipitation feb	2,232	64,121	28.7	0.067	0.036	0.036	0.055
precipitation mar	2,232	64,121	28.7	0.092	0.038	0.046	0.081
precipitation apr	2,232	64,121	28.7	0.095	0.024	0.048	0.086
precipitation may	2,232	64,121	28.7	0.111	0.022	0.050	0.104
precipitation jun	2,232	64,121	28.7	0.109	0.017	0.050	0.101
precipitation jul	2,232	64,121	28.7	0.106	0.023	0.049	0.098
precipitation aug	2,232	64,121	28.7	0.099	0.021	0.047	0.091
precipitation sep	2,232	64,121	28.7	0.094	0.021	0.056	0.082
precipitation oct	2,232	64,121	28.7	0.082	0.019	0.049	0.072
precipitation nov	2,232	64,121	28.7	0.082	0.031	0.044	0.073
precipitation dec	2,232	64,121	28.7	0.078	0.038	0.043	0.067
moderate heat jan	2,232	64,121	28.7	0.018	0.027	0.017	0.004
moderate heat feb	2,232	64,121	28.7	0.027	0.035	0.017	0.011
moderate heat mar	2,232	64,121	28.7	0.070	0.062	0.027	0.051
moderate heat apr	2,232	64,121	28.7	0.138	0.076	0.030	0.125
moderate heat may	2,232	64,121	28.7	0.253	0.082	0.040	0.245
moderate heat jun	2,232	64,121	28.7	0.361	0.071	0.029	0.365
moderate heat jul	2,232	64,121	28.7	0.430	0.061	0.027	0.439
moderate heat aug	2,232	64,121	28.7	0.408	0.068	0.031	0.415
moderate heat sep	2,232	64,121	28.7	0.295	0.082	0.033	0.291
moderate heat oct	2,232	64,121	28.7	0.157	0.079	0.031	0.144
moderate heat nov	2,232	64,121	28.7	0.064	0.055	0.024	0.047
moderate heat dec	2,232	64,121	28.7	0.025	0.033	0.017	0.008
extreme heat jan	2,232	64,121	28.7	0.000	0.000	0.000	0.000
extreme heat feb	2,232	64,121	28.7	0.000	0.001	0.001	0.000
extreme heat mar	2,232	64,121	28.7	0.000	0.003	0.002	0.000
extreme heat apr	2,232	64,121	28.7	0.004	0.010	0.009	0.000
extreme heat may	2,232	64,121	28.7	0.022	0.034	0.026	0.006
extreme heat jun	2,232	64,121	28.7	0.106	0.100	0.071	0.066
extreme heat jul	2,232	64,121	28.7	0.216	0.172	0.119	0.167
extreme heat aug	2,232	64,121	28.7	0.174	0.165	0.108	0.108
extreme heat sep	2,232	64,121	28.7	0.061	0.074	0.054	0.024
extreme heat oct	2,232	64,121	28.7	0.006	0.016	0.011	0.000
extreme heat nov	2,232	64,121	28.7	0.000	0.001	0.001	0.000
extreme heat dec	2,232	64,121	28.7	0.000	0.000	0.000	0.000

Note: An observation is a county-year combination. The panel column indicates the number of counties. The years column gives the average number of observations per county. We also report the between-counties (s.d. B) and within-county (s.d. W) standard deviation. The acres are measured in millions and the nitrogen concentration is measured in mg/L. The precipitation is measured in meters. The moderate heat is measured in 1,000 degree days between 10°C and 29°C. The extreme heat is measured in 100 degree days above 29°C. For additional details, see section 9.5.2.

plays an important role in our assessment of the effects of agriculture on nutrient pollution based on our earlier discussion regarding the tight connection between nitrogen pollution and rainfall.

We have about 64,000 observations and 2,200 counties. On average, we track a county for 29 years during the 48-year period 1970–2017. The mean nitrogen concentration is about 2.5 mg/L and both the between-counties and within-county standard deviation are around 1.65 mg/L. Hence, pollution exhibits similar variation across counties and within a county over time. On average, 38,000 acres of corn are planted per year in a county. Contrary to nitrogen pollution, the variation in acres is much larger across counties (48,000 acres) than within a county over time (11,000 acres). As a benchmark for the acres planted, the mean (median) county land area is 603 (556) square miles or 386,187 (355,969) acres. The total annual precipitation is, on average, close to 1.1 meters and varies more across counties than within a county over time. On average, February and May are the months with the smallest (0.067 meters) and largest (0.111 meters) total precipitation, respectively. July is the month with the largest number of moderate-heat (430) and extreme-heat (21.6) DDs. While monthly precipitation varies more within a county over time than across counties with the exception of January, extreme and moderate heat DDs vary more across counties than within a county over time for most months.

9.5.3   Nitrogen Concentration Across Space and Over Time

The choropleth maps in figure 9.1 offer visualizations of the spatial variation for the variables used in our analyses and provide some descriptive evidence on the spatial correlation between nitrogen concentration and corn acreage. In general, we see higher concentration in watersheds in southern Minnesota, Iowa, Illinois, Indiana, and Ohio that drain large areas of agricultural land. We explore this spatial correlation in more depth using cross-section regressions.

In panel A of figure 9.2, based on monitoring-site level data on average daily nitrogen concentrations (mg/L), we show trends in nitrogen concentration. We also show flow-normalized annual nitrogen concentration in the GoM using data from the USGS National Water Quality Network in panel B. Panels C and D provide information related to fertilizer use and acreage, which are important in understanding the relationship between agriculture and nitrogen concentration.

The use of nitrogen fertilizer increased from about 2.5 million metric tons (mmts) in 1964 to 11.8 mmts in 2015; it reached its peak of about 12 mmts in 2013. Most of the almost fivefold increase took place before the early 1980s (panel C). By 1981, nitrogen use had steadily increased to 10.8 mmts.[15]

15. See table 9 (percent of corn acreage receiving nitrogen fertilizer) in this link: https://www .ers.usda.gov/webdocs/DataFiles/50341/fertilizeruse.xls?v=5014.

Fig. 9.1A–F Nitrogen concentration, corn acreage, and weather-related variables

Note: Panels A–F are read from left to right. In all panels, we show averages for 1970–2017. The shading of the choropleth maps is based on the deciles of the empirical distribution. In panels D, E, and F, we show the months with the highest average values. The acres are in millions and the nitrogen concentration is in mg/L. The precipitation is in meters. The moderate heat is in 1,000 degree days between 10°C and 29°C. The extreme heat is in 100 degree days above 29°C. For additional details, see section 9.5.2.

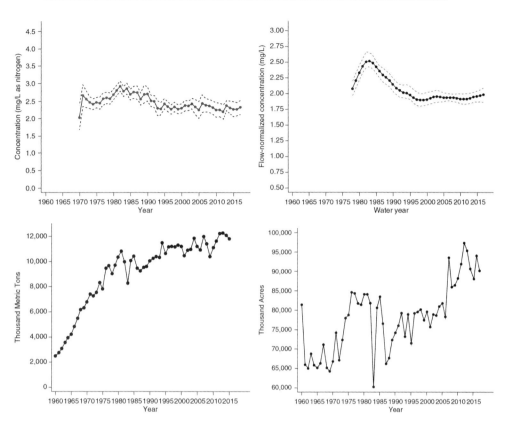

Fig. 9.2A–D   Nitrogen pollution and related factors

Note: Panels A–D are read from top left to bottom right. In panel A, we regress the average daily nitrogen concentration at the USGS monitoring-site level for the CONUS on site fixed effects (FEs), year FEs, day, day squared, day cubed, month, month squared, month cubed, and report the estimated year FEs. The 95% confidence intervals shown are constructed using standard errors clustered by HUC8. Additional details regarding the flow-normalized total nitrogen concentration in the Gulf in panel B are available in the following USGS link: https:// nrtwq.usgs.gov/nwqn/\#/GULF. In panel C, we show US consumption of nitrogen fertilizer from table 9.1 in the USDA ERS report on fertilizer use and price. In panel D, we show corn acres planted from the USDA Historical Track Records. For additional details, see section 9.5.3.

The expansion of nitrogen use during this time was due to expanded acreage (panel D), increase in application rates, and a higher share of acres receiving fertilizer (from 85 percent to 97 percent); the percent of US corn acreage receiving nitrogen fertilizer has been 95 percent, on average, in the last 50 years or so. Since then fertilizer use has fluctuated over time following changes in cropping system implementation and fertilizer crop prices, but has shown no persistent trend (Hellerstein, Vilorio, and Ribaudo 2019). The application rates in the major corn producing states follow similar trends with a notable increase between the mid-1960s and early 1980s. The fertilizer costs have oscillated between 14 percent and 27 percent of the corn gross value of production averaging close to 20 percent.

Overall, there is an increase in nitrogen concentration between the early 1970s and early 1980s from about 2 mg/L to a peak of about 3 mg/L. This pattern is consistent with the increase in corn acreage and nitrogen fertilizer use. Following a downward trend between the mid-1980s and the mid-1990s, nitrogen concentration has plateaued at about 2.3 mg/L in the last 20 years or so. These are roughly the concentration levels in the early 1970s. The flow-normalized annual nitrogen concentration in the GoM exhibits a very similar behavior over time.[16]

9.6   Econometric Estimates

Preamble. Table 9.3 shows detailed results of the panel FE regressions for our (c)ounty-centric analysis. In panel A, we report results from regressing nitrogen pollution on corn acres planted without controlling for weather. In panel B, we control for weather. In particular, we use 12 control variables (one for each month) for precipitation, squared precipitation, moderate-heat DDs, and extreme-heat DDs, for a total of 48 variables. In panel C, we add the interaction of acres with total annual (January–December) precipitation to the set of explanatory variables. The standard errors are clustered at the HUC4 level (124 clusters) accommodating arbitrary correlation of the unobservables across time and space.[17] To explore the implications of climate change for our estimated effects, we also interact corn acreage with moderate- and extreme-heat DDs in a subsequent section.

Baseline estimates. For the models without weather-related controls, the adjusted R-squared ($\bar{R}^2$) is 0.26–0.53 depending on the specification with most of the fit improvement attributed to the county FEs. Apart from the

16. Sprague, Hirsch, and Aulenbach (2011) estimate changes in nitrate concentration and flux during 1980–2008 at eight sites in the MRB using the WRTDS model, which produces flow-normalized (FN) estimates of nitrate concentration and flux. Their results show that little consistent progress had been made in reducing riverine nitrate since 1980, and that FN concentration and flux had increased in some areas. Murphy, Hirsch, and Sprague (2013), who extended the analysis in Sprague, Hirsch, and Aulenbach (2011), show that trends in FN nitrate concentration and flux were increasing or near-level at all sites for 1980–2018. They note, however, that trends at some sites began to exhibit decreases or greater increases during 2000–2008.

17. In Section A.5 of the online appendix (http://www.nber.org/data-appendix/c14692 /appendix.pdf), we discuss results from cross-section regressions. In section A.6, we discuss results from *(h)ydrologic unit-centric* and *(m)onitoring site-centric* analyses. For the h-centric analysis, i denotes an eight-digit hydrologic unit (HUC8), y_{it} is the average nitrogen concentration using sites located in the same HUC8, and a_{it} are acres planted planted in counties that lie in the same HUC8 weighted by their area. For the m-centric analysis, y_{it} is the concentration for monitor i and a_{it} are the acres planted in counties within a 50-mile radius from the site. Regarding the weather-related variables, in the case of the m-centric analysis, p_{it} and $\mathbf{z}_{it}$ are averages across counties within the assumed radius of site i. For the h-centric analysis, we use averages of the same variables weighted by the area of the counties that lie within the HUC8 polygons.

specifications with county-specific linear trends in columns A7 and A8, the acres coefficient is statistically significant at 5 percent level with values between 3.862 (column A5) and 23.581 (column A1). According to these estimates, the implied elasticities are 0.060–0.364 and they are significant at 5 percent level. For the specifications with county-specific linear trends, the elasticities are not significant at conventional levels.[18]

In the presence of weather-related controls, there is a notable change in the acres coefficient from 23.581 (column A1) to 18.458 (column B1) for the specification without county FEs. The model fit improvements, however, are relatively minor. As it was the case for the models without weather-related controls, the acres coefficients fail to be statistically significant at conventional levels for the specifications with county-specific trends (columns B7 and B8). Apart from the specification without county FEs (column B1), the elasticity of nitrogen concentration with respect to corn acreage is between 0.061 (column B5) and 0.093 (column B6).

The interaction of acres with precipitation implies effects that are significant at 5 percent level even in the presence of county-specific trends. Indeed, all but 2 of the 24 elasticities are significant at 5 percent level. Once again, apart from the specification without county FEs that implies elasticities of 0.278 (first precipitation quartile) to 0.395 (third quartile), we see elasticities of up to 0.086, 0.130, and 0.178, depending on the precipitation quartile, all of which are significant at 1 percent level. For the richest specification (column C8) that includes county FEs, county-specific trends, and year FEs, the elasticities are significant at 1 percent level and equal to 0.076 and 0.118 for the second and third precipitation quartiles, respectively; their counterpart for the first quartile is not significant at conventional levels.

Figure 9.3 shows point estimates along with 95 percent CIs for the 48 weather-related controls. Among the 48 coefficients, only the ones associated with January precipitation and its square are statistically significant. Based on multiple-hypotheses testing performed separately for each of the three sets of weather-related controls, the 24 precipitation controls, as well as the 12 extreme-heat controls, are jointly significant at 5 percent. The 12 moderate-heat controls are not jointly significant at conventional levels.[19]

Statistical significance. In all, we see positive and statistically significant effects of corn acreage on nitrogen pollution. The specifications that control for weather and contain an interaction of corn acreage with precipitation

18. Throughout the paper, we refer to statistical significance at ≤ 10 percent as significance at conventional levels.

19. We discuss additional estimates for the panel FE regressions summarized in tables 9.4–6 and figure 4 in section A.6 of the online appendix (http://www.nber.org/data-appendix/c14692/appendix.pdf). A detailed discussion of the motivation behind our additional estimates and any related data sources for the panel FE regressions is available in section A.6.1 and section A.6.2. A similar discussion for the cross-section regressions is available in section A.6.3.

Table 9.3   Panel fixed effect regressions and corn acreage elasticities

	(A1)	(A2)	(A3)	(A4)	(A5)	(A6)	(A7)	(A8)
				A. Acres only				
acres	23.581***	5.146***	4.202**	5.845***	3.862**	6.117***	−0.523	2.741
	(2.032)	(1.714)	(1.659)	(1.902)	(1.596)	(1.941)	(1.987)	(1.955)
$\bar{R}^2$	0.26	0.46	0.47	0.47	0.48	0.48	0.52	0.53
Obs.	64,121	64,121	64,121	64,121	64,121	64,121	64,121	64,121
Clusters	124	124	124	124	124	124	124	124
elast est.	0.364***	0.079***	0.065**	0.090***	0.060**	0.094***	−0.008	0.042
elast s.e.	(0.031)	(0.026)	(0.026)	(0.029)	(0.025)	(0.030)	(0.031)	(0.030)
elast pval	0.000	0.003	0.013	0.003	0.017	0.002	0.793	0.163

	(B1)	(B2)	(B3)	(B4)	(B5)	(B6)	(B7)	(B8)
				B. Acres plus weather				
acres	18.458***	5.212***	4.317***	5.490***	3.941**	6.058***	−0.678	2.477
	(1.872)	(1.669)	(1.640)	(1.864)	(1.525)	(1.970)	(1.941)	(1.942)
$\bar{R}^2$	0.30	0.47	0.47	0.48	0.48	0.49	0.52	0.53
Obs.	64,121	64,121	64,121	64,121	64,121	64,121	64,121	64,121
Clusters	124	124	124	124	124	124	124	124
elast est.	0.285***	0.080***	0.067***	0.085***	0.061**	0.093***	−0.010	0.038
elast s.e.	(0.029)	(0.026)	(0.025)	(0.029)	(0.024)	(0.030)	(0.030)	(0.030)
elast pval	0.000	0.002	0.010	0.004	0.011	0.003	0.727	0.205

	(C1)	(C2)	(C3)	(C4)	(C5)	(C6)	(C7)	(C8)
			C. Acres plus weather and interaction with precipitation					
acres	0.765	−9.793**	−10.379***	−8.444*	−10.699***	−7.947**	−13.730***	−9.566**
	(3.315)	(3.968)	(3.916)	(3.602)	(3.804)	(3.542)	(4.887)	(4.495)
acres × prec	19.486***	16.342***	16.043***	15.355***	15.870***	15.277***	14.409***	13.517***
	(3.381)	(3.899)	(3.877)	(3.667)	(3.845)	(3.715)	(3.927)	(3.805)

$\bar{R}^2$	0.31	0.47	0.47	0.48	0.48	0.49	0.52	0.53
Obs.	64,121	64,121	64,121	64,121	64,121	64,121	64,121	64,121
Clusters	124	124	124	124	124	124	124	124
elast 25 est.	0.278***	0.072**	0.059**	0.079***	0.052**	0.086***	−0.015	0.037
elast 25 s.e.	(0.029)	(0.029)	(0.027)	(0.028)	(0.025)	(0.030)	(0.035)	(0.034)
elast 25 pval	0.000	0.013	0.031	0.006	0.042	0.005	0.666	0.275
elast 50 est.	0.333***	0.119***	0.105***	0.123***	0.097***	0.130***	0.026**	0.076***
elast 50 s.e.	(0.032)	(0.030)	(0.029)	(0.031)	(0.027)	(0.033)	(0.032)	(0.032)
elast 50 pval	0.000	0.000	0.000	0.000	0.000	0.000	0.032	0.001
elast 75 est.	0.395***	0.170*	0.155***	0.171***	0.147***	0.178***	0.071**	0.118***
elast 75 s.e.	(0.037)	(0.036)	(0.035)	(0.037)	(0.034)	(0.040)	(0.033)	(0.035)
elast 75 pval	0.000	0.000	0.000	0.000	0.000	0.000	0.032	0.001
precip 25	0.885	0.885	0.885	0.885	0.885	0.885	0.885	0.885
precip 50	1.070	1.070	1.070	1.070	1.070	1.070	1.070	1.070
precip 75	1.274	1.274	1.274	1.274	1.274	1.274	1.274	1.274
mean acres	0.038	0.038	0.038	0.038	0.038	0.038	0.038	0.038
mean N	2.451	2.451	2.451	2.451	2.451	2.451	2.451	2.451
county FE	✓							
trend		✓	✓					
year FE				✓		✓		
state × trend					✓	✓		
county × trend							✓	✓

Note: We control for weather (precipitation, squared precipitation, moderate-heat degree days, and extreme-heat degree days) in all eight models of panels B and C. The standard errors reported in parentheses are clustered by HUC4. The elasticities in the bottom of each panel are calculated using the means of corn acreage (mean acres) and nitrogen concentration (mean N), and three precipitation quartiles (precip25, precip50, precip75) in panel C. The acreage is in millions, the concentration is in mg/L, and the precipitation is total annual (January–December) in meters. The asterisks denote statistical significance as follows: 1% (***), 5% (**), 10% (*). For additional details, see section 9.6.

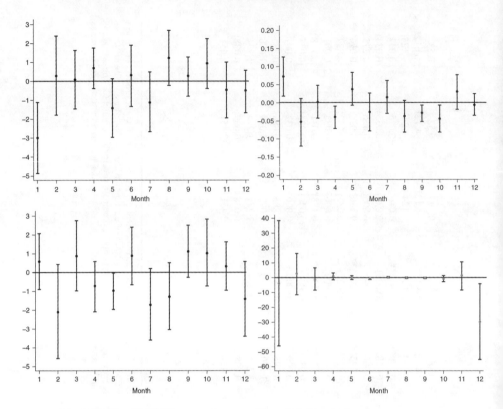

Fig. 9.3A–D   Panel FE regressions, weather related controls

Note: Panels A–D are read from top left to bottom right. The figure shows point estimates and 95% confidence intervals (CIs) for the 48 weather related controls in specification C8 of the panel FE regressions in table 9.3. The CIs are constructed using standard errors clustered by HUC4. The F statistics and the *p*-values in squared brackets for the joint significance of the coefficients shown in the four panels are as follows: 2.06 [0.024] for panel A, 2.94 [0.001] for panel B, 1.39 [0.178] for panel C, and 2.02 [0.027] for panel D. For additional details, see section 9.6.

generally imply larger effects than their counterparts that do not contain such interactions. Spatial FEs matter more than time-related controls for the magnitude of the effects. According to our preferred specification (column C8), the elasticity of nitrogen concentration with respect to corn acreage is 0.076 for the second precipitation quartile and increases to 0.118 for the third quartile. In both instances, the elasticity is significant at 1 percent level.

Economic significance. The statistically significant effects reported above are also economically meaningful according to a back-of-the-envelope calculation that utilizes the (median) potential damage costs of nitrogen due to declines in fisheries and estuarine/marine life of $15.84 per kg ($2008) from table 1 in Sobota et al. (2015). At the third precipitation quartile, a 1 within-county standard deviation increase in corn acres planted implies a 3.3 percent increase in the level of nitrogen concentration. At the average nitrogen concentration of about 2.5 mg/L and the average streamflow of the Mississippi River in the GoM in our sample ($\approx$ 21,500 cubic meters per second), this effect entails close to 50,800 additional metric tons of nitrogen in the GoM. Hence, our estimated increase in nitrogen concentration of 3 percent implies an external cost of $805.5 million per year in $2008, or approximately $805.5 × 1.14 = $918.3 in $2017 (the last year in our sample) using the GDP deflator (FRED GDPDEF series).[20]

Reconciling our baseline estimates. Table 4 in Hendricks, Smith, and Sumner (2014) gives the average nitrogen loss from the edge-of-field (EoF) as predicted by the SWAT model—coupled with an econometric model—for different land uses in Iowa, Illinois, and Indiana for 2000–2010. Nitrogen losses are the sum of nitrate and organic nitrogen loss. Corn after corn generates the largest nitrogen losses (34.7 lb per acre per year [lb/a/y], on average) because more fertilizer is applied to corn after corn since there is no nitrogen carry-over from a previous soybean crop. The mean loss of 34.7 lb/a/y reported by the authors is similar to the average estimate of total nitrogen loss of 34 lb/a/y in the USDA CEAP national nitrogen loss report (NRCS 2017b) we discussed earlier.

Assuming the mean annual EoF loss of about 35 lb/a/y from Hendricks, Smith, and Sumner and 79,384,857 corn acres per year (average of national corn acres planted during the same period according to USDA data), we have 1,260,294 metric tons of total nitrogen per year. This calculation assumes that the EoF losses translate to an equivalent nitrogen loading in the GoM—admittedly a strong assumption, because some nitrogen that leaves the field does not reach the GoM. Note that the average annual total nitrogen flux of the Mississippi and the Atchafalaya Rivers to the GoM between is 1,460,419 metric tons for 1968–2016.[21]

In our case, the average nitrogen concentration is 2.5 mg/L. According to the USGS, the mean annual flow of the Mississippi plus Atchafalaya to the GoM (Ibid) is about 21,376 cubic meters per second for 1968–2016. This mean annual flow implies 1,685,245 metric tons of total nitrogen per year, which translates to 46.8 lb/a/y using the average annual corn acreage for 1968–2016. However, a comparison of 46.8 lb/a/y with 35 lb/a/y from Hendricks, Smith, and Sumner hinges on the assumption that all nitrogen pollution recorded at the USGS monitoring sites is due to fertilizer loss from corn fields, but it is not. A better, albeit imperfect comparison, is to assume that 70 percent of the 1,685,245 metric tons are attributed to agriculture (David, Drinkwater, and McIsaac 2010), in which case we have 32.8 lb/a/y (see Wu and Tanaka 2005 for a similar approach). This loss of 32.8 lb/a/y cal-

20. We use the average flow for years 1970–2016 from column F (Total Mississippi-Atchafalaya River) available in the following link: https://toxics.usgs.gov/hypoxia/mississippi /flux_ests/delivery/Gulf-Annual-2016.xlsx.

21. See the USGS link here: https://toxics.usgs.gov/hypoxia/mississippi/flux_ests/delivery /index.html.

culated using our estimates is similar to the average loss of 34.7 lb/a/y in Hendricks, Smith, and Sumner.

According to our baseline panel FE estimates in column C8 of table 9.3, a 28 percent increase in corn acres planted—assuming an increase equal 1 within-county standard deviation (11,000 acres) and using the mean acreage (38,000 acres) from table 9.2 to calculate the percent increase—implies a a 3.3 percent increase in nitrogen concentration when evaluated at the mean concentration of 2.5 mg/L. A 3.3 percent increase in mean concentration of 2.5 mg/L implies an increase in flux equal to 55,613 metric tons. Assuming that this 3.3 percent increase in concentration is associated with a 28 percent increase in 79,384,857 corn acres, the implied increase is 5.52 lb/a/y.

The effect of additional corn acres on measured nitrogen in waterways is an order of magnitude smaller than agronomic estimates of excess nitrogen applied to those acres assuming EoF losses translate to an equivalent nitrogen loading to streams and rivers. However, we do not interpret our results as evidence that the amount of surplus nitrogen used on crops is much smaller than previously believed. Instead, our findings are consistent with new research in environmental science arguing that there is a large amount of nitrogen stored in subsurface soil and groundwater (e.g., Van Meter, Basu, and Cappellen 2017; Van Meter, Van Cappellen, and Basu 2018; Ilampooranan, Van Meter, and Basu 2019) and contributes to the so-called legacy nitrogen, which may increase loadings in rivers and streams with a long delay.[22] The presence of large quantities of legacy nitrogen has substantive policy implications because it increases the relative efficacy of downstream policies such as fluvial wetlands (i.e., those connected to waterways) and it is a topic we explore in more detail in Metaxoglou and Smith (2022).

Using the elasticity estimate of 0.076 from column C8 of Table 9.3, an additional corn acre generates an average of 3.5 lb/a/y of nitrogen in small (level 4) streams within a 50-mile radius from the country centroids for median precipitation and average streamflow of 362 cubic feet per second (cfs).[23] This estimate is close to 10 percent of the USDA CEAP estimate of 34 lb/a/y of EoF losses. If we instead use 5.52 lb/a/y, per our discussion in the previous paragraph, and a streamflow of 1,997 cfs, which is the average across all streams, an additional corn acre generates an average of 30 lb/a/y in streams and rivers, which is almost 80 percent of the NRCS estimate of surplus nitrogen.

Additional estimates. Panel A of figure 9.4 shows that a more flexible specification for the interaction of corn acreage with precipitation does not have a material effect on our estimated corn acreage elasticities. Similar flexible

22. Van Meter et al. (2016) study soil data from cropland in the Mississippi River basin and find nitrogen accumulation of 25–70 kg per hectare per year (22–62 lb per acre per year).
23. This is the average streamflow based on the Enhanced Unit Runoff Method (EROM) Flow Estimation in the USGS NHD Plus data for years 1971–2000 and is readily available by river segment (COMID).

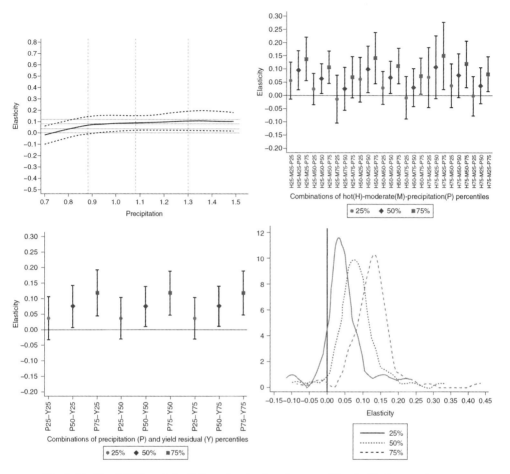

Fig. 9.4A–D Additional elasticity estimates based on panel fixed-effect regressions

Note: Panels A–D are read from top left to bottom right. In panels A–C, we report elasticity estimates along with 95% confidence intervals using standard errors clustered by HUC4. In panels B and C, the legend pertains to the quartiles of total annual precipitation. We use the same set of weather-related controls, county fixed effects (FEs), year FEs, and county-specific trends as in column C8 in table 9.3. In panel A, we use a flexible specification (cubic spline) to model the interaction of corn acreage and precipitation. We use the vertical dashed lines to indicate the precipitation quartiles and the horizontal gray lines to indicate the elasticities from specification C8 in table 9.3. In panel B, we interact corn acreage with total annual precipitation, annual moderate-, and extreme-heat degree days. In panel C, we interact corn acreage with total annual precipitation and corn yield residuals. We obtain the yield residuals by regressing yields on county-specific trends. In panel D, we summarize the elasticity estimates in tables 9.4– 9.6 by precipitation quartile using kernel density plots. For additional details, see section A.6 in the online appendix, http://www.nber.org/data-appendix/c14692/appendix .pdf.

Table 9.4   **Additional estimates of corn acreage elasticities based on panel fixed-effect regressions**

no.	model	sample						coefficients		Mean nitrogen	Mean acres	precipitation			elasticities		
		obs	starts	ends	counties	years	Clusters	β_1	β_2			25%	50%	75%	25%	50%	75%
M1	BEA SAEMP25	64,121	1970	2017	2,232	48	124	−9.561**	13.527**	2.451	0.038	0.885	1.070	1.274	0.037	0.076**	0.118***
M2	BEA SAGDP2	64,121	1970	2017	2,232	48	124	−9.598**	13.502**	2.451	0.038	0.885	1.070	1.274	0.036	0.075**	0.117***
M3	BEA REA CAINC	63,477	1970	2017	2,209	48	124	−7.123	13.477**	2.462	0.038	0.884	1.070	1.274	0.074*	0.113***	0.156***
M4	M1 plus EIA–SEDS	64,121	1970	2017	2,232	48	124	−9.518***	13.619***	2.451	0.038	0.885	1.070	1.274	0.039	0.078***	0.121***
M5	M2 plus EIA–SEDS	64,121	1970	2017	2,232	48	124	−9.547***	13.589***	2.451	0.038	0.885	1.070	1.274	0.038	0.077**	0.120***
M6	M3 plus EIA–SEDS	63,477	1970	2017	2,209	48	124	−7.325	13.564**	2.462	0.038	0.884	1.070	1.274	0.072*	0.111***	0.154***
M7	M4 plus EPA	23,735	1996	2017	1,749	22	117	−12.560*	16.716**	2.710	0.044	0.886	1.063	1.254	0.037	0.085	0.137
M8	M5 plus EPA	23,735	1996	2017	1,749	22	117	−12.428*	16.634**	2.710	0.044	0.886	1.063	1.254	0.038	0.086	0.137
M9	M6 plus EPA	23,488	1996	2017	1,731	22	117	−7.555	16.884***	2.725	0.045	0.884	1.062	1.254	0.121	0.170	0.223**
M10	BLS LAUS	33,496	1990	2017	1,983	28	121	−8.207	10.625**	2.637	0.042	0.894	1.076	1.272	0.020	0.051	0.084
M11	M10 plus EIA–SEDS	33,496	1990	2017	1,983	28	121	−8.112	10.675**	2.637	0.042	0.894	1.076	1.272	0.023	0.053	0.086
M12	M11 plus EPA	23,735	1996	2017	1,749	22	117	−12.664*	16.407**	2.710	0.044	0.886	1.063	1.254	0.030	0.078	0.129
M13	M6 plus WWTPs	11,475	1978	2012	559	35	46	5.738	6.035	2.845	0.042	0.718	0.912	1.124	0.147*	0.164***	0.183***
M14	M6 plus TREND livestock	63,406	1970	2017	2,207	48	124	−7.306*	13.568**	2.459	0.038	0.884	1.070	1.275	0.073*	0.112***	0.155***
M15	M6 plus TREND human	63,406	1970	2017	2,207	48	124	−7.335*	13.581**	2.459	0.038	0.884	1.070	1.275	0.072*	0.111***	0.154***
M16	M6 plus TREND ATMDEP	63,406	1970	2017	2,207	48	124	−7.321*	13.481**	2.459	0.038	0.884	1.070	1.275	0.071*	0.110***	0.153***
M17	M6 plus TREND fertilizer	63,406	1970	2017	2,207	48	124	−7.444*	13.431**	2.459	0.038	0.884	1.070	1.275	0.069*	0.107***	0.150***
M18	M6 plus TREND fixing	63,406	1970	2017	2,207	48	124	−7.561*	13.512***	2.459	0.038	0.884	1.070	1.275	0.068*	0.107***	0.149***
M19	M6 plus TREND uptake	63,406	1970	2017	2,207	48	124	−7.320*	13.649**	2.459	0.038	0.884	1.070	1.275	0.073*	0.113***	0.156***

Note: The coefficients β_1 and β_2 are the ones for corn acres and their interaction with precipitation in equation (1). All regressions are estimated using the 48 weather-related controls, the fixed effects, and the time-related controls, in specification C8 in table 9.3. The corn acres (acres) are in millions and the nitrogen concentration (nitrogen) is in mg/L. The precipitation (measured in meters) and runoff (measured in millimeters) are both total annual. The elasticities in the three rightmost columns are calculated using the precipitation (or runoff), when appropriate quartiles and the mean acreage and nitrogen concentration reported in the table. The standards errors are clustered by HUC4 except for the robustness checks in which we explore alternative clustering schemes. The asterisks denote statistical significance as follows: 1% (***), 5% (**), 10% (*). For additional details, see section A.6 at http://www.nber.crg/data-appendix/c14692/appendix.pdf.

Table 9.5   Additional estimates of corn acreage elasticities based on panel fixed-effect regressions (continued)

model	obs	starts	ends	counties	years	Clusters	β_1	β_2	mean nitrogen	mean acres	precip/runoff 25%	50%	75%	elast 25%	50%	75%
A. Baseline																
Baseline	64,121	1970	2017	2,232	48	124	-9.566**	13.517**	2.451	0.038	0.885	1.070	1.274	0.037	0.076**	0.118***
B. Control for conservation programs, acres of other major crops, and fertilizer sales																
Other acres	64,121	1970	2017	2,232	48	124	-10.440**	14.320**	2.451	0.038	0.885	1.070	1.274	0.034	0.075*	0.120***
CRP acres	64,121	1970	2017	2,232	48	124	-10.135**	13.655**	2.451	0.038	0.885	1.070	1.274	0.030	0.069**	0.112***
Fertilizer sales	58,047	1970	2012	2,230	42	124	-7.842	11.932**	2.408	0.037	0.877	1.071	1.279	0.041	0.076**	0.115***
Other and CRP acres	64,121	1970	2017	2,232	48	124	-11.018**	14.517**	2.451	0.038	0.885	1.070	1.274	0.028	0.070	0.115**
Other and Fertilizer	58,047	1970	2012	2,230	42	124	-7.542	12.540**	2.408	0.037	0.877	1.071	1.279	0.054	0.091**	0.132***
Other, CRP, Fertilizer	58,047	1970	2012	2,230	42	124	-8.049	12.743**	2.408	0.037	0.877	1.071	1.279	0.048	0.087**	0.128***
C. Temporal variation in corn acreage effects																
1970s	12,765	1970	1979	2,079	10	122	-9.481	3.085	2.020	0.034	0.873	1.109	1.341	-0.115	-0.103	-0.091
1980s	17,657	1980	1989	2,058	10	123	-5.666	14.718**	2.416	0.033	0.875	1.040	1.230	0.100*	0.133**	0.172**
1990s	14,878	1990	1999	1,920	10	121	2.534	12.690*	2.568	0.038	0.905	1.095	1.307	0.205	0.241	0.280
2000s	11,204	2000	2009	1,561	10	117	0.568	10.340*	2.630	0.043	0.875	1.064	1.252	0.159*	0.191***	0.223***
2010s	7,095	2010	2017	1,225	8	110	-9.319	15.012	2.785	0.047	0.906	1.056	1.232	0.072	0.109	0.154
D. Spatial variation in corn acreage effects																
MRB	40,906	1970	2017	1,534	48	87	-10.456**	15.178**	2.866	0.048	0.789	1.008	1.230	0.026	0.082**	0.138***
North	26,799	1970	2017	826	48	58	-14.556**	18.324**	3.302	0.065	0.801	0.940	1.087	0.003	0.053	0.106
Middle	15,974	1970	2017	546	48	38	17.912**	-5.230	2.345	0.031	0.881	1.055	1.222	0.174***	0.162***	0.151***
South	21,348	1970	2017	860	48	51	-8.309	4.253	1.462	0.009	1.111	1.279	1.475	-0.022	-0.018	-0.012
E. Alternative time windows for measuring nitrogen concentration, precipitation, and degree days																
January-June	60,177	1970	2017	2,227	48	124	-3.271	11.310**	2.604	0.037	0.417	0.526	0.655	0.021	0.038	0.059**
March-August	61,308	1970	2017	2,227	48	124	-3.487	10.127**	2.528	0.038	0.503	0.600	0.705	0.024	0.038	0.054*
April-September	61,654	1970	2017	2,228	48	124	-5.550*	13.773**	2.461	0.038	0.509	0.599	0.701	0.022	0.042	0.063**
May-October	61,890	1970	2017	2,230	48	124	-9.066**	20.689**	2.435	0.038	0.501	0.588	0.688	0.020	0.048	0.080**
F. Interacting corn acres with runoff instead of precipitation																
Runoff	62,420	1970	2015	2,232	46	124	-0.521	0.002*	2.435	0.038	1752.310	2824.000	4112.086	0.048	0.083**	0.124***
G. Nitrogen concentration accounting for streamflow																
Downstream main	33,335	1970	2015	1,913	46	123	-15.821**	17.665***	2.305	0.038	0.878	1.059	1.261	-0.005	0.047	0.105***
Downstream all	40,145	1970	2017	2,037	48	124	-15.974**	17.006**	2.322	0.038	0.878	1.058	1.262	-0.017	0.033	0.090***

Note: The coefficients β_1 and β_2 are the ones for corn acres and their interaction with precipitation in equation (1). All regressions are estimated using the 48 weather-related controls, the fixed effects, and the time-related controls, in specification C8 in table 9.3. The corn acres (acres) are in millions and the nitrogen concentration (nitrogen) is in mg/L. The precipitation measured in meters is total annual unless it is indicated otherwise (panel E). The runoff (measured in millimeters) is total annual. The models with fertilizer sales in panel B exclude 1986 due to data availability. The elasticities in the three rightmost columns are calculated using the quartiles of precipitation or runoff, and the mean acreage and nitrogen reported in the table. The standard errors are clustered by HUC4 except for the robustness checks in which we explore alternative clustering schemes. The asterisks denote statistical significance as follows: 1% (***), 5% (**), 10% (*). For additional details, see section A.6 at http://www.nber.org/data-appendix/c14692/appendix.pdf.

Table 9.6   Additional estimates of corn acreage elasticities based on panel fixed-effect regressions (continued)

model	obs	starts	ends	years	counties	Clusters	β_1	β_2	mean nitrogen	mean acres	precip 25%	precip 50%	precip 75%	elast. 25%	elast. 50%	elast. 75%
A. Baseline																
Baseline	64,121	1970	2017	48	2,232	124	−9.566**	13.517**	2.451	0.038	0.885	1.070	1.274	0.037	0.076**	0.118***
B. Nitrogen concentration accounting for streamflow and stream levels																
Downstream all L1	13,922	1971	2017	47	753	86	−16.085**	10.893**	1.803	0.026	0.916	1.091	1.290	−0.088	−0.060	−0.029
Downstream all L2	16,581	1970	2017	48	1,155	107	−10.587**	10.669**	1.873	0.038	0.920	1.084	1.277	−0.016	0.020	0.061
Downstream all L3	15,789	1971	2017	47	1,289	117	−26.341**	25.282**	1.813	0.043	0.882	1.061	1.257	−0.096	0.011	0.129**
Downstream all L4	8,667	1970	2017	48	977	112	−8.760	20.881**	1.755	0.045	0.838	1.033	1.231	0.226***	0.331***	0.438***
C. Include lagged acres																
1 lag	58,258	1970	2017	48	2,232	124	−8.307	13.331**	2.451	0.040	0.885	1.070	1.274	0.057*	0.097**	0.142**
2 lags	58,258	1970	2017	48	2,232	124	−6.713	13.516**	2.451	0.040	0.885	1.070	1.274	0.086*	0.126**	0.171**
3 lags	58,258	1970	2017	48	2,232	124	−6.451	13.774**	2.451	0.040	0.885	1.070	1.274	0.093*	0.135*	0.181*
D. Reporting limits in nitrogen concentration																
Reporting limit	64,670	1970	2017	48	2,232	128	−11.043**	15.602**	2.346	0.035	0.884	1.070	1.274	0.041	0.085***	0.133***
Zero	64,676	1970	2017	48	2,232	128	−11.276**	15.128**	2.157	0.035	0.884	1.070	1.274	0.034	0.081***	0.131***
E. Nitrogen concentration using alternative radii																
100 miles	80,612	1970	2017	48	2,242	124	−6.409**	9.943**	2.615	0.040	0.869	1.061	1.271	0.034	0.063***	0.094***
200 miles	85,359	1970	2017	48	2,243	124	−3.426**	6.379**	2.642	0.040	0.863	1.059	1.272	0.031*	0.050***	0.070***
F. Data filtering																
KS filter	64,096	1970	2017	48	2,232	124	−9.533**	13.651**	2.442	0.038	0.885	1.070	1.274	0.039	0.079**	0.122***
G. Alternative datasets and geographic scope (CONUS)																
USGS-NWIS	72,751	1970	2017	48	2,688	188	−10.446**	13.608**	2.418	0.035	0.814	1.038	1.256	0.009	0.053*	0.096***
USGS-NWQN	107,994	1970	2017	48	2,998	205	−10.862**	13.897**	2.200	0.026	0.798	1.049	1.278	0.003	0.043***	0.080***
USGS+EPA	113,372	1970	2017	48	3,027	205	−7.528**	14.386**	2.044	0.025	0.767	1.037	1.270	0.043**	0.091***	0.133***
H. Alternative data aggregation																
Hydrologic unit	17,943	1970	2017	48	986	128	−18.535**	21.895**	2.100	0.030	0.888	1.077	1.280	0.013	0.071*	0.134***
Monitoring site	56,466	1970	2017	48	9,596	129	−1.359**	1.566**	2.356	0.292	0.939	1.108	1.286	0.014	0.046**	0.081***
I. Statistical inference with alternative clustering schemes																
HUC2 × year	64,121	1970	2017	48	2,232	528	−9.566**	13.517**	2.451	0.038	0.885	1.070	1.274	0.037	0.076**	0.118***
HUC4 × year	64,121	1970	2017	48	2,232	5,132	−9.566**	13.517**	2.451	0.038	0.885	1.070	1.274	0.037	0.076**	0.118***
year	64,121	1970	2017	48	2,232	48	−9.566**	13.517**	2.451	0.038	0.885	1.070	1.274	0.037	0.076*	0.118***

Note: The coefficients β_1 and β_2 are the ones for corn acres and their interaction with precipitation in equation (1). All regressions are estimated using the 48 weather-related controls, the fixed effects, and the time-related controls, in specification C8 in table 9.3. The corn acres (acres) are in millions and the nitrogen concentration (nitrogen) is in mg/L. The precipitation (measured in meters) is total annual. The elasticities are calculated using the quartiles of precipitation or runoff and the mean acreage and nitrogen concentration reported in the table. The standards errors are clustered by HUC4 except for the robustness checks in which we explore alternative clustering schemes. The asterisks denote statistical significance as follows: 1% (***), 5% (**), 10% (*). For additional details, see section A.6 at http://www.nber.org/data-appendix/c14692/appendix.pdf.

specifications based on total precipitation for different time windows during the year (March–August and April-September) produced very similar elasticities to the ones shown here. In panels B and C of figure 9.4, we explore the role of crop nutrient uptake. Holding extreme-heat DDs and precipitation constant, additional moderate-heat DDs imply lower elasticities. Holding moderate-heat DDs and precipitation constant, an increase in extreme-heat DDs has no material impact on the magnitude of the acreage elasticities despite the well-documented negative effect of extreme heat on yields. Holding moderate- and extreme-heat DDs constant, an increase in precipitation implies larger elasticities. In all, the elasticity estimates when we interact corn acreage with moderate- and extreme-heat DDs in addition to precipitation are very similar to their baseline counterparts obtained by interacting the corn acreage with precipitation alone. The pattern in the magnitude of the elasticities just described also holds for panel FE regressions estimated using counties in the MRB, and counties in the most northern (coldest) states east of the 100th meridian from Schlenker and Roberts (2009). The elasticity estimates for the most southern (warmest) states from Schlenker and Roberts are generally noisy and indistinguishable from zero at conventional levels. Their counterparts for the middle states exhibit very little variation across the quartiles of precipitation and heat we considered. Yield shocks, calculated as deviations from county-specific yield trends, do not matter for the magnitude of the acreage elasticities either.

The implied corn acreage elasticities for a number of models we estimated performing a series of robustness checks, discussed in detail in section A.6 in the online appendix,[24] are summarized by precipitation quartile using the kernel density plots in panel D of figure 9.4. Similar to the baseline results, the coefficient of the interaction of corn acreage and precipitation (coefficient β_2 in equation [1]) is positive and highly significant in the vast majority of the models we explored. Hence, the amount of precipitation matters for the magnitude of the estimated acreage elasticities. With very few exceptions, the corn acreage elasticities based on the second and third precipitation quartiles are highly significant. Their counterparts based on the first precipitation quartile are not. For the second precipitation quartile, the elasticities that are significant at conventional levels are 0.043–0.331. Their counterparts for the third precipitation quartile are 0.059–0.438. As a reminder, for our preferred baseline specification in column C8 of table 9.3, the acreage elasticities are 0.076 and 0.118 for the second and third precipitation quartiles.

9.7   Climate Change and Nitrogen Pollution

According to our econometric analysis, corn acreage drives nitrogen concentration and the magnitude of the acreage effect depends on precipitation

24. See http://www.nber.org/data-appendix/c14692/appendix.pdf.

with more precipitation implying larger effects for our baseline estimates that pertain to the part of the country east of the 100th meridian. An additional specification in which we also interact corn acreage with moderate and extreme-heat DDs shows that, all else equal, an increase in moderate-heat DDs implies smaller effects, while an increase in extreme-heat DDs has no material impact on the magnitude of the effects.

We now explore the implications of climate change for our findings regarding the relationship between corn acreage and nitrogen concentration. In particular, we use the NASA Earth Exchange Global Daily Downscaled Projections (NEX-GDDP-CMIP6) data set to obtain projections for precipitation, moderate-, and extreme-heat DDs, and, in turn, projections of the marginal effects (MEs) of corn acreage on nitrogen concentration. The NEX-GDDP-CMIP6 data set is comprised of global downscaled climate scenarios derived from the General Circulation Model runs conducted under the Coupled Model Intercomparison Project Phase 6 (Eyring et al. 2016) and across two of the four "Tier 1" greenhouse gas emissions scenarios known as shared socioeconomic pathways (SSPs), namely, SSP2–4.5 and SSP5–8.5.[25]

We use out-of-sample projections from three climate models (CanESM5, UKESM1-0-LL, and GFDL-ESM4) and SSP2-4.5 and SSP5-8.5 for three weather-related variables available at a latitude/longitude resolution of 0.25°, namely, the mean of the daily precipitation rate (pr), the daily minimum near surface air temperature (tasmin), and the daily maximum near surface air temperature (tasmax). Projections of these variables from the climate models based on alternative SSPs allow us to obtain projections of total annual precipitation, moderate-heat, and extreme-heat DDs, which in their turn translate to projections of corn acreage ME on nitrogen concentration. These MEs do not take into account the impacts of climate change on other factors affecting nitrogen concentration and loads (e.g., streamflow, change in farmers' behavior as in Elbakidze et al. 2022, etc.).

Although projections for the three weather-related variables are available until 2100, we obtain ME projections for 2018–2050, as we are skeptical about the use of a model that has been estimated using data for 1970–2017 to project MEs more than 20 to 30 years out of sample. We opt for projections of MEs as opposed to elasticities because the former do not require an assumption about future values of nitrogen concentration and corn acreage while the later do. To the best of our knowledge, projections of both acreage and nitrogen concentration with the spatial and temporal coverage required to obtain projections of elasticities are not available. The MEs discussed are estimated assuming an increase in corn acreage equal to the historical (in-sample) within-county standard deviation and estimating different regressions for five sets of counties. The specification of these regression equations is identical to specification C8 of table 9.3. The sets of counties for which we obtained projections of MEs are as follows: counties east of the 100th meridian excluding Florida (baseline), counties in the MRB, as well as all counties in the northern, middle, and southern states east of the 100th meridian as in Schlenker and Roberts (2009).

The precipitation projections are generally smaller than their historical counterparts across all climate models, SSPs, and quartiles of the precipitation distribution. A notable exception is the median precipitation for the middle counties for which the projections exceed their historical counterpart for all climate models and SSPs. The projected quartiles for moderate-heat DDs are larger than their historical counterparts for all climate models and SSPs for all sets of counties and all three quartiles of precipitation considered. The projected quartiles for extreme-heat DDs, on the other hand, are generally smaller than their historical counterparts, especially for the lower quartiles of the extreme-heat distribution. It is also the case that the differences between projected and historical quartiles are generally larger for the moderate- and extreme-heat DDs than for precipitation.

For the discussion that follows, it important to keep in mind that for the panel FE regressions in which we interact acreage only with precipitation, the coefficient on the interaction is significant at conventional levels for the MRB and northern counties, in addition to the baseline counties. For the regressions in which we interact corn acreage with precipitation and DDs, in addition to the baseline counties, the coefficient on the interaction of corn acreage with precipitation is significant at conventional levels in the MRB and northern counties. The coefficients on the interaction of the corn acreage with moderate-heat DDs, as well as those on the interaction of the corn acreage with extreme-heat DDs, are indistinguishable from zero at conventional levels.

For the baseline counties—depending on the climate model and SSP—the projected median precipitation is 1.047–1.078 meters (panel A, table 9.7). Its third-quartile counterpart is 1.242–1.289 meters. The implied MEs based on the projected median precipitation are 0.048–0.053 mg/L, which are similar in magnitude to the ME of 0.051 mg/L based on the historical median precipitation. For the MRB counties, an area of particular interest for policies aiming to address the GoM HZ areal extent, the median precipitation projections are 0.945–0.980 meters implying MEs of 0.045–0.051 mg/L, the lower end of which is slightly smaller than their historical counterpart of 0.056 mg/L but similar to their baseline counterparts. For the northern counties, the median precipitation projections are 0.875–0.937 meters implying MEs of 0.017–0.031 mg/L, respectively. Their historical ME counterpart is 0.032 mg/L. For the middle counties, the median precipitation projections are 1.057–1.079 meters implying MEs of 0.133–0.134 mg/L, which are essentially identical to their historical counterpart, noting that the coefficient of the interaction of corn acreage with precipitation is statistically indistinguishable from zero. Finally, for the southern coun-

Table 9.7 Projections of precipitation, moderate, and extreme heat degree days

source	year	precipitation			moderate heat			extreme heat		
		25%	50%	75%	25%	50%	75%	25%	50%	75%
A. Baseline										
Historical	1970–2017	0.885	1.070	1.274	1.700	2.149	2.695	0.150	0.415	0.854
CANESM5 SSP245	2018–2050	0.848	1.070	1.269	2.309	2.877	3.496	0.061	0.244	0.739
CANESM5 SSP585	2018–2050	0.852	1.078	1.289	2.395	2.960	3.593	0.071	0.301	0.812
GFDL-ESM4 SSP245	2018–2050	0.870	1.073	1.275	1.985	2.557	3.232	0.048	0.198	0.559
GFDL-ESM4 SSP585	2018–2050	0.866	1.062	1.251	2.016	2.590	3.265	0.052	0.213	0.615
UKESM1-0-LL SSP245	2018–2050	0.834	1.073	1.282	2.360	2.923	3.568	0.070	0.362	0.968
UKESM1-0-LL SSP585	2018–2050	0.838	1.047	1.242	2.411	2.989	3.633	0.116	0.488	1.342
B. Mississippi River Basin										
Historical	1970–2017	0.789	1.008	1.230	1.699	2.043	2.402	0.177	0.401	0.756
CANESM5 SSP245	2018–2050	0.695	0.963	1.203	2.278	2.705	3.151	0.055	0.234	0.716
CANESM5 SSP585	2018–2050	0.691	0.967	1.218	2.356	2.798	3.263	0.072	0.292	0.834
GFDL-ESM4 SSP245	2018–2050	0.720	0.980	1.208	1.928	2.369	2.818	0.033	0.146	0.452
GFDL-ESM4 SSP585	2018–2050	0.726	0.975	1.191	1.965	2.408	2.880	0.034	0.160	0.523
UKESM1-0-LL SSP245	2018–2050	0.673	0.960	1.199	2.308	2.725	3.170	0.058	0.296	0.813
UKESM1-0-LL SSP585	2018–2050	0.686	0.945	1.163	2.343	2.789	3.258	0.098	0.395	1.092
C. Northern states east of the 100th meridian										
Historical	1970–2017	0.801	0.940	1.087	1.421	1.642	1.887	0.061	0.142	0.286
CANESM5 SSP245	2018–2050	0.761	0.905	1.068	1.936	2.206	2.478	0.015	0.070	0.207
CANESM5 SSP585	2018–2050	0.767	0.907	1.074	1.979	2.281	2.590	0.018	0.084	0.243

GFDL-ESM4 SSP245	2018–2050	0.797	0.937	1.089	1.621	1.876	2.139	0.008	0.048	0.140
GFDL-ESM4 SSP585	2018–2050	0.793	0.927	1.069	1.649	1.911	2.176	0.008	0.050	0.137
UKESM1-0-LL SSP245	2018–2050	0.725	0.875	1.031	1.967	2.256	2.535	0.008	0.068	0.218
UKESM1-0-LL SSP585	2018–2050	0.731	0.883	1.019	1.992	2.305	2.612	0.021	0.109	0.296
D. Middle states east of the 100th meridian										
Historical	1970–2017	0.881	1.055	1.222	2.005	2.196	2.379	0.302	0.494	0.730
CANESM5 SSP245	2018–2050	0.848	1.066	1.215	2.622	2.856	3.058	0.106	0.307	0.711
CANESM5 SSP585	2018–2050	0.845	1.072	1.225	2.681	2.935	3.187	0.150	0.401	0.830
GFDL-ESM4 SSP245	2018–2050	0.868	1.076	1.244	2.305	2.542	2.752	0.089	0.249	0.536
GFDL-ESM4 SSP585	2018–2050	0.878	1.064	1.204	2.329	2.572	2.786	0.105	0.283	0.565
UKESM1-0-LL SSP245	2018–2050	0.866	1.079	1.216	2.650	2.894	3.107	0.156	0.445	0.817
UKESM1-0-LL SSP585	2018–2050	0.857	1.057	1.189	2.682	2.960	3.197	0.249	0.617	1.160
E. Southern states east of the 100th meridian										
Historical	1970–2017	1.111	1.279	1.475	2.673	3.000	3.368	0.636	0.991	1.414
CANESM5 SSP245	2018–2050	1.072	1.253	1.438	3.354	3.693	4.058	0.232	0.628	1.377
CANESM5 SSP585	2018–2050	1.082	1.275	1.468	3.427	3.788	4.177	0.270	0.676	1.372
GFDL-ESM4 SSP245	2018–2050	1.047	1.243	1.433	3.090	3.435	3.813	0.235	0.535	1.139
GFDL-ESM4 SSP585	2018–2050	1.047	1.233	1.401	3.110	3.471	3.853	0.256	0.604	1.370
UKESM1-0-LL SSP245	2018–2050	1.123	1.298	1.460	3.435	3.758	4.120	0.503	1.024	1.675
UKESM1-0-LL SSP585	2018–2050	1.074	1.253	1.436	3.480	3.808	4.184	0.601	1.382	2.227

Note: We report quartiles of total annual precipitation, moderate-heat, and extreme-heat degree days based on projections from three climate models (UKESM1-0-LL, CANESM5, and GFDL-ESM4) and two shared socioeconomic pathways (SSPs), namely 245, 585, for 2018–2050. We also report quartiles based on historical data for 1970–2017. The precipitation is total annual and it is measured in meters. The moderate heat is measured in 1,000 degree days between 1°C and 29°C. The extreme heat is measured in 100 degree days above 29°C. We use baseline to refer to counties east of the 100th meridian excluding Florida. We classify states as northern, middle, and southern following Schlenker and Roberts (2009). For additional details, see section 9.7.

Table 9.8   **Marginal effects of corn acreage on nitrogen concentration alternative climate models & SSPs**

Model & SSP	Year	P25%	P50%	P75%	ME25%	ME50%	ME75%
A. Baseline							
Historical	1970–2017	0.885	1.070	1.274	0.025	0.051	0.080
CANESM5 SSP245	2018–2050	0.848	1.070	1.269	0.020	0.051	0.080
CANESM5 SSP585	2018–2050	0.852	1.078	1.289	0.020	0.053	0.083
GFDL-ESM4 SSP245	2018–2050	0.870	1.073	1.275	0.023	0.052	0.081
GFDL-ESM4 SSP585	2018–2050	0.866	1.062	1.251	0.023	0.050	0.077
UKESM1 0 LL SSP245	2018–2050	0.834	1.073	1.282	0.018	0.052	0.082
UKESM1-0-LL SSP585	2018–2050	0.838	1.047	1.242	0.018	0.048	0.076
B. Mississippi River Basin							
Historical	1970–2017	0.789	1.008	1.230	0.018	0.056	0.095
CANESM5 SSP245	2018–2050	0.695	0.963	1.203	0.001	0.048	0.090
CANESM5 SSP585	2018–2050	0.691	0.967	1.218	0.000	0.048	0.092
GFDL-ESM4 SSP245	2018–2050	0.720	0.980	1.208	0.005	0.051	0.091
GFDL-ESM4 SSP585	2018–2050	0.726	0.975	1.191	0.006	0.050	0.088
UKESM1-0-LL SSP245	2018–2050	0.673	0.960	1.199	−0.003	0.047	0.089
UKESM1-0-LL SSP585	2018–2050	0.686	0.945	1.163	−0.001	0.045	0.083
C. Northern states east of the 100th meridian							
Historical	1970–2017	0.801	0.940	1.087	0.002	0.032	0.063
CANESM5 SSP245	2018–2050	0.761	0.905	1.068	−0.007	0.024	0.059
CANESM5 SSP585	2018–2050	0.767	0.907	1.074	−0.006	0.024	0.060
GFDL-ESM4 SSP245	2018–2050	0.797	0.937	1.089	0.001	0.031	0.064
GFDL-ESM4 SSP585	2018–2050	0.793	0.927	1.069	−0.000	0.029	0.059
UKESM1-0-LL SSP245	2018–2050	0.725	0.875	1.031	−0.015	0.017	0.051
UKESM1-0-LL SSP585	2018–2050	0.731	0.883	1.019	−0.014	0.019	0.049
D. Middle states east of the 100th meridian							
Historical	1970–2017	0.881	1.055	1.222	0.144	0.134	0.125
CANESM5 SSP245	2018–2050	0.848	1.066	1.215	0.146	0.133	0.125
CANESM5 SSP585	2018–2050	0.845	1.072	1.225	0.146	0.133	0.124
GFDL-ESM4 SSP245	2018–2050	0.868	1.076	1.244	0.145	0.133	0.123
GFDL-ESM4 SSP585	2018–2050	0.878	1.064	1.204	0.144	0.133	0.126
UKESM1-0-LL SSP245	2018–2050	0.866	1.079	1.216	0.145	0.133	0.125
UKESM1-0-LL SSP585	2018–2050	0.857	1.057	1.189	0.145	0.134	0.126
E. Southern states east of the 100th meridian							
Historical	1970–2017	1.111	1.279	1.475	−0.030	−0.024	−0.017
CANESM5 SSP245	2018–2050	1.072	1.253	1.438	−0.031	−0.025	−0.018
CANESM5 SSP585	2018–2050	1.082	1.275	1.468	−0.031	−0.024	−0.017
GFDL-ESM4 SSP245	2018–2050	1.047	1.243	1.433	−0.032	−0.025	−0.018
GFDL-ESM4 SSP585	2018–2050	1.047	1.233	1.401	−0.032	−0.025	−0.020
UKESM1-0-LL SSP245	2018–2050	1.123	1.298	1.460	−0.029	−0.023	−0.017
UKESM1-0-LL SSP585	2018–2050	1.074	1.253	1.436	−0.031	−0.025	−0.018

Note: For each climate model and SSP combination, we report precipitation (P) quartiles and marginal effects (MEs) calculated assuming an increase in corn acreage equal to 1 within-county standard deviation using the appropriate set of counties in each panel. For comparison, we show MEs calculated using data for 1970–2017. The precipitation is total annual and it is measured in meters. In panel A, the MEs are in mg/L and they are calculated using specification C8 of the panel fixed-effect (FE) regressions in table 9.3. In panels B–E, the MEs are also in mg/L and they are calculated for the same specification of the panel FE regressions estimated using counties in the Mississippi River Basin, and the northern, middle, and southern states following the classification in Schlenker and Roberts (2009). For additional details, see section 9.7.

Fig. 9.5A–L   Corn acreage marginal effects with GFDL-ESM4 precipitation projections

Note: Panels A–L are read from top left to bottom right. We show corn acreage marginal effects (MEs) in mg/L for specification C8 of the panel fixed-effect (FE) regressions in table 9.3. We use baseline to refer to counties east of the 100th meridian excluding Florida. We define the northern, middle, and southern states following Schlenker and Roberts (2009). For the MEs based on the historical data, we use precipitation averages for 1970–2017. For the MEs based on the projections from two SSPs of the GFDL-ESM4 climate model, we use precipitation averages for 2018–2050. The shading of the choropleth maps is based on deciles of the ME empirical distribution. For additional details, see section 9.7.

ties, the median precipitation projections are 1.233–1.298 meters implying MEs of −0.025 to −0.023 mg/L, which are also essentially identical to their historical counterpart. Similar to the middle counties, the coefficient of the interaction of corn acreage with precipitation is statistically indistinguishable from zero for the southern counties.

Figure 9.5 shows the spatial variation of the MEs when we interact corn acres with precipitation projections for the two SSPs of the GFDL-ESM4 climate model. For comparison, we also show MEs based on historical precipitation. For each county, we calculate MEs using the average precipitation for either 1970–2017 (historical) or 2018–2050 (projected) and the appropriate coefficients of the estimated panel FE regression. For the baseline counties, we see some of the largest MEs in counties in the South

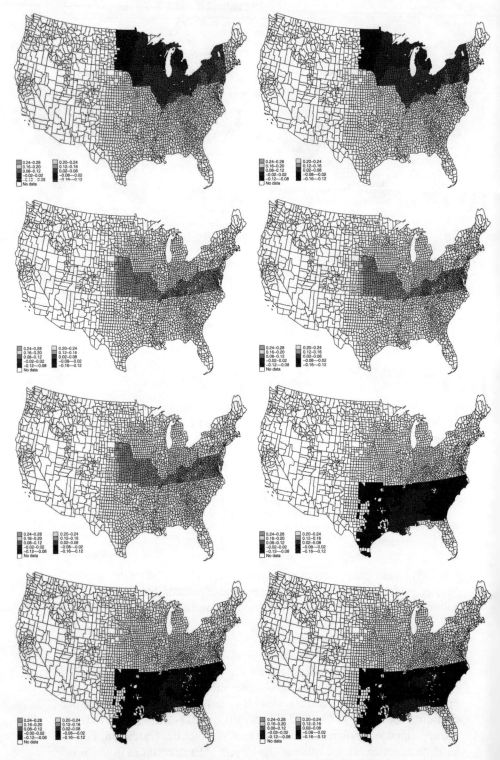

Fig. 9.5 (cont.)

Fig. 9.6A–L   Corn acreage marginal effects with GFDL-ESM4 precipitation and heat projections

Note: Panels A–L are read from left to right. We show corn acreage marginal effects (MEs) in mg/L for the panel fixed-effect (FE) regressions in which we interact corn acreage with precipitation, moderate-heat DDs, and extreme-heat DDs. In the regressions, we use the same set of weather-related controls, county fixed effects (FEs), year FEs, and county-specific trends as in column C8 in table 9.3. We use baseline to refer to counties east of the 100th meridian excluding Florida. We define the northern, middle, and southern states following Schlenker and Roberts (2009). For the MEs based on the historical data, we use precipitation, moderate-, and extreme-heat DD averages for 1970–2017. For the MEs based on the projections from two SSPs of the GFDL-ESM4 climate model, we use averages for 2018–2050. The shading of the choropleth maps is based on deciles of the ME empirical distribution. For additional details, see section 9.7.

(e.g., Louisiana, Mississippi, Alabama, Arkansas) and some of the smallest effects in the Plains (e.g., northern Texas, Oklahoma) and in the upper Midwest (e.g., Michigan, Wisconsin). We see a very similar spatial pattern in the MEs for the MRB counties. The lack of variation across the middle and southern counties is because of the coefficients on the interaction of corn acreage with precipitation being indistinguishable from zero. For the northern counties, we see negative MEs in North and South Dakota, and some of the larger positive MEs in Pennsylvania and New Jersey. The negative MEs are due to a combination of a large negative coefficient on corn acreage and very low precipitation.

Figure 9.6 shows the spatial variation of MEs when we interact corn

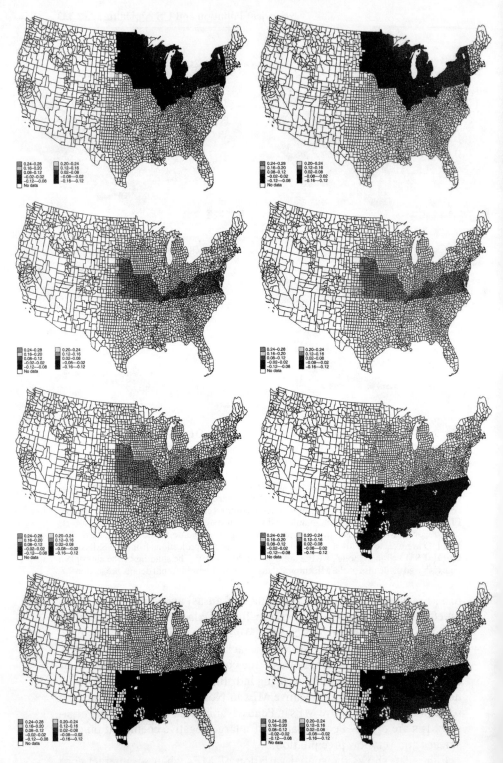

Fig. 9.6 (cont.)

acres with precipitation, moderate-heat DDs, and extreme-heat DDs for the two SSPs of the GFDL-ESM4 climate model. For each county, we calculate MEs using the average precipitation, extreme-heat, and moderate heat DDs for either 1970–2017 (historical) or 2018–2050 (projected) and different panel FE regressions for each of the five sets of counties. Across the baseline set of counties, the median ME based on the historical data is 0.049. Its projections-based counterparts are 0.030 for SSP 245 and 0.027 SSP 585. All three median MEs are smaller than their counterparts based on the panel FE regression in which we interact corn acreage with precipitation only. This is especially true for the projected MEs. In terms of the spatial pattern of the MEs, we see some of the largest effects in Tennessee, and in the northern parts of Alabama and Mississippi. Some of the smallest MEs are those for counties along the 100th meridian, as well as in Georgia and South Carolina. Across the MRB counties, we also see smaller median MEs when we interact corn acres with precipitation and the DDs and more so when we use the 2018–2050 projections. The same is true for the middle and northern counties. For the southern counties, the median historical and projected MEs are negative and larger in magnitude than their counterparts based on the interaction of corn acreage with precipitation alone.

9.8   Conclusion and Policy Implications in an Era of Climate Change

We study the relationship between water nutrient pollution and US agriculture using data from 1970–2017 documenting a causal positive effect of corn acreage on nitrogen concentration in the country's water bodies east of the 100th meridian using alternative empirical approaches. According to our baseline estimates, a 10 percent increase in corn acreage increases nitrogen concentration in water by up to 1 percent. Annual precipitation plays an important role in the magnitude of the estimated effects with higher precipitation exacerbating the acreage effect on nitrogen concentration. Temperature also matters for the magnitude of the acreage effect. An increase in moderate-heat degree leads to smaller effects due to its beneficial effect on the crop nutrient uptake. Extreme-heat degree days do not seem to matter for the magnitude of the effect. The 1 percent increase in the average level of nitrogen concentration in the Midwest coupled with the average streamflow of the Mississippi River at the Gulf of Mexico during this period and damages of about $16 per ton of nitrogen implies an annual external cost of $800 million.

Our estimated effect of additional corn acres on measured nitrogen in waterways is an order of magnitude smaller than agronomic estimates of excess nitrogen applied to those acres assuming edge-of-field losses translate to an equivalent nitrogen loading to streams and rivers. Our findings regarding the magnitude of the effect are consistent with a new line of research showing that large amounts of nitrogen stored in subsurface soil and groundwater give rise to the so-called legacy nitrogen, which may contribute to loadings in rivers and streams with a long delay, a topic we explore in more detail in Metaxoglou and Smith (2022).

Given the role of precipitation and temperature on the magnitude of the estimated effect of corn acreage on nitrogen concentration, we explore the implications of climate change for our findings. We use the NASA Earth Exchange Global Daily Downscaled Projections data set to obtain precipitation and temperature projections for 2018–2050, which we translate to projections of marginal effects of corn acreage on nitrogen concentration. The marginal effects based on precipitation projections from the NASA GFDL-ESM4 climate model and two shared socioeconomic pathways are very similar in magnitude to their counterparts calculated using historical data. The marginal effects based on temperature projections are slightly smaller than those using historical data. These estimated effects do not account for the impacts of climate change on acreage, nitrogen fertilizer use, legacy nitrogen, runoff, and streamflow, all of which contribute to nutrient pollution.

Based on recent work identifying wetlands as a powerful weapon in the war against nutrient pollution, especially due to their efficacy in also removing legacy nitrogen, we ought to emphasize their vulnerability to changes in landscapes and weather patterns impacted by climate change. Increased flooding, drought spells, extreme heat, and frequency of severe storms due to climate change all can negatively affect wetlands (Salimi, Almuktar, and Scholz 2021). Taking into consideration other ecosystem services that wetlands also provide, such as absorbing floodwaters, providing habitat for wildlife, and acting as net carbon sinks, adds to the case for policy discussion of these issues, especially in the light of recent developments in redefining the *Waters of the United States* that are protected by the Clean Water Act.

References

Alexander, R., and R. Smith. 1990. "County-Level Estimates of Nitrogen and Phosphorus Fertilizer Use in United States, 1945–1985." *U.S. Geological Survey Open File Report 90–130*.

Alexander, R., R. Smith, G. Schwarz, E. Boyer, J. Nolan, and J. Brakebill. 2008. "Differences in Phosphorous and Nitrogen Delivery to the Gulf of Mexico from the Mississippi River Basin." *Environmental Science and Technology* 42: 822–30.

Anderson, D., P. Hoagland, Y. Kaoru, and A. White. 2000. "Estimated Annual Economic Impacts from Harmful Algal Blooms (HABs) in the United States." *Woods Hole Oceanographic Institution Technical Report*.

Annan, F., and W. Schlenker. 2015. "Federal Crop Insurance and the Disincentive to Adapt to Extreme Heat." *American Economic Review Papers and Proceedings* 105: 262–66.

Birch, M. B. L., B. M. Gramig, W. R. Moomaw, O. C. Doering, III, and C. J. Reeling. 2011. "Why Metrics Matter: Evaluating Policy Choices for Reactive Nitrogen in the Chesapeake Bay Watershed." *Environmental Science & Technology* 45: 168–74.

Blottnitz, H. V., A. Rabl, D. Boiadjiev, T. Taylor, and S. Arnold. 2006. "Damage Costs of Nitrogen Fertilizer in Europe and their Internalization." *Journal of Environmental Planning and Management* 49: 413–33.

Brakeball, J., and J. Gronberg. 2017 "County-Level Estimates of Nitrogen and Phosphorus from Commercial Fertilizer for the Conterminous United States, 1987–2012." U.S. Geological Survey data release, https://doi.org/10.5066/F7H41PKX.

Byrnes, D., K. Van Meter, and N. Basu. 2020. "Long-Term Shifts in US Nitrogen Sources and Sinks Revealed by the New TREND-Nitrogen Data Set (1930–2017)." *Global Biogeochemical Cycles* 34: e2020GB006626.

Capel, P., K. McCarthy, R. H. Coupe, K. Grey, S. Amenumey, N. T. Baker, and R. Johnson. 2018. "Agriculture–A River Runs Through It–The Connections Between Agriculture and Water Quality." *U.S. Geological Survey Circular 1433*.

CENR. 2000. "Integrated Assessment of Hypoxia in the Northern Gulf of Mexico." *National Science and Technology Council Committee on Environment and Natural Resources*.

Claassen, R., M. Bowman, J. McFadden, D. Smith, and S. Wallander. 2018. "Tillage Intensity and Conservation Cropping in the United States." *U.S. Department of Agriculture Economic Information Bulletin Number 197*.

Compton, J. E., J. A. Harrison, R. L. Dennis, T. L. Greaver, B. H. Hill, S. J. Jordan, H. Walker, and H. V. Campbell. 2011. "Ecosystem Services Altered by Human Changes in the Nitrogen Cycle: A New Perspective for US Decision Making." *Ecology Letters* 14: 804–15.

David, M. B., L. E. Drinkwater, and G. F. McIsaac. 2010. "Sources of Nitrate Yields in the Mississippi River Basin." *Journal of Environmental Quality* 39: 1657–1667.

Deschênes, O., and M. Greenstone. 2007. "The Economic Impacts of Climate Change: Evidence from Agricultural Output and Random Fluctuations in Weather." *American Economic Review* 97: 354–85.

Diaz, R. J., and R. Rosenberg. 2008. "Spreading Dead Zones and Consequences for Marine Ecosystems." *Science* 321: 926–29.

Dodds, W. K., W. W. Bouska, J. L. Eitzmann, T. J. Pilger, K. L. Pitts, A. J. Riley, J. T. Schloesser, and D. J. Thornbrugh. 2009. "Eutrophication of U.S. Freshwaters: Analysis of Potential Economic Damages." *Environmental Science and Technology* 43: 12–19.

Doering, O., F. Diaz-Hermelo, C. Howard, R. Heimlich, F. Hitzhusen, R. Kazmierczak, J. Lee, L. Libby, W. Milon, T. Prato, and M. Ribaudo. 1999. "Evaluation of the Economic Costs and Benefits of Methods for Reducing Nutrient Loads to the Gulf of Mexico: Topic 5 Report for the Integrated Assessment on Hypoxia in the Gulf of Mexico." National Oceanic and Atmospheric Administration.

Dubrovsky, N., K. Burow, G. Clark, J. Gronberg, P. Hamilton, K. Hitt, D. Mueller, M. Munn, B. Nolan, L. Puckett, M. Rupert, T. Short, N. Spahr, L. Sprague, and W. Wilber. 2018. "The Quality of Our Nation's Water: Nutrients in the Nation's Streams and Groundwater, 1992–2004." *U.S. Geological Survey Circular 1350*.

Elbakidze, L., Y. Xu, J. Arnold, and H. Yen. 2022. "Climate Change and Downstream Water Quality in Agricultural Production: The Case of Nutrient Runoff to the Gulf of Mexico." In *American Agriculture, Water Resources, and Climate Change*, edited by G. Libecap and A. Dinar. Chicago, IL: University of Chicago Press. This volume.

EPA. 2007. "Hypoxia in the Northern Gulf of Mexico: An Update by the EPA Science Advisory Board." U.S. Environmental Protection Agency.

————. 2015. "A Compilation of Cost Data Associated with the Impacts and Control of Nutrient Pollution." U.S. EPA Office of Water EPA 820-F-15-096.

————. 2016. "National Rivers and Streams Assessment 2008–2009: A Collborative Survey." U.S. Environmental Protetcion Agency Office of Research and Development.

Eyring, V., S. Bony, G. A. Meehl, C. A. Senior, B. Stevens, R. J. Stouffer, and K. E. Taylor. 2016. "Overview of the Coupled Model Intercomparison Project Phase 6 (CMIP6) Experimental Design and Organization." *Geoscientific Model Development* 9: 1937–1958.

Falcone, J. 2018. "Changes in Anthropogenic Influences on Streams and Rivers in the Conterminous U.S. over the Last 40 Years, Derived for 16 Data Themes: U.S. Geological Survey data release." https://doi.org/10.5066/F7XW4J1J.

Garcia, A. M., R. B. Alexander, J. G. Arnold, L. Norfleet, M. J. White, D. M. Robertson, and G. Schwarz. 2016. "Regional Effects of Agricultural Conservation Practices on Nutrient Transport in the Upper Mississippi River Basin." *Environmental Science & Technology* 50: 6991–7000.

GOMNTF. 2013. "Reassessment 2013: Assessing Progress Made Since 2008." Mississippi River Gulf of Mexico Watershed Nutrient Task Force.

————. 2015. "Mississippi River Gulf of Mexico Watershed Nutrient Task Force: 2015 Report to the Congress." U.S. Environmental Protection Agency.

Goolsby, D., W. Battaglin, G. Lawrence, R. Artz, B. Aulenbach, R. Hooper, D. Keeney, and G. Stensland. 1999. "Flux and Sources of Nutrients in the Mississippi-Atchafalaya River Basin: Topic 3 Report for the Integrated Assessment on Hypoxia in the Gulf of Mexico." National Oceanic and Atmospheric Administration.

Gourevitch, J., B. Keeler, and T. Ricketts. 2018. "Determining Socially Optimal Rates of Nitrogen Fertilizer Application." *Agriculture Ecosystems and Environment* 254: 292–99.

Hellerstein, D., D. Vilorio, and M. Ribaudo. 2019. "Agricultural Resources and Environmental Indicators 2019." *U.S. Department of Agriculture Economic Information Bulletin Number 208*.

Hendricks, N., A. Smith, and D. Sumner. 2014. "Crop Supply Dynamics and the Illusion of Partial Adjustment." *American Journal of Agricultural Economics* 96: 1469–1491.

Ho, J. C., A. M. Michalak, and N. Pahlevan. 2019. "Widespread Global Increase in Intense Lake Phytoplankton Blooms since the 1980s." *Nature* 574: 667–70.

Ilampooranan, I., K. J. Van Meter, and N. B. Basu. 2019. "A Race Against Time: Modeling Time Lags in Watershed Response." *Water Resources Research* 55: 3941–3959.

Jägermeyr, J., C. Müller, A. C. Ruane, J. Elliott, J. Balkovic, O. Castillo, B. Faye, I. Foster, C. Folberth, J. A. Franke, et al. 2021. "Climate Impacts on Global Agriculture Emerge Earlier in New Generation of Climate and Crop Models." *Nature Food* 2: 873–85.

Keeler, B. L., J. D. Gourevitch, S. Polasky, F. Isbell, C. W. Tessum, J. D. Hill, and J. D. Marshall. 2018. "The Social Costs of Nitrogen." *Science Advances* 2.

Keiser, D., and J. Shapiro. 2018. "Consequences of the Clean Water Act and the Demand for Water Quality." *Quarterly Journal of Economics* 134: 349–96.

Kling, C. 2011. "Economic Incentives to Improve Water Quality in Agricultural Landscapes: Some New Variations on Old Ideas." *American Journal of Agricultural Economics* 93: 297–309.

Kling, C. L., Y. Panagopoulos, S. S. Rabotyagov, A. M. Valcu, P. W. Gassman, T. Campbell, M. J. White, J. G. Arnold, R. Srinivasan, M. K. Jha, J. J. Richardson, L. M. Moskal, R. E. Turner, and N. N. Rabalais. 2014. "LUMINATE: Linking

Agricultural Land Use, Local Water Quality and Gulf Of Mexico Hypoxia." *European Review of Agricultural Economics* 41: 431–59.

Kudela, R., E. Berdalet, S. Bernard, M. Burford, L. Fernand, S. Lu, S. Roy, P. Tester, G. Usup, R. Magnien, et al. 2015. "Harmful Algal Blooms. A Scientific Summary for Policy Makers. IOC." UNESCO, IOC/INF-1320, 20pp.

Marshall, E., M. Aillery, M. Ribaudo, N. Key, S. Sneeringer, L. Hansen, S. Malcolm, and A. Riddle. 2018. "Reducing Nutrient Losses From Cropland in the Mississippi/Atchafalaya River Basin: Cost Efficiency and Regional Distribution." *U.S. Department of Agriculture Economic Research Report Number 258.*

McCabe, G. J., and D. M. Wolock. 2011 "Independent Effects of Temperature and Precipitation on Modeled Runoff in the Conterminous United States." *Water Resources Research* 47: W11522.

McLellan, E., J. Proville, and M. Monast. 2016. "Restoring the Heartland: Costs and Benefits of Agricultural Conservation in the U.S. Corn Belt." Environmental Defense Fund Draft Report.

McLellan, E., D. Robertson, K. Schilling, M. Tomer, J. Kostel, D. Smith, and K. King. 2015. "Reducing Nitrogen Export from the Corn Belt to the Gulf of Mexico: Agricultural Strategies for Remediating Hypoxia." *JAWRA Journal of the American Water Resources Association* 51: 263–89.

Mendelsohn, R., W. Nordhaus, and D. Shaw. 1994. "The Impact of Gloabl Warming on Agriculture: A Ricardian Analysis." *American Economic Review* 84: 753–71.

Metaxoglou, K., and A. Smith. 2022. "Agriculture's Nitrogen Legacy." Working Paper.

Miao, R., M. Khanna, and H. Huang. 2015. "Responsiveness of Crop Yield and Acreage to Prices and Climate." *American Journal of Agricultural Economics* 98: 191–211.

Mitsch, W., J. W.Day, J. Gilliam, P. Groffman, D. Hey, G. Randall, and N. Wang. 1999. "Reducing Nutrient Loads, Especially Nitrate Nitrogen, to Surface Water, Ground Water, and the Gulf of Mexico: Topic 5 Report for the Integrated Assessment on Hypoxia in the Gulf of Mexico." National Oceanic and Atmospheric Administration.

Murphy, J., R. Hirsch, and L. Sprague. 2013. "Nitrate in the Mississippi River and Its Tributaries, 1980 2010: An Update." *U.S. Geological Survey Scientific Investigations Report 2013 5169.*

Nakagaki, N., and M. Wieczorek. 2016. "Estimates of Subsurface Tile Drainage Extent for 12 Midwest States, 2012: U.S. Geological Survey data release" http://dx.doi.org/10.5066/F7W37TDP.

Nakagaki, N., M. Wieczorek, and S. Qi. 2016. "Estimates of Subsurface Tile Drainage Extent for the Conterminous United States, Early 1990s: U.S. Geological Survey data release." http://dx.doi.org/10.5066/F7RB72QS.

NRCS. 2017a. "Assessment of the Effects of Conservation Practices on Cultivated Cropland in the Upper Mississippi River Basin." U.S. Department of Agriculture National Resources Conservation Services Conservation Effects Assessment Project.

————. 2017b. "Effects of Conservation Practices on Nitrogen Loss from Farm Fields." U.S. Department of Agriculture National Resources Conservation Services Conservation Effects Assessment Project.

Olmstead, S. 2010. "The Economics of Water Quality." *Review of Environmental Economics and Policy* 4: 44–62.

Olmstead, S. M., L. A. Muehlenbachs, J.-S. Shih, Z. Chu, and A. J. Krupnick. 2013. "Shale Gas Development Impacts on Surface Water Quality in Pennsylvania." *Proceedings of the National Academy of Sciences* 110: 4962–4967.

Paudel, J., and C. L. Crago. 2020. "Environmental Externalities from Agriculture: Evidence from Water Quality in the United States." *American Journal of Agricultural Economics 103 (1).*

Rabotyagov, S., C. Kling, P. Gassman, N. Rabalais, and R. Turner. 2014a. "The Economics of Dead Zones: Causes, Impacts, Policy Challenges, and a Model of the Gulf of Mexico Hypoxic Zone." *Review of Environmental Economics and Policy* 8: 58–79.

Rabotyagov, S. S., T. D. Campbell, M. White, J. G. Arnold, J. Atwood, M. L. Norfleet, C. L. Kling, P. W. Gassman, A. Valcu, J. Richardson, R. E. Turner, and N. N. Rabalais. 2014b. "Cost-Effective Targeting of Conservation Investments to Reduce the Northern Gulf of Mexico Hypoxic Zone." *Proceedings of the National Academy of Sciences* 111: 18530–18535.

Read, E. K., L. Carr, L. De Cicco, H. A. Dugan, P. C. Hanson, J. A. Hart, J. Kreft, J. S. Read, and L. A. Winslow. 2017. "Water Quality Data for National-Scale Aquatic Research: The Water Quality Portal." *Water Resources Research* 53: 1735–1745.

Ribaudo, M., J. Delgado, L. Hansen, M. Livingston, R. Mosheim, and J. Williamson. 2011. "Nitrogen in Agricultural Systems: Implications for Conservation Policy." *U.S. Department of Agriculture Economic Research Report Number 127.*

Ribaudo, M. O., R. Heimlich, R. Claassen, and M. Peters. 2001. "Least-Cost Management of Nonpoint Source Pollution: Source Reduction versus Interception Strategies for Controlling Nitrogen Loss in the Mississippi Basin." *Ecological Economics* 37: 183–97.

Roberts, M., W. Schlenker, and J. Eyer. 2012. "Agronomic Weather Measures in Econometric Models of Crop Yield with Implications for Climate Change." *American Journal of Agricultural Economics* 95: 236–43.

Robertson, D., and D. Saad. 2006. "Spatially Referenced Models of Streamflow and Nitrogen, Phosphorus, and Suspended-Sediment Loads in Streams of the Midwestern United States." *U.S. Geological Survey Scientific Investigations Report 2019 5114.*

Salimi, S., S. A. Almuktar, and M. Scholz. 2021. "Impact of Climate Change on Wetland Ecosystems: A Critical Review of Experimental Wetlands." *Journal of Environmental Management* 286: 112160.

Schlenker, W., and M. Roberts. 2009. "Nonlinear Temperature Effects Indicate Severe Damages to U.S. Crop Yields under Climate Change." *Proceedings of the National Academy of Sciences* 106: 15594–15598.

Sinha, E., A. M. Michalak, and V. Balaji. 2017. "Eutrophication Will Increase during the 21st Century as a Result of Precipitation Changes." *Science* 357: 405–8.

Sobota, D. J., J. E. Compton, M. L. McCrackin, and S. Singh. 2015. "Cost of Reactive Nitrogen Release from Human Activities to the Environment in the United States." *Environmental Research Letters* 10: 025006.

Sprague, L., R. Hirsch, and B. Aulenbach. 2011. "Nitrate in the Mississippi River and Its Tributaries, 1980 to 2008: Are We Making Progress?" *Environmental Science & Technology* 45: 7209–7216.

Sprague, L., G. Oelsner, and D. Argue. 2017. "Challenges with Secondary Use of Multi-source Water Quality Data in the United States." *Water Research* 110: 252–61.

Sugg, Z. 2007. "Assessing U.S. Farm Drainage: Can GIS Lead to Better Estimates of Subsurface Drainage Extent?" Water Resources Institute.

Tallis, H., S. Polasky, J. Hellmann, N. P. Springer, R. Biske, D. DeGeus, R. Dell, M. Doane, L. Downes, J. Goldstein, T. Hodgman, K. Johnson, I. Luby, D. Pennington, M. Reuter, K. Segerson, I. Stark, J. Stark, C. Vollmer-Sanders, and S. K.

Weaver. 2019. "Five Financial Incentives to Revive the Gulf of Mexico Dead Zone and Mississippi Basin Soils." *Journal of Environmental Management* 233: 30–38.

Taylor, C., and G. Heal. 2021. "Algal Blooms and the Social Cost of Fertilizer." Working Paper.

UCS. 2020. "Reviving the Dead Zone: Solutions to Benefit Both Gulf Coast Fishers and Midwest Farmers." Union of Concerned Scientists.

USEPA. 2001. "The National Costs of the Total Maximum Daily Load Program."

Valayamkunnath, P., M. Barlage, F. Chen, D. J. Gochis, and K. J. Franz. 2020. "Mapping of 30-meter Resolution Tile-Drained Croplands Using a Geospatial Modeling Approach." *Scientific Data* 7: 1–10.

Van Grinsven, H. J. M., M. Holland, B. H. Jacobsen, Z. Klimont, M. a. Sutton, and W. Jaap Willems. 2013a. "Costs and Benefits of Nitrogen for Europe and Implications for Mitigation." *Environmental Science & Technology* 47: 3571–3579.

———. 2013b. "Costs and Benefits of Nitrogen for Europe and Implications for Mitigation." *Environmental Science & Technology* 47: 3571–3579.

Van Meter, K., P. Van Cappellen, and N. Basu. 2018. "Legacy Nitrogen May Prevent Achievement of Water Quality Goals in the Gulf of Mexico." *Science* 360: 427–30.

Van Meter, K. J., N. B. Basu, and P. Van Cappellen. 2017. "Two Centuries of Nitrogen Dynamics: Legacy Sources and Sinks in the Mississippi and Susquehanna River Basins." *Global Biogeochemical Cycles* 31: 2–23.

Van Meter, K. J., N. B. Basu, J. J. Veenstra, and C. L. Burras. 2016. "The Nitrogen Legacy: Emerging Evidence of Nitrogen Accumulation in Anthropogenic Landscapes." *Environmental Research Letters* 11: 035014.

Whittaker, G., B. L. Barnhart, R. Srinivasan, and J. G. Arnold. 2015. "Cost of Areal Reduction of Gulf Hypoxia through Agricultural Practice." *Science of The Total Environment* 505: 149–53.

Wolock, D., and G. McCabe. 2018. "Water Balance Model Inputs and Outputs for the Conterminous United States, 1900–2015." https://doi.org/10.5066/F71V5CWN.

Wu, J., and K. Tanaka. 2005. "Reducing Nitrogen Runoff from the Upper Mississippi River Basin to Control Hypoxia in the Gulf of Mexico: Easements or Taxes." *Marine Resource Economics* 20: 121–44.

Xu, Y., L. Elbakidze, H. Yen, J. Arnold, P. Gassman, J. Hubbart, and M. Strager. 2021. "Integrated Assessment of N Runoff in the Gulf of Mexico." Working Paper.

10

The Political Economy of Groundwater Management
Descriptive Evidence from California

Ellen M. Bruno, Nick Hagerty, and Arthur R. Wardle

10.1 Introduction

Water pumped from underground aquifers contributes to agricultural production worldwide with particular importance in times of drought. When surface water flows are lower than expected, groundwater resources provide an important reserve capable of decoupling agricultural production from year-to-year variation in precipitation (Tsur and Graham-Tomasi 1991). With warming temperatures and resulting changes to precipitation and surface water storage due to climate change, agriculture will increasingly rely on groundwater to make up shortfalls in surface water supplies. As a result, both the demand for and buffer value of groundwater will increase. Despite the growing need for groundwater as a tool to adapt to climate change, pumping in excess of recharge threatens the sustainability of groundwater

Ellen M. Bruno is an assistant professor of Cooperative Extension in the Department of Agricultural and Resource Economics at the University of California, Berkeley.

Nick Hagerty is an assistant professor in the Department of Agricultural Economics and Economics at Montana State University.

Arthur R. Wardle is a PhD student in the Department of Agricultural and Resource Economics at the University of California, Berkeley.

This project has benefited from the excellent research assistance of Ezana Anley, Paige Griggs, and Lindsay McPhail. We also thank Andrew Ayres, Ariel Dinar, Michael Hanemann, Katrina Jessoe, Gary Libecap, and Richard Sexton for helpful comments and discussions. This research was supported by the Giannini Foundation of Agricultural Economics and the Foundation for Food & Agriculture Research under award number—Grant ID: NIA21-0000000036. The content of this publication is solely the responsibility of the authors and does not necessarily represent the official views of the Foundation for Food & Agriculture Research or the National Bureau of Economic Research. For data appendix and disclosure of the authors' material financial relationships, if any, please see https://www.nber.org/books-and -chapters/american-agriculture-water-resources-and-climate-change/political-economy -groundwater-management-descriptive-evidence-california.

aquifers worldwide (Edwards and Guilfoos 2021). Persistent drawdown is a particularly acute problem in many of the world's major food-producing regions, including California's Central Valley.

Despite a broad range of available regulatory solutions—from the formalization of property rights to pumping restrictions and volumetric fees—groundwater regulation remains rare. Examples of groundwater management do exist, ranging from quantity controls in parts of Kansas (Drysdale and Hendricks 2018) to price controls in small parts of Colorado (Smith et al. 2017) and California (Bruno and Jessoe 2021). Groundwater basins that have instituted rules that bear resemblance to first-best policies have enjoyed greater economic returns from their water as a result (Hornbeck and Keskin 2014; Edwards 2016; Ayres, Meng, and Plantinga 2021). But as with any common-pool resource dilemma, groundwater overdraft often continues despite its resulting economic losses due to the difficulty of replacing open-access management with institutions designed to preserve aquifers' value. Challenges arise due to the high transaction costs associated with collective action and the political economic forces that influence policy choice.

Classic characterizations of the groundwater commons dilemma often oversimplify both the problems and remedies facing real-world basins. The tragedy of the commons can be overcome through collective action, but the ability to do so is determined by myriad factors affecting the magnitude and distribution of the gains from management and the costs of bargaining. What prompts groundwater pumpers to attempt collective management, which factors influence the success of those attempts, and what determines policy choices are all central questions in the political economy of groundwater management. Understanding the political economic forces that give rise to collective action and first- or second-best policies is critical to the sustained economic viability of groundwater-dependent regions in the face of climate change.

This paper sets forth a framework for determining the likelihood for collective action and uses this to outline and test five hypotheses in the context of groundwater management in California. Our case study is California's Sustainable Groundwater Management Act (SGMA) of 2014, a landmark statewide mandate for local institutional transition. SGMA required the formation of hundreds of local Groundwater Sustainability Agencies (GSAs), formed by coalitions of preexisting water and land management agencies like water districts, cities, and counties, and charged each GSA to develop management actions to meet sustainability criteria. By mandating sustainability, SGMA forces parties to negotiate and therefore reduces barriers to collective action that would have persisted in the absence of the law. We construct a novel data set on the management choices, environmental conditions, and governing structures of 343 groundwater agencies subject to the legislation and use it to characterize cross-sectional trends in collective action and management strategies across the state.

California's SGMA offers a unique real-world setting for describing changing institutions and investigating determinants of collective action and policy instrument choice. Affecting hundreds of groundwater agencies simultaneously, the legislation covers all major agricultural areas, which account for over 90 percent of the state's groundwater pumping, in the nation's largest agricultural state.[1] Second, SGMA provides a statewide framework with local authority and flexibility while requiring that groundwater agencies engage with the public and document their governance structures and intended management actions. In essence, SGMA reduces the transaction costs to bargaining over collective action, empowers local water agencies to manage groundwater with new authorities, and requires that their processes and actions be recorded publicly. While it is still early in the process, California's initial implementation of SGMA offers a rare look at the barriers to collective action and the drivers of policy instrument choice.

Our assessment reveals a significant departure from the prior status quo of open-access groundwater use. We find that two features are positively correlated with an increased likelihood of collective action: more severe groundwater depletion and less heterogeneity among resource users in a locality. Contrary to expectations, a higher number of bargaining parties is not associated with a decreased likelihood of active groundwater management. Additionally, we find that agencies are approaching groundwater sustainability with substantial policy heterogeneity across the state by proposing a mixture of price and quantity instruments as well as a suite of other conservation incentive programs and ad hoc pumping restrictions. The most common proposed policy change is an introduction of taxes or fees, for which 60 percent of management plans stated a plan to implement or consider such a change. Almost half of the submitted plans include allocations to determine individual pumping limits, and two-thirds of the agencies setting allocations are considering trade of those individual allocations. Using constructed measures of local political power and local heterogeneity in groundwater demand, we find that proposals to allow trade of allocations are more likely when plans are governed by a board with a greater share of representation by agricultural interests (through special districts) and that this appears to better predict planned trading programs than a proxy for the available gains from this instrument.

This paper contributes to the literature on how groundwater management institutions develop in light of new empirical evidence offered by SGMA. The literature describing the political economy of this type of institutional transition is thick, but many open questions remain due to the inherent difficulty of collecting adequate data in these contexts. Most closely related to this work are Leonard and Libecap (2019) and Ayres, Edwards, and Libe-

---

1. Groundwater makes up 40 percent of the agricultural water supply on average in California but can be a much larger portion during dry years.

cap (2018), both of which study institutional transitions of common-pool water resources and attempt to explain the economic characteristics that lead to institutional change. By compiling a new data set on the management choices, governance, and economic and hydrologic features of hundreds of agencies following the passing of a statewide legislation that substantially altered the bargaining environment over collective action, we are able to present new evidence on where, how, and why groundwater management is occurring in practice.

California's approach to groundwater management through SGMA provides lessons for other regions facing similar common-pool resource issues. Our analysis reveals a state-led process to empower local agencies to collectively take action via a unifying framework that served to reduce bargaining costs. The state was effective at reducing the transaction costs associated with collective bargaining by improving access to information, altering the policy default, influencing the number and composition of bargaining parties, and providing direct financial support. Other regions looking to incentivize collective action at the local level could look to California's SGMA. The fact that we see substantial heterogeneity in policy instrument choice across space suggests that a uniform top-down approach may not allow local agencies to adopt preferred policies that reflect diverse regional hydrologic and economic conditions.

The rest of the paper proceeds as follows. We first provide background on SGMA. In section 10.3, we develop a conceptual framework for overcoming both the open-access problem and the criteria for policy instrument choice. We provide an overview of the political economy literature regarding what stands in the way of effective management and outline five testable hypotheses. Section 10.4 describes the data. Section 10.5 presents empirical results that document patterns in how the GSAs are planning to meet the sustainability requirements and in local characteristics that predict the chosen strategies. The final section concludes.

10.2   The Sustainable Groundwater Management Act

Groundwater serves as a critical buffer during periods of surface water scarcity, with average use in California increasing from 40–80 percent of the water supply during drought years. Groundwater reserves in California's Central Valley have been declining over the last several decades, raising fears about the long-term availability of the resource.

The passing of California's Sustainable Groundwater Management Act (SGMA) in 2014 provides an ideal opportunity to study the political economy of a groundwater regulation in its early implementation stage. Passing during the peak of the state's last major drought, SGMA provides a statewide framework for local agencies to manage groundwater and bring their basins into balance. It requires stakeholders in overdrafted basins throughout California to reach and maintain long-term stability in their groundwater levels by either 2040 or 2042, depending on their priority status. Local management authority is assigned to new groundwater sustainability agencies (GSAs) that were required to be formed in each basin or subbasin by 2017.[2] GSAs are given the authority and flexibility to manage the resource however they see fit, as long as their approach is documented in a "Groundwater Sustainability Plan" (GSP) outlined and approved by the state.

The timeline to adhere to SGMA is determined by a state-designated level of priority. Based on current conditions of groundwater overdraft, the California Department of Water Resources (DWR) assessed whether each basin was experiencing "critical overdraft," which bumps major SGMA deadlines up by two years. Based on a much wider suite of variables, such as expected future population growth, DWR also separated each groundwater basin into High, Medium, Low, or Very Low priority. Only high- and medium-priority basins face most of SGMA's mandates. All GSPs for the 94 high- and medium-priority basins were required to be adopted by January 31, 2022. GSAs managing groundwater in high- and medium-priority basins subject to critical conditions of overdraft had to adopt a GSP two years earlier, by January 31, 2020. Once they adopt, the plan goes into effect. The state provided both advisory and monetary resources for the development of plans. Failure to comply will result in top-down state regulation as a backstop.

SGMA created substantial variation in regulatory stringency, since basins with more overdraft must adopt greater pumping restrictions in order to achieve sustainability. Figure 10.1 shows the state-designated priority level, including which basins were deemed to be in conditions of critical overdraft.

Recognizing the institutional and policy path dependence in which SGMA emerged is important for characterizing the local developments and management strategies we observe. While historic in its nature to mandate groundwater management statewide, SGMA naturally built upon decades of previous water policies designed to support and encourage groundwater management (Ayres, Edwards, and Libecap 2018; Dennis et al. 2020). Its emphasis on local control, giving the newly formed GSAs the authority to leverage fees and facilitate trade, reflects a history of groundwater measurement and management at the local level. Prior to SGMA, the state provided funds to local water agencies to monitor groundwater and conduct studies,

---

2. We will use *basins* to refer to the basic spatial units of management under SGMA, which California Department of Water Resources (DWR) calls *basins* and *subbasins*. A basin is an entire aquifer that is relatively physically isolated from other groundwater resources; many large basins are further divided into subbasins. Basin and subbasin boundaries were developed by DWR for other purposes prior to SGMA and are not influenced by local choices. Each basin (i.e., or subbasin) is required to form at least one GSA; GSAs never contain parts of more than one basin, but one basin may have more than one GSA. In some cases, multiple GSAs within a basin joined together to collaboratively develop one GSP; our analysis treats them together as one unit.

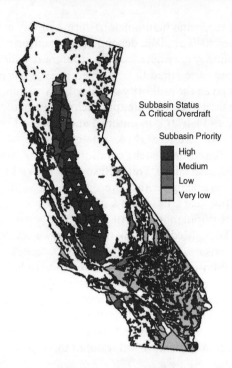

Fig. 10.1   Priority and overdraft designation of California groundwater basins
Note: High- and medium-priority basins are subject to SGMA and must write GSPs. These are concentrated in the Central Valley. Critically overdrafted basins are subject to an earlier compliance timeline.

entrenching this idea of local control (Dennis et al. 2020). Some water agencies and irrigation districts took advantage of these incentives and others did not, placing agencies at different starting points when SGMA was passed.

10.2.1   Changes to the Bargaining Landscape

The passing of SGMA changed the bargaining landscape in several important ways that are relevant for the emergence of collective action. Prior to the passing of SGMA, active groundwater management was only occurring in a small number of adjudicated basins (Ayres, Edwards, and Libecap 2018), implying that in most cases the transaction costs of bargaining outweighed the gains from management, despite stark declines in groundwater reserves in many regions.[3] We see SGMA serving to enable less costly institutional transitions, pulling some basins into collective action and

3. Given that court adjudication is often a decades-long and highly litigious process, this may not be surprising.

active groundwater management. We anticipate that GSAs will introduce meaningful groundwater management where the transaction costs associated with bargaining over collective action are now smaller than the gains from management (Demsetz 1967).

SGMA has altered the bargaining environment in four key ways. First, SGMA serves to lessen information asymmetries and incomplete information by requiring hydrologic modeling and the development of a detailed water budget that must be consistent with other GSPs in the same basin. It also requires the establishment of a monitoring network of wells to track key sustainability indicators. Combined with its requirements to conduct public outreach and stakeholder engagement, this likely reduced information barriers to collective action.

Second, SGMA generates a new role for the state to act as a backstop if plans are insufficient, altering the policy default, and reducing the likelihood of management plans that lack teeth. By imposing a 2040 sustainability mandate, SGMA restricts the set of potential collective agreements, eliminating the possibility that parties come together and decide that business-as-usual is in their mutual best interest.[4]

Third, SGMA broadens the jurisdiction and power of local agencies by giving them the new authority to monitor and meter wells, levy taxes, and facilitate groundwater trade. It empowers GSA board members to agree on management actions as representatives of the interests in the region, thereby limiting the number of bargaining parties directly involved, and bolstering their ability to conduct effective monitoring and enforcement. Even some "very low priority" basins are forming GSAs and writing GSPs, even though they are not required to do so, implying that these shifts in the bargaining environment have been significant even in instances where there is no new binding sustainability mandate.

Finally, SGMA sinks many transaction costs by mandating the development of plans, which forces negotiation among GSA board members, and by providing direct financial support for plan development.

10.3 Conceptual Framework

When does effective groundwater management occur, and how? We next review the open-access problem, casting the outcome as an equilibrium result of balancing the gains from management and the costs of collective action. Based on prior literature, we characterize the conditions under which

4. We note that DWR cannot perfectly observe or predict whether a given plan will actually achieve sustainability, meaning the state is only likely to reject plans that fail to target sustainability by a large and apparent margin. For this reason, we may expect basins where bargaining costs continue to outweigh the gains from management to propose only a minimal set of actions to appease state regulators.

we might expect certain management strategies to emerge. We obtain five testable hypotheses that we take to the early data on SGMA.

10.3.1   Overcoming the Open-Access Problem

The basic problem facing groundwater management is a tragedy of the commons. With unrestricted authority to pump from underlying aquifers, individual pumpers choose groundwater extraction based on their own private costs and benefits and ignore the external costs imposed on other basin pumpers through reduced aquifer storage. Choosing to extract additional water today imposes negative externalities on other users, reducing the amount available in the future, increasing pumping costs for neighboring pumpers, affecting groundwater quality, and inducing other spatial environmental effects. In the face of significant costs for bargaining over new management among users, economic theory predicts that individual pumpers will pump individually optimal but socially excessive amounts, leading to long-run drawdown of the aquifer.

The tragedy of the commons can be overcome through collective action, but the ability to do so is determined by myriad factors affecting the magnitude and distribution of the gains from management and the costs of bargaining. Textbook treatments of the commons problem facing groundwater users elegantly describe how individually optimal extraction decisions can be socially suboptimal but oversimplify both the problems and remedies facing real-world basins. What prompts groundwater pumpers to attempt collective management, the factors influencing the success of those attempts, and what determines the choice of management instruments are all central questions in the political economy of groundwater management. Here, we start by describing the commons problem and characterizing the gains from optimal management. We then discuss the drivers of bargaining costs that together determine the likelihood of collective action. We use this framework to outline five testable hypotheses.

10.3.1.1   Gains from Management

To formalize this notion of the gains from management, consider a basin with many pumpers i, each of whom have a profit function $\pi_i(w_i(t), h(t))$ describing their profit from groundwater use as a function of the volume of water $w(t)$ pumped at time t and the height of the water table $h(t)$. The equation of motion for the height of the water table is $\dot{h}(t) = r(t) - \sum_i w_i(t)$, where $r(t)$ describes recharge.

A benevolent social planner would solve the following problem:

(1)
$$\max_{\{w_i(t)\}} \int_0^\infty \sum_i \pi_i(w_i(t), h(t)) \, dt$$

$$\text{s.t. } \dot{h}(t) = r(t) - \sum_i w_i(t), \, h(0) = H_0,$$

which maximizes collective profits of pumpers on the basin subject to the constraint determining the rate of change in the height of the groundwater table. This is an extremely simplified model, often referred to as the "bathtub" model of groundwater, which abstracts away from the concept of conductivity and other important spatial aspects of the groundwater hydrology that translate to differences in individual net returns from management.

The seminal paper by Gisser and Sanchez using the "bathtub" model found the gains from optimal management to be negligible when extraction was small relative to the size of the aquifer, suggesting small stock externalities (Gisser and Sanchez 1980). This approach assumes the absence of cones of depression around wells and the sizeable spatial pumping externalities that exist in many aquifers, which increase the gains from coordination and management (Brozović, Sunding, and Zilberman 2010). It has been shown that high hydraulic conductivity and lower recharge are associated with higher relative land value increases when groundwater management is implemented (Edwards 2016).

All else equal, we expect basins that would experience greater gains from management given a certain set of aquifer conditions to be more likely to experience active demand management under SGMA. We can proxy for the expected gains from management with the degree of current overdraft.

HYPOTHESIS 1 *Basins with greater overdraft will be more likely to adopt a demand management policy.*

10.3.1.2   Costs of Collective Action

While the gains from management help to determine the likelihood of successful bargaining to end open access, a complete accounting includes the costs of bargaining as well. In principle, transitioning to a more efficient groundwater management policy should produce enough value to compensate any potential losers in the transition; this is the very definition of what it means to be efficiency improving. In practice, determining exactly how new property rights to groundwater ought to work and who should receive the gains and in what shares is a costly process that can spur deep disagreements among bargaining participants.

Both the size and distribution of bargaining costs among users influence the likelihood of institutional change. Once at the negotiating table, users are constrained in what actions they can implement both by their ability to reconcile their heterogeneous preferences and the enforceability of their agreements.[5]

First, resource users need to agree upon baseline information about the nature of the groundwater resource and the value of individuals' claims.

---

5. A complete accounting of the variables influencing the endogenous management process would be beyond the scope of this paper—Ostrom (2009) identifies 53 unique variables important to understanding socio-ecological systems like groundwater.

With imperfect scientific understanding of the groundwater resource, sub-stantial disagreement over the rate of recharge, interactions with surface water flows, or the extent of hydraulic conductivity can easily spill over to disagreement over the best course of management action (Wiggins and Libecap 1985; Ostrom 1990, 33–34). Imperfect and asymmetric informa-tion regarding the value of water to different participants can also inhibit defining appropriate compensating transfers to smooth over disagreements (Wiggins and Libecap 1985; Sallee 2019). Outright deception in an asym-metric information bargaining environment (for the purpose of securing a larger allocation, for example) further aggravates these problems (Libecap 1989, 26).

The number of bargaining parties also naturally raises the difficulty of reaching agreement. With few participants, norms of interpersonal conduct (Ellickson 1991) or Coasean bargaining (Coase 1960) can reliably encour-age effective resource management. With larger groups, the complexity of negotiations increases and the scope of potential compensating transfer opportunities shrinks. In the context of settling disputed American Indian water claims, Sanchez, Edwards, and Leonard (2020) show that the number of bargaining parties increases the duration of negotiations. In the context of oil field unitization, which is highly similar to groundwater management in terms of relevant bargaining characteristics, Libecap and Wiggins (1984) find that only relatively concentrated fields are capable of reaching unitiza-tion agreements; fields with multitudes of smaller operators fail to reach agreement and continue overproducing.

HYPOTHESIS 2   *A greater number of bargaining parties increases the costs of collective action and reduces the likelihood of active demand management.*

Heterogeneity among resource users has a more contested influence over bargaining for collective action. Early treatments tended to treat heteroge-neity as an unambiguous drag on the bargaining process (Libecap 1989, 22–23). When some users gain substantially from the status quo, disputes between incumbents seeking to maintain their privileges and burgeoning users desiring more equitable resource allocations can derail negotiations. Heterogeneity in terms of identity can also inhibit agreement—where nego-tiators bring existing socio-cultural resentments to the bargaining table, dis-trust further narrows the scope of achievable agreements. Varughese and Ostrom (2001) synthesize this literature and find that heterogeneity need not be a barrier to collective action. According to Ruttan (2008), heterogeneity in benefits of management can even facilitate transition to efficient manage-ment when "economically advantaged individual(s) gain from providing the collective good, and are thus willing to pay a greater share of the costs [and/or] where the actions of one or a few individuals provide sufficient positive externalities to provide the good for all."

HYPOTHESIS 3 *Greater heterogeneity among bargaining parties can both increase and decrease the costs of collective action, resulting in an ambiguous effect on the likelihood of active management.*

Finally, the broader legal and political environment can both impose limitations and enable further progress on potential collective action agreements. While organizing for collective action completely outside the auspices of government is possible, recognition and support from formal authorities enables a broader suite of monitoring and enforcement possibilities.

10.3.2   Determinants of Instrument Choice

For basins in which the gains from management exceed the costs of collective action, the question becomes how to manage. The choice of policy instrument will depend on several political and economic factors. Major evaluation criteria discussed in the literature include the relative cost-effectiveness of different policies, the distribution of benefits and costs among users, and the minimization of risk associated with missing the policy target in the face of uncertainty (Baumol and Oates 1988). The optimal policy instrument for a given basin will depend on the subjective weight placed on each dimension and the political feasibility of implementing a given strategy.

HYPOTHESIS 4 *Interests of governing board members may influence policy instrument choice.*

The cost-effectiveness advantage of incentive-based policies depends on the heterogeneity among regulated firms (Goulder and Parry 2008). In the context of groundwater, we may expect to see markets emerge in places where variation in demand for groundwater is greatest. Heterogeneity in groundwater demand may stem from differences across users in the marginal value product of groundwater and the marginal cost to extract. Marginal value product will vary with the crops grown in the region and the presence or absence of urban water consumers while the marginal costs to extract will vary with the depth to groundwater. We can proxy for local heterogeneity in groundwater demand by considering the variation in crops grown in a given region.

HYPOTHESIS 5 *Trading is more likely to occur where greater heterogeneity exists in demand for groundwater.*

10.4   Data

The early implementation of SGMA provides an opportunity to compile data that characterize trends in groundwater management, including where, how, and why groundwater management is occurring in the state. To characterize groundwater management under SGMA, we collected publicly available data on basins, GSAs, and their GSPs from DWR. First, we created a GSP-level data set on policy instrument choice, including all demand-side programs under consideration and which GSAs are involved in which GSP, by reviewing all 107 groundwater sustainability plans that had been submitted to DWR by May 2022.[6] Second, we assembled data on the control of governing board seats by inspecting GSA formation documents and websites. Third, we link these to a basin-level data set on priority status, including whether or not each basin is critically overdrafted, and informa- tion on crop acreages. These data were collected for all 343 groundwater agencies and all 107 groundwater plans that were submitted to DWR. The 94 medium- and high-priority basins, on which 236 GSAs formed to col- lectively write and submit 102 GSPs, were the only areas mandated to do so under the law. An additional five low- and very-low-priority basins volun- tarily submitted plans which were also included in our analysis.

Table 10.1 provides a descriptive overview of the variables collected, including the unit, number of observations, interpretation, and source. The data comprise a cross-sectional snapshot of how SGMA is unfolding.

GSPs include lists of management actions that the GSA is considering to achieve sustainability. These vary a great deal in terms of specificity and certainty. Though the majority of management actions listed in GSPs are supply augmentation and conservation projects conducted by GSAs them- selves, we focus exclusively on management actions that alter the pumping incentives of groundwater end users. We characterized these management strategies in each GSP by recording the intentions of GSAs to set allocations or pumping restrictions, allow trade, set taxes or fees, or provide incentives for conservation and efficiency improvements. The presence or absence of a given strategy was characterized as "Yes," "No," or "Maybe" to reflect the natural uncertainty at this early stage of SGMA development. If plans stated that a given strategy would be developed or implemented, regardless of the degree of detail described, we marked them as a "Yes." Plans were given a "Maybe" designation with language such as "we may adopt" or "we may consider implementing" a certain strategy.[7] Even in GSPs with highly certain language, plans are subject to change, especially where litigation prevents immediate action. Despite these drawbacks, management plans in GSPs offer the most complete description of management actions on the table at this nascent stage of SGMA implementation.

Each category of management action that we record is an abstraction that

6. In cases where smaller GSAs joined to form a larger GSA (e.g., the Northern Delta GSA), we count only the smaller, individual GSAs.

7. Levels of both specificity and certainty vary substantially between GSPs; where one plan may include a throwaway line about potentially considering a pumping charge, another may set out a multi-page plan for a specific groundwater allocation and market development scheme, perhaps even with results from a pilot.

Table 10.1   **Variables collected**

Data	Interpretation	Primary Source
GSA Level (343 obs, 236 signatory to a "High" or "Medium" Priority GSP)		
Board Seats	List of districts, cities, etc. with board representation	GSA Formation Documents, GSA websites
Single Agency	Is GSA a single agency? (As opposed to MOU, JPA)	From Board Seats
GSP Level (107 obs, 102 "High" or "Medium" Priority)		
GSA Participants	List of GSAs included in GSP	GSP Submissions
Allocations	Does GSP include making an "allocation"? Y/M/N	GSP Submissions
Trading	Does GSP allow trading of allocations? Y/M/N	GSP Submissions
Taxes or Fees	Does GSP impose taxes or fees? Y/M/N	GSP Submissions
Tax Base	What is the tax based on? Acreage, extraction, not specified	GSP Submissions
Rate Structure	How is the tax structured? Tiered, flat, not specified	GSP Submissions
Pumping Restrictions	Does the GSP impose other restrictions on pumping? Y/M/N	GSP Submissions
Restriction Description	Open field describing pumping restrictions	GSP Submissions
Efficiency Incentives	Does GSP offer incentives for conservation/efficiency? Y/M/N	GSP Submissions
Incentive Description	Open field describing conservation/efficiency incentives	GSP Submissions
Crop Acreages	Acres harvested in specific crops in 2014	Land IQ
Subbasin Level (515 obs, 94 "High" or "Medium" Priority)		
Priority	DWR-assigned priority for SGMA compliance	DWR
Critical Overdraft	Is basin in critical overdraft?	DWR
Prioritization Data	All data used for prioritization by DWR	DWR

Notes: The table summarizes the variables collected with source information. Gaps in district-level data were filled manually. GSA formation documents and GSP submissions are accessible at https://sgma .water.ca.gov/portal/.

captures many varied management responses. Here, we give further detail about the definitions of each management action variable recorded.

10.4.1   Allocations

Adjudication has long been an available but costly option for California groundwater basins seeking to establish formalized property rights to water. Without undergoing adjudication, California law prevents a clear, simple groundwater entitlement allocation; therefore, policies suggesting allocations often avoid using that word directly or function as allocations only in roundabout ways. For example, it is not uncommon to see a two-tier block rate structure where the first rate is basically free and the second rate is prohibitively expensive. In this way, the initial tier basically constitutes an allocation. Other plans discuss allowing farmers to generate groundwater "credits" by pumping below some expected/allowable level which can be sold to other users. Not all GSPs that discuss allocations specify how the allocations will be made; among those that do, allocations based on either historic pumping or owned acreage are common.

10.4.2   Trading

This variable is only relevant for GSAs making (or at least considering making) allocations and includes any procedure whereby allocation owners can trade their allocations to other groundwater users in cash sales. Trading schemes often come with restrictions, including bans on exporting water outside the basin or volumetric limits. We do not include individual banking and borrowing (trading across time periods rather than across users) in this variable.

10.4.3   Taxes or Fees

New authority to levy taxes on groundwater extraction is a major new power bestowed on GSAs by SGMA. This variable includes new taxes that affect agricultural production decisions on some margin (i.e., it excludes completely flat fees imposed on every property owner). For taxes and fees that specify their tax basis (groundwater extraction, irrigated acreage, or acreage), we record this as well. This variable does include the tiered extraction taxes that make up some of the allocation schemes as described earlier. Among the GSPs that specify a tax structure, all plans involve tiered (as opposed to flat) rates. Most plans leave the specific monetary level of the tax to future determination.

10.4.4   Ad Hoc Pumping Restrictions

While all of the above can be considered "pumping restrictions" in some sense, we reserve this variable for outright bans on pumping in certain circumstances or geographies. These restrictions generally take the form of conditional restrictions that are triggered in event of a drought declaration, for example. Many GSPs receive a "Maybe" in this category for the inclusion of a vague sentence alluding to the potential need to consider outright pumping restrictions in the event that the remainder of the GSP management actions are insufficient for achieving sustainability. Other examples include geographic pumping bans to prevent specific undesirable outcomes like seawater intrusion or impacts to groundwater-dependent ecosystems.

10.4.5   Efficiency Incentives

Of all the variables, this captures the broadest diversity of policies. Examples include payments for fallowing, switching to less water-intensive crops, investments in more water-efficient irrigation infrastructure, and payments for residential rainwater harvesting, lawn removal, or appliance efficiency. Importantly, this variable does not include descriptions of existing water utility efficiency programs (they must be new), programs offering merely education or technological support without direct monetary incentives, or efficiency improvements made only to the infrastructure directly controlled by the agencies forming the GSA, e.g., canal lining.

10.5   Results

We first document the broad trends of how the GSAs are planning to meet the sustainability requirements of SGMA, with a focus on the demand-side strategies, and then identify patterns and characteristics that predict the proposed strategies.

10.5.1   Policy Instrument Choice

A breakdown of the number of plans that suggest a given policy is reported in figure 10.2. Our count of reported management strategies reveals both substantial variation in the approaches taken by local agencies and a substantial departure from pre-SGMA management strategies. Notably, 17 plans report the establishment of individual groundwater pumping allocations, with another 33 plans considering setting such allocations. Prior to SGMA, this type of quantification was only achieved through a costly adjudication process. A smaller subset of these plans are developing groundwater markets (5) or considering the development of markets (26) to facilitate trade of these newly defined allocations.

The establishment of taxes or fees on groundwater extraction or land use represents another departure from the previous status quo in which groundwater pumpers faced only the energy costs to extract groundwater from below. Taxes and fees represent one of the most common demand side management actions proposed by GSAs with 18 GSPs outlining definite plans and another 46 with possible plan to institute a tax, together representing 60 percent of the plans in our data.

Of the 107 GSPs submitted to DWR, 19 of them exclude mention of

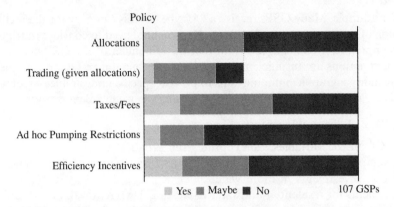

Fig. 10.2   Breakdown of proposed management actions

Note: The figure displays a summary of proposed demand-side policies. Data were collected manually from 107 Groundwater Sustainability Plans submitted to the Department of Water Resources, available on the online SGMA Portal. Detailed definitions of each variable are provided in section 10.4.

any demand-side strategy, and are likely relying exclusively on supply-side strategies to correct overdraft and achieve sustainability. These supply-side strategies include importing additional surface water supplies for in-lieu groundwater recharge, artificial groundwater recharge with excess winter flood flows, and recycled water programs. While these programs may help achieve the goal of slowing or stopping groundwater drawdown, they also impose costs on the district that must be recuperated. Rather than aligning the individual and social costs of pumping, these projects drive a larger wedge by socializing the costs of finding additional water sources when groundwater is overextracted.

Figures 10.3 and 10.4 show the spatial distribution of (1) allocations and trading and (2) taxes and fees, respectively, with definite and potential proposals shown separately. A look at the spatial spread reveals a concentration of these policies in the Tulare Lake region of the southern Central Valley where the majority of critically overdrafted basins reside.

10.5.2   Determinants of Collective Action and Instrument Choice

We next explore how collective action and policy instrument choice correlate with different features of the localities in which they emerge to shed light on the hypotheses laid out in section 10.3. Table 10.2 reports these associations, restricting the sample to only GSPs that report definite plans to proceed with a given management strategy.

HYPOTHESIS 1 (Gains from Management). We expect that collective action is more likely to occur where the gains from management are greatest. Not surprisingly, the presence of each demand management strategy (allocations, taxes, pumping restrictions, or efficiency incentives) being imple-

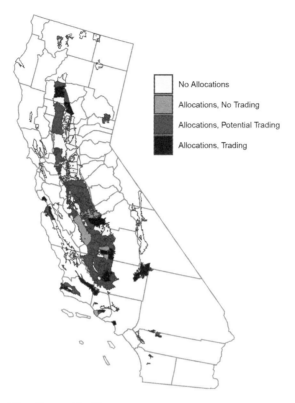

Fig. 10.3   Allocation and trading programs

Note: The map shows the spatial distribution of Groundwater Sustainability Plans that proposed allocations and trading. Both certain and potential allocations are included in this map. Data were collected manually from Groundwater Sustainability Plans submitted to the Department of Water Resources, available on the online SGMA Portal.

mented with confidence is positively correlated with a basin being designated as critically overdrafted.

HYPOTHESIS 2 (Number of Bargaining Parties). The next two variables—the number of GSAs coordinating on one GSP and the number of board seats governing GSAs involved in the GSP—proxy for the number of bargaining actors. We anticipated that a larger number of players reduces the likelihood of collective action. However, column (6) reveals a positive correlation between these proxies and the likelihood of any demand management action being proposed. Comparing across columns, plans with a larger number of coordinating GSAs are more likely to propose pumping restrictions and efficiency incentives than allocations, trading, or taxes. However, these correlations are fairly weak and have inconsistent signs across types of policy instruments, so there is not strong evidence to support nor to oppose the hypothesis.

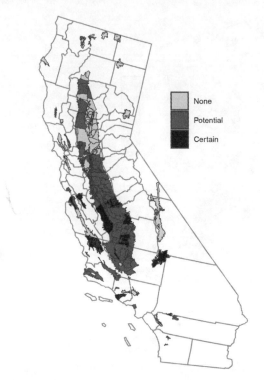

Fig. 10.4   Proposed fees

Note: The map shows the spatial distribution of Groundwater Sustainability Plans that include proposals for fees on extraction, irrigated acreage, or some other measure of water intensity. Data were collected manually from Groundwater Sustainability Plans submitted to the Department of Water Resources, available on the online SGMA Portal.

HYPOTHESIS 3 (Heterogeneity of Bargaining Parties). To shed light on this hypothesis, we construct a Herfindahl-Hirschman Index (HHI) of the concentration of GSA board seats held by different interest types and show the correlation between this and policy outcomes in row 4 of table 10.2. This variable captures the concentration of voting power within a certain type of groundwater user. Large values of the index suggest a high concentration of interests and less diversity. Likewise, small values of the HHI imply more diversity of interests represented on the governing board. Overall, we find a negative correlation between the board seat HHI and the presence of any demand management, meaning that greater diversity of bargaining parties correlates with an increased likelihood of collective action.

HYPOTHESIS 4 (Balance of Power). The next set of attributes, which describes the representation on the board, proxy for whose interests are dominating. Many local water and land use agencies elected to partner with other organizations and form multi-agency GSAs. GSAs pursuing this route formed

Table 10.2   Correlation coefficients between policy choice and GSP attributes ("Yes" Only)

	Allocations (1)	Trading* (2)	Taxes or Fees (3)	Pumping Restrictions (4)	Efficiency Incentives (5)	Any (6)
GSP in Critically Overdrafted	0.331	0.323	0.406	0.208	0.155	0.323
Subbasin	(0.008)	(0.018)	(0.008)	(0.009)	(0.009)	(0.009)
Number of GSAs in GSP	−0.071	−0.033	−0.107	0.035	0.131	0.03
	(0.009)	(0.02)	(0.009)	(0.01)	(0.009)	(0.01)
Number of Seats in GSAs in	0.036	−0.219	−0.172	−0.067	0.24	0.133
GSP	(0.01)	(0.019)	(0.009)	(0.009)	(0.009)	(0.009)
HHI of GSA Board Seats by	−0.112	0.165	0.095	0.026	−0.224	−0.117
Category	(0.009)	(0.02)	(0.009)	(0.01)	(0.009)	(0.009)
Share of Seats Held by	0.01	0.173	0.121	−0.013	−0.123	−0.077
Special Districts	(0.01)	(0.02)	(0.009)	(0.01)	(0.009)	(0.009)
Share of Seats Held by Cities	−0.106	−0.064	−0.089	0.107	0.009	−0.006
and Counties	(0.009)	(0.02)	(0.009)	(0.009)	(0.01)	(0.01)
HHI of Area Harvested	−0.185	−0.24	−0.153	−0.117	−0.23	−0.29
Among Top 12 CA Crops	(0.009)	(0.019)	(0.009)	(0.009)	(0.009)	(0.009)

Note: The table presents correlation coefficients between management actions and GSP attributes. We focus here on management plans that are considered definite and exclude management plans that are simply under consideration ("Yes" only). Standard errors are reported in parentheses. For counting seats, single-agency GSAs are considered to have a single seat controlled by the forming agency.

*When considering how trading correlates with GSP attributes, we restrict the sample set to only plans that are setting allocations.

boards, with substantial leeway to design board size and representation. Some GSAs granted board seats to non-agency partners, like water companies, private well stakeholders, or environmental organizations. The majority of GSA board seats are held by special districts and local water agencies. Special districts, including reclamation, water, and irrigation districts, are local government entities created under state law to administer specific public services. An irrigation district, for instance, maintains irrigation canals and distributes surface water. We largely anticipate that special districts are aligned with the incentives and priorities of farmers and agribusiness in their jurisdictions.

Cities and counties are also common board seat holders in collaborative GSAs that are motivated to maintain groundwater supplies for community water systems. Counties have an extra role under SGMA implementation to fill in as the GSA representative for any basin areas left unmanaged by the formation of other GSAs.

A look at the share of seats held by different agency representatives in table 10.2 shows that GSPs where the governing boards feature a higher share of seats held by special districts are more likely to propose allocations and taxes and less likely to impose pumping restrictions and efficiency incentive programs. The opposite is true for GSPs with a greater fraction of seats held by cities and counties. These results are suggestive of the hypothesis that unobserved interests of governing parties plays a role in policy instrument choice.

HYPOTHESIS 5 (Heterogeneity of Groundwater Demand). The final variable is meant to capture the heterogeneity in groundwater demand across the GSP-managed area. We construct a Herfindahl-Hirschman Index (HHI) of the concentration of planted acreage among California's top 12 crops in each basin since crop type is a major determinant of water demand. Not all GSPs have values for this since some GSPs cover exclusively non-agricultural land. It is not necessarily clear ex ante how this should be associated with management decisions. Highly homogenized groundwater uses may make finding agreement easier, but this also limits the potential gains from trade. A look at column (2) shows that conditional on setting allocations, trading programs are negatively correlated with a higher HHI of crop types. A higher HHI implies more concentration and less heterogeneity in groundwater demand. Contrary to hypothesis 5, this proxy suggests that heterogeneity in demand is correlated with a decreased likelihood of allowing trade of allocations.

Table 10.3 presents the same set of correlations but this time inclusive of potential plans to implement a given policy. Results are consistent between these two definitions in terms of both direction and magnitude when con-

Table 10.3   Correlation coefficients between policy choice and GSP attributes ("Yes" and "Maybe")

	Allocations (1)	Trading* (2)	Taxes or Fees (3)	Pumping Restrictions (4)	Efficiency Incentives (5)	Any (6)
GSP in Critically	0.229	0.257	0.581	0.17	0.107	0.273
Overdrafted Subbasin	(0.009)	(0.019)	(0.006)	(0.009)	(0.009)	(0.009)
Number of GSAs in GSP	−0.095	0.321	−0.013	−0.07	0.092	−0.015
	(0.009)	(0.018)	(0.01)	(0.009)	(0.009)	(0.01)
Number of Seats in GSAs	0.074	0.142	0.042	−0.048	0.079	−0.055
in GSP	(0.009)	(0.02)	(0.01)	(0.01)	(0.009)	(0.009)
HHI of GSA Board Seats	−0.085	0.223	0.138	0.033	0.024	0.081
by Category	(0.009)	(0.019)	(0.009)	(0.01)	(0.01)	(0.009)
Share of Seats Held by	0.107	0.373	0.205	−0.048	0.178	0.092
Special Districts	(0.009)	(0.018)	(0.009)	(0.01)	(0.009)	(0.009)
Share of Seats Held by	−0.137	−0.178	−0.169	0.046	−0.233	−0.04
Cities and Counties	(0.009)	(0.02)	(0.009)	(0.01)	(0.009)	(0.01)
HHI of Area Harvested	−0.231	−0.301	−0.376	−0.039	−0.057	−0.175
Among Top 12 CA Crops	(0.009)	(0.019)	(0.008)	(0.01)	(0.009)	(0.009)

Note: The table presents correlation coefficients between management actions and GSP attributes. Here we consider management plans that are both definite and potential ("Yes" and "Maybe"). Standard errors are reported in parentheses. For counting seats, single-agency GSAs are considered to have a single seat controlled by the forming agency. *When considering how trading correlates with GSP attributes, we restrict the sample set to only plans that are setting allocations.

sidering prioritization of the basin and the share of seats held by different entities. Differences emerge when considering associations between management policies and the number of GSAs or number of board seats.

10.6 Concluding Remarks

In many ways, the Sustainable Groundwater Management Act builds naturally on historical institutional developments that have occurred in the use and management of both surface water and groundwater over time. Prior appropriation laws, which emerged shortly after westward expansion began, enabled landowners without direct access to river water to construct ditches for diversions, which could be shared among nearby landowners. The high costs of constructing water storage and delivery infrastructure led to the development of mutual water companies and irrigation districts to solve the collective action problem of making necessary infrastructure investments (Libecap 2011; Hanemann 2014; Leonard and Libecap 2019).

By contrast, groundwater extraction does not require expensive capital for storage and distribution, meaning groundwater use required far less neighborly cooperation to develop. US irrigation in the 20th century remained fairly flat over the first four decades of the century, with slow gains in irrigated acreage driven mostly by the formation of new surface water districts. After 1940, technological innovations such as center-pivot irrigation led to an explosion in irrigated acreage supplied by groundwater pumping, managed by individuals and operating outside the auspices of irrigation districts or cooperatives (Edwards and Smith 2018). As groundwater pumping advanced and losses from open-access management became apparent, collective action (or costly adjudication) began to emerge, although collective action over groundwater has remained difficult in many settings.

In this paper, we studied the early and ongoing implementation of California's Sustainable Groundwater Management Act through a political economy lens. While the regulation is still in its infancy, proposed policies shed light on how agencies are tackling sustainability mandates, revealing which governance and basin characteristics are correlated with various policy outcomes. We compiled a novel data set that documents the proposed strategies in 107 management plans and used it to test hypotheses about collective action and policy instrument choice. Consistent with expectations, we found collective action to be more likely in basins characterized by a greater degree of overdraft, likely indicative of greater returns from management. We found evidence to suggest that homogeneity in groundwater users and board representatives was associated with decreased likelihood of demand management. We found no evidence of trading schemes being more likely to occur where greater heterogeneity of demand exists, despite potential for greater cost-effectiveness relative to other instruments.

Open-access issues around groundwater will become even more critical to resolve as climate change causes higher temperatures, alters the frequency and severity of droughts, and shifts the precipitation regime. Our assessment of California's Sustainable Groundwater Management Act has shown that efforts by a centralized government to reduce transaction costs over bargaining can drive local management. In other groundwater-stressed regions of the world characterized by many competing actors and large transaction costs, policy changes that reduce information asymmetries and force negotiation may be fruitful avenues for collective action in the face of climate change.

References

Ayres, Andrew B., Eric C. Edwards, and Gary D. Libecap. 2018. "How Transaction Costs Obstruct Collective Action: The Case of California's Groundwater." *Journal of Environmental Economics and Management* 91: 46–65.

Ayres, Andrew B., Kyle C. Meng, and Andrew J. Plantinga. 2021. "Do Environmental Market Improve on Open Access? Evidence from California Groundwater Rights." *Journal of Political Economy* 129 (10): 2817–2860.

Baumol William J., and Wallace E. Oates. 1988. *The Theory of Environmental Policy.* Cambridge, UK: Cambridge University Press.

Brozović, Nicholas, David L. Sunding, and David Zilberman. 2010. "On the Spatial Nature of the Groundwater Pumping Externality." *Resource and Energy Economics* 32 (2): 154–64.

Bruno, Ellen M., and Katrina Jessoe. 2021. "Missing Markets: Evidence on Agricultural Groundwater Demand from Volumetric Pricing." *Journal of Public Economics* 196: 104374.

Coase, Ronald H. 1960. "The Problem of Social Cost." *Journal of Law & Economics* 3: 1–44.

Demsetz, Harold. 1967. "Toward a Theory of Property Rights." *American Economic Review* 57 (2): 347–59.

Dennis, Evan M., William Blomquist, Anita Milman, and Tara Moran. 2020. "Path Dependence, Evolution of a Mandate and the Road to Statewide Sustainable Groundwater Management." *Society & Natural Resources* 33 (12): 1542–1554.

Drysdale, Krystal M., and Nathan P. Hendricks. 2018. "Adaptation to an Irrigation Water Restriction Imposed through Local Governance." *Journal of Environmental Economics and Management* 91: 150–65.

Edwards, Eric C. 2016. "What Lies Beneath? Aquifer Heterogeneity and the Economics of Groundwater Management." *Journal of the Association of Environmental and Resource Economists* 3 (2): 453–91.

Edwards, Eric C., and Todd Guilfoos. 2021. "The Economics of Groundwater Governance Institutions across the Globe." *Applied Economics Perspectives and Policy* 43 (4): 1571–1594.

Edwards, Eric C., and Steven M. Smith. 2018. "The Role of Irrigation in the Development of Agriculture in the United States." *Journal of Economic History* 78 (4): 1103–141.

Ellickson, Robert C. 1991. *Order without Law: How Neighbors Settle Disputes.* Cambridge, MA: Harvard University Press.

Gisser, Micha, and David A. Sanchez. 1980. "Competition versus Optimal Control in Groundwater Pumping." *Water Resources Research* 16 (4): 638–42.

Goulder, Lawrence H., and Ian W. H. Parry. 2008. "Instrument Choice in Environmental Policy." *Review of Environmental Economics and Policy 2 (2)*.

Hanemann, Michael. 2014. "Property Rights and Sustainable Irrigation—A Developed World Perspective." *Agricultural Water Management* 145: 5–22.

Hornbeck, Richard, and Pinar Keskin. 2014. "The Historically Evolving Impact of the Ogallala Aquifer: Agricultural Adaptation to Groundwater and Drought." *American Economic Journal: Applied Economics* 6 (1): 190–219.

Leonard, Bryan, and Gary D. Libecap. 2019. "Collective Action by Contract: Prior Appropriation and the Development of Irrigation in the Western United States." *Journal of Law & Economics* 62 (1): 67–115.

Libecap, Gary D. 1989. *Contracting for Property Rights*. Cambridge, UK: Cambridge University Press.

Libecap, Gary D. 2011. "Institutional Path Dependence in Climate Adaptation: Coman's 'Some Unsettled Problems of Irrigation.'" *American Economic Review* 101 (1): 64–80.

Libecap, Gary D., and Steven N. Wiggins. 1984. "Contractual Responses to the Common Pool: Prorationing of Crude Oil Production." *American Economic Review* 74 (1): 87–98.

Ostrom, Elinor. 1990. *Governing the Commons: The Evolution of Institutions for Collective Action*. Cambridge, UK: Cambridge University Press.

Ostrom, Elinor. 2009. "A General Framework for Analyzing Sustainability of Social-Ecological Systems." *Science* 325 (5939): 419–22.

Ruttan, Lore M. 2008. "Economic Heterogeneity and the Commons: Effects on Collective Action and Collective Goods Provisioning." *World Development* 36 (5): 969–85.

Sallee, James M. 2019. "Pigou Creates Losers: On the Implausibility of Achieving Pareto Improvements from Efficiency-Enhancing Policies." Technical report, National Bureau of Economic Research.

Sanchez, Leslie, Eric C. Edwards, and Bryan Leonard. 2020. "The Economics of Indigenous Water Claim Settlements in the American West." *Environmental Research Letters* 15.

Smith, Steven M., Krister Andersson, Kelsey C. Cody, Michael Cox, and Darren Ficklin. 2017. "Responding to a Groundwater Crisis: The Effects of Self-Imposed Economic Incentives." *Journal of the Association of Environmental and Resource Economists* 4 (4): 985–1023.

Tsur, Yacov, and Theodore Graham-Tomasi. 1991. "The Buffer Value of Groundwater with Stochastic Surface Water Supplies." *Journal of Environmental Economics and Management* 21 (3): 201–24.

Varughese, George, and Elinor Ostrom. 2001. "The Contested Role of Heterogeneity in Collective Action: Some Evidence from Community Forestry in Nepal." *World Development* 29 (5): 747–65.

Wiggins, Steven N., and Gary D. Libecap. 1985. "Oil Field Unitization: Contractual Failure in the Presence of Imperfect Information." *American Economic Review* 75 (3): 368–85.

Estimating the Demand for In Situ Groundwater for Climate Resilience
The Case of the Mississippi River Alluvial Aquifer in Arkansas

Kent F. Kovacs and Shelby Rider

11.1 Introduction

Groundwater systems are connected to climate change and variability both through natural recharge and through changes in the use of groundwater. Those impacts depend on human choices such as changes in land use. Since groundwater is a common source of high-quality fresh water, there is frequent development of the resource which can easily scale to meet local needs without a major need for infrastructure (Giordano 2009). Throughout the world, groundwater supplies a third of freshwater for domestic use, more than a third for agriculture use, and nearly a third for industrial use (Döll et al. 2012). In periods of low or absent rainfall, the groundwater will naturally replenish the baseflow of waterbodies such as streams and wetlands. While certainly crucial to natural and human systems, there is general lack of studies on the relationship between climate, groundwater, and its monetary value that restricts how well the Intergovernmental Panel on Climate Change (IPCC) can assess human impacts related to climate change. The value of in situ groundwater is difficult to measure because there is no market for the resource, and this complicates the evaluation of climate impacts on groundwater value. We examine how agricultural property value changes

Kent F. Kovacs is a research scientist in the Department of Agricultural Economics and Agribusiness at the University of Arkansas.

Shelby Rider is a program associate in the Department of Agricultural Economics and Agribusiness at the University of Arkansas.

The findings and conclusions in this manuscript are those of the authors and should not be construed to represent any official USDA or US government determination or policy. For acknowledgments, sources of research support, and disclosure of the authors' material financial relationships, if any, please see https://www.nber.org/books-and-chapters/american -agriculture-water-resources-and-climate-change/estimating-demand-situ-groundwater -climate-resilience-case-mississippi-river-alluvial-aquifer.

with climate using the relationship between agricultural land values and saturated thickness in the Lower Mississippi River basin in Arkansas, US.

Decision makers seeking to understand the value of land through the underlying groundwater resource face uncertainties in the hydrologic, economic, and institutional aspects of groundwater management. There is uncertainty in the problem of predicting the consequence of the future climate. The challenge stems from the difficult evaluation of groundwater benefits in the future and irreversible nature of groundwater management impacts. A central distinction in groundwater value is between extractive value, which occurs from the extraction of groundwater and use, and in situ values that occur by keeping the water in the aquifer. Examples of in situ value include values associated with subsidence, buffer values, recreational values, ecological values, and existence values.

Groundwater problems receive ever greater attention because greater withdrawals cause problem like destruction of wildlife, habitat, subsidence, and saltwater intrusion. In addition, groundwater is important as a buffer, or emergency supply, and this has become more widely acknowledged. The importance of this value was evident in California during the drought in the early 1990s and the 2010s, when the surface water demand greatly exceeded the supply available. The use of effluent to restore groundwater is frequent in the southern region of California. The aquifer is converted into an adaptively managed storage receptacle, and the supply of the groundwater is replenished by surface water imports, treated effluents, and flood flows. Over a relatively short period, the water travels through the material of the aquifer and then provides a buffer against surface water shortages.

We make several contributions to the literature on climate and groundwater. First, we provide empirical evidence for the change in agricultural land value as overdraft intensifies due to the heating and drying beyond the current levels. Second, we estimate a non-marginal WTP for groundwater using the revealed preference hedonic property value method. Using the consumer surplus from the uncompensated demand, the loss in property value from a decrease in average precipitation is $160 and $202 depending on the severity of the climate change, assuming that the current saturated thickness is between 100 to 120 feet.

11.2   Theoretical Model

Suppose M is identical agricultural landowners, and parcels of land overlie a portion of the aquifer area. The profit of the landowners is given by $\pi = (p, h)$, where p is the pumping rate and h is the height of water table. The water table height (or saturated thickness) is the distance between water level and the bottom of the aquifer. We assume an open access regime with profit maximizing landowners ignoring the effects of their pumping on the water table. All users pump at the same rate with open access, and the height of the water table is the same for all landowners. The water table height changes over time as

$$\dot{h}(t) = RE - Mp(h(t)),$$

where RE is natural recharge and $Mp(h(t))$ is aggregate pumping. If aggregate pumping exceeds the recharge, there is a water table decline and the pumping rate falls. When the pumping rate and recharge are equal, then there is a steady state. The price of a parcel of land is equal to the present discounted value of the stream of profits.

$$V = \int_0^\infty \pi(p(s),h(s))e^{-\delta s}ds,$$

with a discount rate δ and the time frame includes the declining water table and the steady state. Climatic change diminishes the natural recharge and leads to a greater decline in the water table. Our expectation is that lower profits accrue to the agricultural landowners, and the price of the parcel of land falls. We test this hypothesis with the empirical setting of agricultural landowners in the Arkansas Delta to examine whether declines in the natural recharge of an aquifer due to a drier and hotter climate affect the value of land.

11.3 Data

Arkansas is the largest user of the Mississippi River Valley Alluvial (MRVA), which is the third most used aquifer in the US (Konikow 2013). Much of the region has experienced declines in groundwater levels to half of those before settlement (Clark, Westerman, and Fugitt 2013). County land records for Arkansas are the basis for agricultural land sale information (DataScout, LLC 2020). The 4,071 agricultural land transactions occur from 1993 to 2019 for parcels greater than 10 acres in size. We remove transactions where the total assessed value exceeds the land assessed value and where the price per acre is greater than the 95th percentile or below the 5th percentile. We use a geographic information system to link a parcel identification number to a spatial coordinate for each property. Daily gridded climate data merged to the parcels come from the PRISM to understand how the climate affects the parcel sale (table 11.1). Average growing season precipitation is for the past ten years and for the previous thirty years. Also, we use the average number of degree days between 10°C and 32°C in the past thirty years, and the average number of degree days when heat harms crop growth (i.e., above 32°C) in the past thirty years (Schlenker, Hanemann, and Fisher 2005).

The calculation of the saturated thickness is the difference between the depth to the bottom of the aquifer from the US Geologic Survey (USGS) and the three-year rolling average depth to the saturated region of the aqui-

Table 11.1   Variable summary statistics for the first-stage hedonic equation

Variable	Well on parcel (n=890)		No well on parcel (n=3,811)	
	Mean	Std. Dev.	Mean	Std. Dev.
Price per acre ($/acre)	3,146.5	2,165.2	2,689.9	2,473.1
Growing season precipitation: ten-year average (inches)	26.2	6.9	24.2	6.7
Growing season precipitation: thirty-year average (inches)	24.5	4.9	23.2	5.2
Degree days between 10 and 32 Celsius: thirty-year average (degrees*days)	2,414.2	315.2	2,427.2	378.6
Degree days over 32 Celsius: thirty-year average (degrees*days)	0.25	0.42	0.3	0.3
Well within quarter mile (Binary)	—		0.5	0.5
Well within half mile (Binary)	—		1.0	0.2
Saturated thickness (ft)	119.5	57.8	119.1	54.1
Hydraulic Conductivity (ft/day)	141.1	92.4	142.0	94.7
Intermittent stream within quarter mile (Binary)	0.6	0.5	0.6	0.5
Reservoir within half mile (Binary)	0.01	0.1	0.01	0.1
Root zone available water storage (inches)	10.2	1.6	10.3	1.7
Soil organic matter (kg per square meter)	1.50	0.4	1.49	0.4
Acidic soils (percent of land pH<5.3)	3.1	12.5	3.5	13.6
Commute time to 5,000 population (minutes)	26.3	11.7	27.2	12.6
Commute time to 40,000 population (minutes)	50.1	25.5	54.6	29.5

fer from the Arkansas Department of Agriculture, Division of Natural Resources. Figure 11.1 shows the saturated thickness largely declined over the time frame of the analysis, but some sub-regions have seen a recovery. Lateral hydro-conductivity for the alluvial aquifer depends on the slug tests by the USGS for 42 wells. We use irrigation well dummy variables for parcels that have a well on the property, within a quarter mile of the property, and within a half mile of the property. The presence of an irrigation well comes from the Arkansas water well construction commission (WWCC), and the information on the well includes the location coordinates, pumping capacity, and designated use.

Additional control variables for the first-stage hedonic analysis include proximity to streams or rivers from the National Hydrology Dataset or proximity to on-farm reservoirs or tail-water recovery systems (West and Kovacs 2018). Soil characteristics such as the root available water storage, the soil organic matter, and percentage of the parcel land with a soil pH less the 5.3 come from the on-line SSURGO soil survey with the USDA Natural Resources Conservation Service. Urban influence controls include ArcGIS network analyst derived commute times to towns with greater than 5,000 in population and greater than 40,000 in population.

The estimation of in situ value of groundwater with the inverse demand

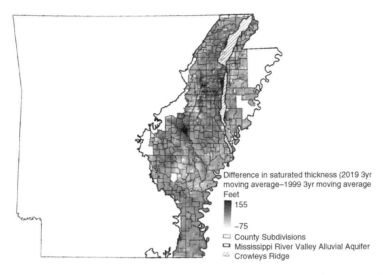

Fig. 11.1   Change in saturated thickness between the three-year moving average for 1999 and the three-year moving average for 2019

equation uses survey responses from farm landowners. A 2016 questionnaire through a phone survey had more than 100 questions, and 199 producers completed the survey in full for a response rate of 32 percent. There were 182 survey responses from farm landowners in the Arkansas Delta used for estimation of the demand equation. The features of the farm that enter as explanatory variables in the demand equation include the climatic variables, the number of irrigated acres, and socioeconomic characteristics such as income and education (table 11.2). The climatic variables from the hedonic equation are matched to survey responses based on the county.

11.4   Empirical Model

The hedonic price function has the specification in equation 1. The natural log of the price per acre of parcel i sold during period t is $\ln P_{it}$, and the saturated thickness of the MRVA aquifer is

$$(1) \quad \ln P_{it} = \beta_{0j} + \beta_{1j}S_{it} + \beta_{2j}S_{it}^2 + \beta_{3j}S_{it}^3 + \beta_{4j}W_{it} + \boldsymbol{\eta}_j' \mathbf{z_{it}} + \boldsymbol{v}_j' x_i + \tau + \theta_{c,t,q}$$

$$+ W_{it}(\beta_{5j}S_{it} + \beta_{6j}S_{it}^2 + \beta_{7j}S_{it}^3 + \beta_{8j}H_i + \beta_{9j}R_{it} + \beta_{10j}PR_{it}) + \varepsilon_{it}.$$

We avoid bias in the OLS estimation of a log-linear model by using a generalized linear model with the average of the dependent variable transformed rather than all observations of the dependent variable (Sampson, Hendricks, and Taylor 2019). Using the Box-Cox functional form to examine the appropriate functional form for the hedonic model, we find the log of

Table 11.2   Definitions and summary statistics of the farm operation characteristics for the second-stage groundwater inverse demand equation

Variable	Definition	Sample Mean	Sample standard deviation	2017 Census of Agriculture Mean
SATTHICK	Saturated thickness (feet)	84.01	38.25	
Demand shifters				
LMKT1	=1 if respondent live in the land market one	0.14	0.35	
LMKT2	=1 if respondent live in the land market two	0.39	0.49	
LMKT3	=1 if respondent live in the land market three	0.16	0.36	
PRECIP_10	Growing season (April to October) precipitation: ten year average (inches)	27.21	2.42	
PRECIP_30	Growing season (April to October) precipitation: thirty year average (inches)	26.52	2.51	
DHARM_30	Degree days over 32 Celsius: thirty-year average (degrees*days)	0.23	0.32	
ACRES	Acres irrigated	2,308	2,716	1459.1
INC	Household income in 2015 from all sources ($ thousands)	104.9	105.5	152.2
INC_NA	=1 if household income not reported	0.23	0.42	
EDU	=1 if no formal education and =8 if beyond Master's degree	4.95	1.55	
Excluded instruments				
SI	Index of the average saturated thickness for a county. =1 for the lowest saturated thickness, =2 for the next lowest saturated thickness, and so forth.	12.31	6.99	
LMKT2_PCTCOT	LMKT2*Percentage of irrigated cropland in cotton	1.27	7.77	
LMKT3_PCTCOT	LMKT3*Percentage of irrigated cropland in cotton	0.86	7.58	

Note: GMM Instruments include all the demand shifters and excluded instruments.

price provides the best fit statistically (Cropper, Deck, and McConnell 1988; Kuminoff, Parmeter, and Pope 2010). A cubic form for saturated thickness provides flexibility in the examination of the non-linear marginal value of the groundwater stock. The dummy variable W_{it} takes on the value of 1 if there is an irrigation well on the parcel i in period t. The vector $\mathbf{z}_{it}$ comprises climatic and other time-varying characteristics (e.g., precipitation, number of degree days, proximity to on-farm reservoirs), and time invariant characteristics are in the vector $\mathbf{x}_i$ (e.g., commute time to population centers). Spatial fixed effects from no controls to county subdivision controls, τ, account for unobserved heterogeneity in land prices that do not vary over time. All specifications have critical groundwater area (CWA) by year by quarter dummies, $\theta_{c,t,q}$, to control for commodity price movements and water management rule changes that could affect CWAs differently over time (ADA 2021).

The price per acre of a parcel may be affected differently by the explanatory variables in equation 1 if a well is present on the parcel. We examine this through interaction variables between W_{it} and climatic features such as precipitation (PR_{it}), aquifer features (S_{it} and lateral hydro-conductivity H_i), and irrigation infrastructure like reservoirs (R_{it}). The subscript j on β, $\boldsymbol{\eta}$, and $\boldsymbol{\nu}$ indicate that these coefficients, which determine the shape of hedonic price function, are estimated for several land markets. Coefficient estimates from several land markets are necessary to properly estimate the demand equation for in situ groundwater (Zhang, Boyle, and Kuminoff 2015). We classify four different agricultural land markets with the Mid-South Land Values and Lease Trend Reports (ASFMRA 2021). We account for heteroscedasticity from spatially correlated errors by allowing for intragroup correlation using counties for the clusters.

The implicit price of saturated thickness specific in each land market comes from the derivative of the hedonic price equation with respect to saturated thickness, and the second stage analysis uses the implicit prices associated with agricultural parcels that have a well on the property. The demand function for saturated thickness with the implicit price for the dependent variable is

$$(2) \quad p_{SAT} = \alpha_0 + \alpha_1 SATTHICK + \alpha_2 LMKT1 + \alpha_3 LMKT2$$
$$+ \alpha_4 LMKT3 + \alpha_5 PRECIP\_10 + \alpha_6 PRECIP\_30$$
$$+ \alpha_7 DHARM\_30 + \alpha_8 ACRES + \alpha_9 INC$$
$$+ \alpha_{10} INC\_NA + \alpha_{11} EDU + \mu.$$

SATTHICK is the saturated thickness estimate associated with each survey respondent's farm, and LMKT1, LMKT2, LMKT3 are land market dummies corresponding to the agricultural land market. PRECIP_10 and PRECIP_30 are measures of precipitation in the past 10 and 30 years, respectively, and DHARM_30 is the average number of degree days that heat harms crop growth over the past 30 years. A negative coefficient on PRE-CIP_10 (α_5) or PRECIP_30 (α_6) implies that producers who receive greater rainfall have a lower shadow price of groundwater. A positive coefficient on DHARM_30 implies that producers who experience a greater number of high temperature degree days have a higher shadow price of groundwater. ACRES is the acres of cultivated land on the farm; INC and INC_NA represent the household income and a dummy if income not reported; EDU is an index for the years of education attained; μ is an error term, and the vector α are preference parameters to estimate.

We use a set of instruments inspired by the literature of residential sorting (Klaiber and Kuminoff 2014) and land market/demand shifter interaction terms (Bartik 1987). An index for the average level of saturated thickness in a county, SI, is a sorting instrument which takes a value of one for a county with lowest saturated thickness, a value of two in the county with the second lowest saturated thickness, and so on. The land market dummies (LMKT2 and LMKT3) interacted with the percentage of farmland in cotton are valid instruments under the assumption that the hedonic function varies across land market but unobserved tastes do not. The percentage of farmland in cotton proxies as a natural recharge demand shifter in LMKT2 and LMKT3 because cotton is principally grown in a region with more natural recharge.

11.5   Results and Discussion

The hedonic model on the left (table 11.3) has spatial controls for 23 counties in the study area, and the column on the right has the estimates for a hedonic model using spatial controls for 235 county subdivisions defined by the US Census Bureau. The coefficients on the saturated thickness variables interacted with well on parcel are significant. The presence of a well means that a parcel of land increases in value as groundwater abundance rises. Based on the cubic relationship between land value and saturated thickness, the land value increases at a decreasing rate with greater saturated thickness, and the land value is largely unaffected by saturated thickness after the thickness is 160 feet or greater. The complete set of coefficient estimates for the first-stage hedonic model is shown in table 11A.1.

The growing season precipitation has a positive influence on the land value, and the 30-year average of precipitation has a greater influence on land value than the 10-year average of precipitation. An average increase in degree days over 32°C over the last 30 years decreases land value for the hedonic model. Climatic variables interacted with the dummy for well on a parcel are also statistically significant. Parcels with a well sold for more if the precipitation in the past 30 years was higher because buyers presumably have a lower cost of irrigation. Also, parcels with a well and a greater number of degree days over 32°C over the last 30 years have lower agricultural land value since the greater heat stress on the crop lowers the crop productivity or

Table 11.3   **Coefficient estimates for the first-stage hedonic model**

	County spatial fixed effect	County subdivision fixed effects
Well on parcel interacted with saturated thickness	0.0128[c]	0.026[b]
	(0.005)	(0.008)
Well on parcel interacted with square of saturated thickness	–9.85E-05[c]	–1.63E-04[b]
	(5.81E-05)	(6.74E-05)
Well on parcel interacted with cube of saturated thickness	2.38E-07	3.49E-07[b]
	(1.94E-07)	(1.82E-07)
Acidic soils	–5.29E-04	–1.25E-03[c]
	(9.92E-04)	(1.61E-03)
Growing season precipitation: ten year average	1.10E-03[b]	3.17E-03[b]
	(6.95E-04)	(6.62E-04)
Growing season precipitation: thirty year average	1.64E-03[b]	3.84E-03[b]
	(6.16E-04)	(6.49E-04)
Degree days over 32 Celsius: thirty year average	–0.03[b]	–0.041[b]
	(0.006)	(0.004)
Commute time to 40,000 population	–0.007[b]	–0.0129[a]
	(0.003)	(0.003)
Other variables interacted with well on parcel		
Growing season precipitation: thirty year average	5.94E-04[a]	7.35E-04[a]
	(2.21E-04)	(2.62E-04)
Degree days over 32 Celsius: thirty year average	–2.45E-03[a]	–4.01E-03[a]
	(1.31E-04)	(2.89E-04)
Spatial fixed effects (#)	23	235
BIC	85,400	84,879

Note: Number of observations: 4,701. Standard errors clustered at counties in parentheses. All models have controls for groundwater region by year by quarter dummy variables.
[a] $p < 0.01$. [b] $p < 0.05$. [c] $p < 0.1$.

increases the irrigation costs. Other variables in the hedonic model, though not our main interest, have significant coefficients. Very acidic soils (pH less than 5.3) can harm crops, although rice prefers slightly acidic soil, and this lowers the land values. An increase in commute time to a city with more than 40,000 people lowers the agricultural land value, but greater commute time to a city with more than 5,000 people has no effect.

The implicit prices from the first-stage hedonic property price equation represent the dependent variable in equation (2) for the estimation of the saturated thickness demand parameters. We assign a value of zero to observations with a negative implicit price in the baseline model for the second stage (Netusil, Chattopadhyay, and Kovacs 2010; Day, Bateman, and Lake 2007). Estimation of IV Model 1 and IV Model 2 is through a two-step instrumental variable (IV) generalized method of moments (GMM) estimator with a saturated thickness sorting index (SI) used for instrumental variables in IV Model 1 and additional demand shifter IVs (LMKT2_PCTCOT and LMKT3_PCTCOT) used in IV Model 2 (table 11.4).

Table 11.4   **Coefficient estimates for GMM estimation of the second-stage groundwater inverse demand equation**

Variable	IV Model 1	IV Model 2
SATTHICK	−0.242[a] (0.027)	−0.183[a] (0.016)
LMKT1	19.1[a] (6.82)	12.11 (9.31)
LMKT2	16.18[a] (2.16)	17.82[a] (3.83)
LMKT3	21.38[a] (2.32)	19.72[a] (3.41)
PRECIP_10	−2.11[a] (0.081)	−2.25[a] (0.113)
PRECIP_30	−3.47[a] (0.012)	−3.22[a] (0.013)
DHARM_30	3.11 (1.011)	4.56 (0.921)
ACRES	0.001 (0.0004)	−0.0003 (0.001)
INC	0.004 (0.005)	0.008 (0.006)
INC_NA	−1.26 (3.03)	0.636 (3.17)
EDU	−1.57[a] (0.443)	−1.99[a] (0.445)
Constant	68.62[a] (19.91)	59.73[a] (15.70)
Instruments	SI	SI; LMKT2_PCTCOT; LMKT3_PCTCOT
R^2	0.34	0.37
First stage F-statistic (p-value)	201.2[a] (0.00)	816.40[a] (0.00)
Overidentification Hansen J (p-value)	0.55 (0.29)	2.26 (0.41)

Note: Robust standard errors clustered at counties in parentheses.

[a] $p<0.01$. [b] $p < 0.05$. [c] $p < 0.1$.

The negative implicit prices from the first stage are adjusted to zero.

The negative coefficient on SATTHICK across all models indicates that landowners' WTP for saturated thickness decreases as aquifer conditions improve. Instrumenting for the endogenous quantity variable suggests that the slope of the demand function is either −0.183 or more negative, given the positive bias expected even in IV estimation (Nevo and Rosen 2012). Weak instruments can lead to even more bias in the coefficients than OLS (Stock, Wright, and Yogo 2002). The first-stage F-statistic is greater than 200 for IV Model 1 and greater than 816 for IV Model 2, suggesting that the instruments are sufficiently strong. Another concern is that the IVs are correlated with the error term, but the Hansen J statistic for GMM estimation is not significant in either model.

Several of the covariates in equation (2) are statistically significant, providing evidence that farmers living in areas with higher average precipitation (PRECIP_10 and PRECIP_30) are willing to pay less for saturated thickness. Farmers living in areas with more degree days over 32°C over the last 30 years (DHARM_30) have a stronger preference for saturated thickness. These coefficient signs match expectations as farmers have preferences for precipitation rather than costly irrigation inputs, and the number of degree days over 32°C increases the crop need for groundwater resources for irriga-

Table 11.5   Change in the value per acre of agricultural land for a 20-foot decline in saturated thickness for alternative changes in the 10- and 30-year average precipitation based on the inverse demand equation for groundwater

Change in average precipitation (inches)	Loss of agricultural land value from a twenty-foot decline in saturated thickness	
	10-year average	30-year average
−0.5	−160	−165
−1	−171	−180
−1.5	−181	−192
−2	−189	−202

Note: We assume that the initial saturated thickness is 120 feet. The first stage is the cubic specification for saturated thickness and county subdivision fixed effects while the second stage is the linear specification for inverse groundwater demand.

tion. The number of years of education a farmer has (EDU) is significant and negative, indicating that education makes farmers less willing to pay for groundwater.

The welfare implications of a decrease in average precipitation from 0.5 inches to 2 inches due to climate change are shown in table 11.5. The value of agricultural land declines in a drier climate because the value of groundwater lost to overdraft is greater. A 20-foot decline in saturated thickness leads to lower property values of $148 per acre if the initial saturated thickness is 120 feet (table 11A.2). A decrease in average precipitation by 0.5 inches would decrease the per acre value of land by $160 for the 10-year average and $165 for the 30-year average, respectively. If the 10-year average precipitation falls a further half inch, the value of the agricultural land would decrease by $171 per acre. The value of land falls by $189 per acre if the average precipitation declines by 2 inches. Slightly greater decreases in the value of land occur when using the 30-year average for precipitation.

11.6   Conclusion

Empirical measurement of the in situ value of groundwater in response to climate change is a challenge because in situ groundwater is a non-market good. One approach available to the practitioner is the second-stage hedonic analysis to estimate an inverse demand for natural capital. Shifters of the demand equation include measures of precipitation and heat because those influence farmers' management of their natural resources. Our empirical analysis of groundwater in the Arkansas Delta provides evidence for losses of agricultural property value as the level of precipitation declines due to climate change. Groundwater overdraft is a chronic challenge for agricultural and urban communities alike as populations increase, agriculture intensifies, and the climate changes. Proper groundwater management requires comparing private benefits of agricultural producers versus the conservation of natural resources.

Policy interventions are often created with the aim of increasing groundwater as illustrated by the recent development of California's Sustainable Groundwater Management Act (Kiparsky et al. 2017). However, estimation of a non-marginal change in the value of groundwater through the use a groundwater demand curve is a challenge since landowners choose how much groundwater to purchase and the price paid for the groundwater simultaneously. We contribute to the hedonic literature on groundwater using a panel data set with two decades of agricultural land sales to determine the welfare implications associated with a non-marginal increase in saturated thickness. We predict that farm landowners in Arkansas will lose \$160 to \$202 in property value per acre in the next 30 years as saturated thickness declines faster due to a drier climate.

Our application to groundwater shows that precipitation can have a vital role in systems with interacting natural capital stocks. Our approach could be extended to include a greater array of climate measures (e.g., growing degree days, heat stress) and natural capital (e.g., water and soil quality). Policy makers and natural resource managers may use the empirically measured relationship among the groundwater and climatic indicators to assess trade-offs with scarce budgets.

Appendix

Table 11A.1 has the full set of coefficient estimates for the first-stage hedonic model. Table 11A.2 indicates the per acre property value benefit from changes in saturated thickness using second-stage welfare measures.

Table 11A.1   Coefficient estimates for the first-stage hedonic model

	No spatial fixed effects	County spatial fixed effect	County subdivision fixed effects
Saturated thickness	−6.06E-03	−4.13E-03	2.99E-03
	(5.63E-03)	(5.89E-03)	(8.25E-03)
Square of saturated thickness	3.47E-05	2.68E-05	−2.16E-05
	(4.33E-05)	(4.37E-05)	(6.11E-05)
Cube of saturated thickness	−4.73E-08	−4.45E-08	4.38E-08
	(9.98E-08)	(9.83E-08)	(1.34E-07)
Root zone available water storage	0.019	0.013	0.012
	(0.013)	(0.013)	(0.010)
Soil organic matter	0.009	0.047	0.0532
	(0.049)	(0.054)	(0.047)
Acidic soils	1.16E-04	−5.52E-04	−1.89E-03[c]
	(7.94E-04)	(9.37E-04)	(1.04E-03)
Degree days between 10 and 32 Celsius:	1.08E-04	1.05E-04	1.17E-04
five year average	(8.00E-05)	(1.30E-04)	(1.17E-04)
Commute time to 5,000 population	−8.67E-04	−2.73E-03	−1.59E-03
	(2.78E-03)	(3.88E-03)	(7.35E-03)
Well on parcel	−0.715[a]	−0.747[a]	−0.972[a]
	(0.227)	(0.269)	(0.305)
Well within quarter mile	−0.101[b]	−0.115[b]	−0.133[a]
	(0.044)	(0.050)	(0.051)
Well within half mile	0.189[b]	0.201[b]	0.213
	(0.090)	(0.097)	(0.132)
Reservoir within half mile	0.057	0.038	0.021
	(0.199)	(0.180)	(0.110)
Intermittent stream within quarter mile	0.046	0.053	0.058
	(0.041)	(0.048)	(0.048)
Other variables interacted with well on parcel			
Hydraulic Conductivity	4.06E-04[c]	4.09E-04[c]	4.79E-04[c]
	(2.20E-04)	(2.37E-04)	(2.91E-04)
Reservoir within half mile	0.335[b]	0.325[c]	0.332[b]
	(0.145)	(0.134)	(0.187)
Implicit price if well on parcel			
80 feet	1.47 (3.46)	4.58 (3.35)	8.96[b] (3.92)
110 feet (average)	−0.88 (2.38)	0.62 (2.23)	−3.28 (3.42)
Spatial fixed effects (#)	0	23	235
BIC	85,498	85,400	84,879
Number of observations	4,701	4,701	4,701

Note: Standard errors clustered at counties in parentheses. All models have controls for groundwater region by year by quarter dummy variables.

[a] $p < 0.01$. [b] $p < 0.05$. [c] $p < 0.1$.

^ Rice parcels include any parcel with rice in the last five years.

Table 11A.2   Per acre property value benefit from changes in saturated thickness using
                     second-stage welfare measures

Change in saturated thickness (feet)	Baseline (No change in average precipitation)
20 to 40	362 ± 304
60 to 80	255 ± 281
100 to 120	148 ± 257
140 to 160	41 ± 233

Note: 95% confidence intervals shown beside each estimate of the per acre property value benefit. The first stage is the cubic specification for saturated thickness and township fixed effects while the second stage is the IV Model 2 for the inverse groundwater demand.

References

American Society of Farm Managers and Rural Appraisers (ASFMRA). 2021. *2021 Mid-South Land Values and Lease Report*. Jonesboro, AR. https://www.glaubfm .com/blog/2021-mid-south-land-values-report-now-available.

Arkansas Department of Agriculture (ADA). 2021. Natural Resources Division. *2020 Arkansas Groundwater Protection and Management Report*. Little Rock, AR, May.

Bartik, T. 1987. "The Estimation of Demand Parameters in Hedonic Price Models." *Journal of Political Economy* 95 (1): 81–88.

Clark, B. R., D. A. Westerman, and D. T. Fugitt. 2013. "Enhancements to the Mississippi Embayment Regional Aquifer Study (MERAS) Groundwater-Flow Model and Simulations of Sustainable Water-Level Scenarios." The U.S. Geological Survey Scientific Investigations Report 2013–5161, 29.

Cropper, M. L., L. B. Deck, and K. E. McConnell. 1988. "On the Choice of Functional Form for Hedonic Price Functions." *Review of Economics and Statistics* 70: 668–75.

DataScout, LLC. 2020: County Land Record for Eastern Arkansas. DataScout, LLC. Data Archive, received June 5, 2020. http://datascoutllc.com/.

Day, B., I. Bateman, and I. Lake. 2007. "Beyond Implicit Prices: Recovering Theoretically Consistent and Transferable Values for Noise Avoidance from a Hedonic Property Price Model." *Environmental and Resource Economics* 37 (1): 211–32.

Döll, P., et al. 2012. "Impact of Water Withdrawals from Groundwater and Surface Water on Continental Water Storage Variations." *Journal of Geodynamics* 59: 143–56.

Giordano, M. 2009. "Global Groundwater? Issues and Solutions." *Annual Review of Environment and Resources* 34: 153–78.

Kiparsky, M., A. Milman, D. Owen, and A. Fisher. 2017. "The Importance of Institutional Design for Distributed Local-Level Governance of Groundwater: The Case of California's Sustainable Groundwater Management Act." *Water* 9 (10): 755.

Klaiber, H., and N. Kuminoff. 2014. "Equilibrium Sorting Models of Land Use and Residential Choice." *The Oxford Handbook of Land Economics*, 352–79. New York: Oxford University Press.

Konikow, L. F. 2013. Groundwater Depletion in the United States (1900–2008): U.S. Geological Survey Scientific Investigations Report 2013–5079, 63.

Kuminoff, N. V., C. F. Parmeter, and J. C. Pope. 2010. "Which Hedonic Models Can We Trust to Recover the Marginal Willingness to Pay for Environmental Amenities?" *Journal of Environmental Economics and Management* 60 (3): 145–60.

Netusil, N. R., S. Chattopadhyay, and K. F. Kovacs. 2010. "Estimating the Demand for Tree Canopy: A Second-Stage Hedonic Price Analysis in Portland, Oregon." *Land Economics* 86 (2): 281–93.

Nevo, A., and A. M. Rosen. 2012. "Identification with Imperfect Instruments." *Review of Economics and Statistics* 94 (3): 659–71.

Sampson, G. S., N. P. Hendricks, and M. R. Taylor. 2019. "Land Market Valuation of Groundwater." *Resource and Energy Economics* 58: 101120.

Schlenker, W., W. M. Hanemann, and A. C. Fisher. 2005. "Will US Agriculture Really Benefit from Global Warming? Accounting for Irrigation in the Hedonic Approach." *American Economic Review* 95: 395–406.

Stock, J. H., J. H. Wright, and M. Yogo. 2002. "A Survey of Weak Instruments and Weak Identification in Generalized Method of Moments." *Journal of Business & Economic Statistics* 20: 518–29.

West, G., and K. Kovacs. 2018. "Tracking the Growth of On-site Irrigation Infrastructure in the Arkansas Delta." Arkansas Bulletin of Water Research: A Publication of the Arkansas Water Resources Center. https://agcomm.uark.edu/awrc/publications/bulletin/Issue-2018-Kovacs-and-West-Growth-of-onsite-irrigation-infrastructure-in-Arkansas-delta.pdf.

Zhang, C., K. J. Boyle, and N. V. Kuminoff. 2015. "Partial Identification of Amenity Demand Functions." *Journal of Environmental Economics and Management* 71: 180–97.

Author Index

Subject Index